Turbulence, Atmosphere and Climate Dynamics (Obukhov 2020)

IOP Conference Series: Earth and Environmental Science
Volume 1040

Moscow, Russia
10-12 November 2020

ISBN: 978-1-7138-5818-8
ISSN: 1755-1307

Printed from e-media with permission by:

Curran Associates, Inc.
57 Morehouse Lane
Red Hook, NY 12571

Some format issues inherent in the e-media version may also appear in this print version.

Additional copies of this publication are available from:

Curran Associates, Inc.
57 Morehouse Lane
Red Hook, NY 12571 USA
Phone: 845-758-0400
Fax: 845-758-2633
Email: curran@proceedings.com
Web: www.proceedings.com

TABLE OF CONTENTS

IOP Conf. Series: Earth and Environmental Science 1040 (2022) 011002 doi:10.1088/1755-1315/1040/1/011002

Peer Review Statement

All papers published in this volume have been reviewed through processes administered by the Editors. Reviews were conducted by expert referees to the professional and scientific standards expected of a proceedings journal published by IOP Publishing.

- **Type of peer review:** Double Anonymous
- **Conference submission management system:** Morressier
- **Number of submissions received:** 64
- **Number of submissions sent for review:** 59
- **Number of submissions accepted:** 44
- **Acceptance Rate (Submissions Accepted / Submissions Received × 100):** 68.8
- **Average number of reviews per paper:** 2
- **Total number of reviewers involved:** 38
- **Contact person for queries:**
  Name: Irina Repina
  Email: repina@ifaran.ru
  Affiliation: A.M. Obukhov Institute of Atmospheric Physics RAS

IOP Conf. Series: Earth and Environmental Science 1040 (2022) 011001 doi:10.1088/1755-1315/1040/1/011001

Turbulence, Atmosphere and Climate Dynamics

I I Mokhov, O G Chkhetiani, I A Repina

A.M. Obukhov Institute of Atmospheric Physics RAS, 119017, Moscow, Russia

E-mail: repina@ifaran.ru

Abstract. The conference "Turbulence, Atmosphere and Climate Dynamics" dedicated to the memory of the Alexander M. Obukhov was held in Moscow from November 10 to 12, 2020. The topics of the conference covered the following scientific areas: turbulence; geophysical hydrodynamics; atmospheric and climate system dynamics; physics and composition of the atmosphere; air-sea interaction; wave propagation. The conference showed a high scientific level of almost all the presentations. Studies of turbulent, climatic and atmospheric processes are traditionally conducted in our country at the highest level, as evidenced by the publication in high-ranking scientific journals and the active participation of Russian scientists in international programs.

The conference "Turbulence, Atmosphere and Climate Dynamics" dedicated to the memory of the outstanding scientist, Academician of the USSR Academy of Sciences Alexander M. Obukhov (05.05.1918–03.12.1989) was held in 2020 for the third time. The founder and director of the Institute of Atmospheric Physics, A.M. Obukhov, together with A.N. Kolmogorov, created the foundations of turbulence theory, which determined one of the main vectors of development in many branches of scientific research, both natural phenomena of different scales, and various applied problems in the second half of the last century and up to the present time.

The wide interest of the scientific community and active participation in the first and second conferences "Turbulence, Atmosphere and Climate Dynamics" in 2013 and 2018 showed the need to organize these conferences regularly. At the third conference from 10 to 12 November 2020 in Moscow 86 oral reports and 70 posters were presented. The conference was attended by scientists from Moscow, St. Petersburg, Ufa, Perm, Tomsk, Novosibirsk, Irkutsk, Kazan, Nizhny Novgorod, Obninsk, Sevastopol with representatives from Great Britain, Finland, Cyprus. Significant number of reports made by young scientists. The topics of the third conference covered the following scientific areas: turbulence in the atmosphere and ocean, geophysical hydrodynamics, general circulation of the atmosphere and dynamics of the climate system, structure and composition of the atmosphere and transport of impurities, propagation and interaction of waves in the atmosphere, physics of the atmospheric boundary layer.

The key topics of the conference were related to the development of scientific studies initiated under the leadership of A.M. Obukhov, including the first section “Turbulence in the atmosphere and ocean”. Turbulence plays a very significant and diverse role in the physical processes that determine the regime of the surface and boundary layers of the atmosphere, in particular. Its theoretical study is necessary to solve the problems of atmospheric diffusion, momentum transfer and eddy transport, transport of impurities and assessment of atmospheric pollution, remote sensing, signal propagation (scattering and transformation), flight safety, climate modeling. The main presentations were related with fundamental topics of the theory of turbulence; observations, instrumentation and theoretical problems in atmospheric turbulence; laboratory and numerical experiments; model and practical approaches in the research of turbulence.

The section "Geophysical hydrodynamics" was devoted to theoretical problems, numerical and laboratory modeling arising in the description of various geophysical processes, including those using the theory of turbulence. The greatest interest was aroused by the reports devoted to quasilinear and nonlinear mechanisms of flow formation, processes in a stratified rotating fluid, baroclinic and barotropic instabilities at large and meso scales, convection, development of geophysical flow modeling in the shallow water approximation on the sphere. A number of reports was presented at this section, demonstrating the experience of applying theoretical constructions to forecasting different meteorological phenomena.

A large number of reports were presented at the sections "General circulation of the atmosphere and dynamics of the climate system" and "Structure and composition of the atmosphere and transport of impurities". Problems of atmospheric structure and composition and climate change are attracting the attention of specialists in various scientific fields. The main tool for reproducing the climate conditions and their changes, including extreme weather and climate situations, is climate modeling. A number of reports of the section "General circulation of the atmosphere and dynamics of the climate system" were devoted to the development, improvement and application of modern weather forecast and climate models. Model approaches developed for operational weather forecasting systems, the results of modeling of future climate changes using global climate models, analysis of the response of the climate system to external influences, as well as natural climate fluctuations, analysis the current state of the climate system, including polar regions, were presented. Series of reports were devoted to the diagnosis and forecasting of dangerous mesoscale meteorological phenomena - tornadoes, squalls, polar cyclones. In a number of presentations, the problems of interaction of phenomena of various scales were considered. Particular attention was paid to the urban climate. The presented reports once again confirmed that the to improve the predictability of extreme atmospheric phenomena and climate anomalies at various spatial scales it is necessary to combine results of ground-based measurements, satellite observations and mesoscale modeling.

The section "The structure and composition of the atmosphere and the transport of impurities" represents the direction of atmospheric physics and chemistry that has been most actively developing in recent decades. This is due to the fact that the dynamics of the composition of the atmosphere is not only an indicator of climatic changes and the anthropogenic impact, but also an indicator of the quality of the human environment. Processes and changes in state in different layers of the atmosphere were considered - from the troposphere to the thermosphere. The results of experimental studies, the results of model simulations and new theoretical models of the propagation of atmospheric pollutants were presented. A number of reports considered the aerosol composition of the atmosphere and the influence of external (volcanic activity) and internal (turbulent transfer) processes. A number of reports was devoted to aeroelectric and aerophysical observations focused on the development of numerical models of electricity in the atmospheric boundary layer. Special reports were devoted to the change in the composition of the atmosphere of the Moscow region during the quarantine measures in spring 2020.

One of the most important applications of the theory of turbulence is to describe the processes of propagation of waves of various ranges in the atmosphere. in the reports of the section " Propagation and interaction of waves in the atmosphere" the propagation of different waves in the atmosphere, including radio waves, acoustic waves, infrasound, was considered. Results of lidar observations of atmospheric structure, theoretical and laboratory modeling of wave propagation in various media were presented. Several reports were devoted to the physical foundations of the acoustic sounding method using the scattering of sound waves by small-scale turbulent inhomogeneities, initiated by A.M. Obukhov.

The section "Physics of the boundary layer atmosphere" was organized for the first time. The reports presented the results of the development of parametrizations of processes in the atmospheric boundary layer, theoretical and experimental studies on various landscapes, the results of modeling of mesoscale circulations. One of reports was related with the studies of the foehn effect at the Spitsbergen archipelago. The section included a special subsection "Study of the structure and dynamics of the atmospheric boundary layer using unmanned aerial vehicles". This subsection was attended by representatives of the Moscow Institute of Physics and Technology, Moscow Aviation Institute, Central Aerological Observatory, A.M. Obukhov Institute of Atmospheric Physics RAS (Moscow), Moscow State University, Arctic and Antarctic Research Institute (St. Petersburg), as well as representatives of the V.E. Zuev Institute of Atmospheric Optics SB RAS (Tomsk), Institute for Monitoring of Climatic and Ecological Systems SB RAS (Tomsk) and SA Khristianovich Institute of Theoretical and Applied Mechanics SB RAS (Novosibirsk). As part of the sectional discussion, possible ways to combine the technical and methodological efforts of various initiative groups and options for future joint field experiments using unmanned aerial vehicles were discussed.

A distinctive feature of the conference in 2020 was a significant poster section. Posters were presented in all areas of the conference and were actively discussed both in person and online. All sections of the conference were interconnected, taking into account the topics of the reports presented. The organizers even had problems with what section to place a number of reports. This demonstrates the close connection between the various topics of the conference. The work of the conference showed the general high scientific level of the reports presented. A number of topical problems were specially discussed at the conference. In particular, to improve the description of interaction of the atmosphere with the surface and the transfer of impurities it is necessary to improve corresponding parameterizations for the atmospheric boundary layer based on the theory of similarity and the coefficients of turbulent transport. New approaches should be adequate for simulations in extreme conditions of stratification and in conditions of complex surfaces, in particular in the Arctic. A wide range of scientific issues discussed at the conference and the attention of the scientific community showed the importance of topics of the "Turbulence, Atmosphere and Climate Dynamics".

Program committee

Chairman: Georgy Golitsyn, RAS academecian, Dr.Sci., Professor, Obukhov Institute of Atmospheric Physics RAS, Moscow
Deputy: Igor Mokhov, RAS academecian, Dr.Sci., Professor, A.M. Obukhov Institute of Atmospheric Physics RAS, Moscow

Members

Roman Vilfand, Dr.Sci, Hydrometeorological Research Center of Russian Federation, Moscow
Valentin Dymnikov, RAS academician, Marchuk Institute for numerical mathematics RAS, Moscow
Nikolay Elanskiy, RAS corresponding member, Dr.Sci, Obukhov Institute of Atmospheric Physics RAS, Moscow
Margarita Kallistratova, Dr.Sci, professor, Obukhov Institute of Atmospheric Physics RAS, Moscow
Alexander Kislov, Dr.Sci, RAS professor Lomonosov Moscow State University, Moscow
Sergei Kulichkov, Dr.Sci. Obukhov Institute of Atmospheric Physics RAS, Moscow
Mikhail Kurganskiy, Dr.Sci. Obukhov Institute of Atmospheric Physics RAS, Moscow
Alexander Makosco, RAS corresponding member, Dr.Sci. Obukhov Institute of Atmospheric Physics RAS, Moscow
Vladimir Semenov, RAS corresponding member, Dr.Sci. Obukhov Institute of Atmospheric Physics RAS, Moscow
Otto Chkhetiani, Dr.Sci, Obukhov Institute of Atmospheric Physics RAS, Moscow

Organizing committee

Chairman: Sergei Kulichkov, Dr.Sci. Obukhov Institute of Atmospheric Physics RAS, Moscow
Deputies: Otto Chkhetiani, Dr.Sci, Obukhov Institute of Atmospheric Physics RAS, Moscow
Irina Repina, Dr.Sci, Obukhov Institute of Atmospheric Physics RAS, Moscow

Members

Artamonov Arseniy, Obukhov Institute of Atmospheric Physics RAS, Moscow
Oksana Koval, Obukhov Institute of Atmospheric Physics RAS, Moscow

IOP Conf. Series: Earth and Environmental Science 1040 (2022) 011001 doi:10.1088/1755-1315/1040/1/011001

Natalia Pankratova, PhD, Obukhov Institute of Atmospheric Physics RAS, Moscow
Maria Parfenova, Obukhov Institute of Atmospheric Physics RAS, Moscow
Vladimir Semenov, RAS corresponding memeber, Dr.Sci., A.M. Obukhov Institute of Atmospheric Physics RAS, Moscow
Andrei Skorokhod, PhD, Obukhov Institute of Atmospheric Physics RAS, Moscow
Victoria Falaleeva, Obukhov Institute of Atmospheric Physics RAS, Moscow
Evgeniya Fedorova, Obukhov Institute of Atmospheric Physics RAS, Moscow

IOP Conf. Series: Earth and Environmental Science 1040 (2022) 012001 doi:10.1088/1755-1315/1040/1/012001

Lunar Observatory for monitoring the radiation balance of the Earth and climate

H I Abdussamatov[1]

[1]Pulkovo observatory of the RAS, St. Petersburg, RF

Abstract.
Continuous uniform high-precision data on the Earth's radiation balance over many decades and hundreds of years are essential for reliable research and prediction of climate change. This necessarily requires the use of long-term fixed, stable space platforms at a sufficient distance from Earth. However, measurements by all low-orbit and geostationary satellites cannot provide high-precision identifications of the radiation balance, since they observe its limited areas. Two identical especial optical robotic telescopes, of which the Lunar Observatory (LO) consists of, are a single system of sequentially functioning in the automatic mode of Earth observations. Telescopes will be installed along the equatorial zone at opposite edges of the Moon, 80.9±0.1° distant from its visible center. LO provides measurements of the variations in the magnitude of the specific power reflected by the planet into the space of the total solar irradiance in the spectral range Δλ=0.2–4 μm and outgoing intrinsic thermal radiation of the Earth in the ranges Δλ=4–50 μm and the atmospheric transparency window Δλ=8–13 μm. The system of two telescopes, replacing each other, will sequentially survey the entire Earth's surface continuously for more than 94% of the time of a lunar day only at lunar night time. LO will make it possible to determine the parameters of the average annual values of total energies departing from the Earth into space with an error of the order of 0.1%, which is almost an order of magnitude more accurate than the errors in the measurement of these parameters by instruments of any orbital spacecraft. Monitoring of energy variations in the range Δλ=8–13 μm will make it possible to control and determine the relative contribution of the variation in the concentration of greenhouse gases in the atmosphere to climate change.

1. Introduction

One of the most important scientific problems is the search for a physical mechanism responsible for global climate change. A study of significant climate fluctuations over the past 800,000 years shows that quasi-bicentennial and one hundred-thousand-year cyclic variations in solar radiation are the main reasons for the alternation of warming and cooling. The only constant feature of climate change for millennia and hundreds thousands years has been its natural cyclical variability. The Sun is the main factor controlling the Earth's climate system, and even minor long-term changes of the total solar irradiance (TSI) can have serious consequences for the climate [1-3]. For the successful study of the most important climatic parameters, which are of great importance in creation the most reliable methods for predicting future deep climate changes, operational and continuous information on the temporal variations of their values is required. Continuous homogeneous high-precision data are needed on the average annual radiation balance of the Earth

$$E = \frac{(S_\odot + \Delta S_\odot)}{4} - \frac{(A_{BE} + \Delta A_{BE})(S_\odot + \Delta S_\odot)}{4} - \varepsilon\sigma(T_p + \Delta T_p)^4 \quad (1)$$

for many decades and hundreds of years for reliable climate research and prediction of future changes. Here, $S_\odot$ is TSI; $\Delta S_\odot$ is the TSI increment; A_{BE} is the Earth's global albedo (Bond albedo); ΔA_{BE} is the

IOP Conf. Series: Earth and Environmental Science 1040 (2022) 012001 doi:10.1088/1755-1315/1040/1/012001

increment of the Bond albedo; ε is the irradiating capacity of the surface – atmosphere system; σ is the Stefan – Boltzmann constant; T_p is the planetary thermodynamic temperature; ΔT_p is the increment of the planetary thermodynamic temperature; E is the specific power of change in the enthalpy of the active oceanic and atmospheric layer (W m^{-2}), which can be treated as the energy balance of our planet's average annual budget in terms of debit and credit of thermal energy. The planetary thermodynamic temperature is the average temperature over the entire planetary surface (the surface of the Earth and atmosphere). The Earth's Bond Albedo A_{BE} is difficult to measure, and astronomers derived it from measurements of the corresponding geometric albedo (A_{gE}) by the relation [4]:

$$A_{BE} = QA_{gE}, \tag{2}$$

where Q is a phase function

$$Q = \int_{-\pi}^{\pi} f(\alpha) d\alpha , \tag{3}$$

and $f(\alpha)$ is the phase function under angle α. The phase function (the ratio of the illumination observed by the spacecraft created by the planet in this direction α to that created towards the Sun $\alpha = 0°$), changing over the entire range of phase angles α from -180° to + 180°, taking on a value from 0 to 1, expressing the coefficient with which the intensity of the reflected and scattered from the entire spherical system surface-atmosphere of solar radiation changes back into outer space depending on the phase angle α. The normalization is chosen so that at a phase angle of 0° (fully illuminated hemisphere of the Earth) the phase function reaches a maximum value of 1. The exact average annual value of the Earth's Bond albedo is valuable in itself. Only knowing its exact value using known TSI data can we determine the exact average annual value of the solar energy absorbed by the planet. The value of the phase integral can be obtained by calculating the area under the phase curve, graph $f(\alpha)$ [4]. The sign of the deviation of the Earth's energy balance equation (1) from the equilibrium state for a long (more than 20 years) period of time is determined the corresponding changes in the thermal state of the Earth's surface – atmosphere system and, therefore, the forthcoming climate changes and their amplitude. The Earth's climate change should obviously be considered as a function of the long-term deviation of the Earth's energy balance from the equilibrium state [1-3,5-8]. Therefore, long-term monitoring of the thermal state of the planet is required. It is necessary to continuously and reliably monitor the magnitude of the difference between the amounts of TSI entering the outer layers of the Earth's atmosphere and the energies of the TSI fraction reflected by the planet and its own thermal radiation going back into space. This certainly requires the use of long-term fixed stable space platforms at a sufficient distance from the Earth, allowing observing the entire surface of the planet, for continuous extraction of homogeneous information about changes in the average annual energy balance of the Earth. From 1990 until about 2043±11, our planet will receive less and less solar energy, but due to the thermal inertia of the Ocean, it will radiate much greater energy throughout this entire period of time. As a result, over a period of ~53 to ~75 years, the Earth will have a negative energy balance. Changes in the parameters of the earth's surface and atmosphere (an increase in the Bond albedo, a decrease in the concentration of water vapor in the atmosphere, etc.) caused by cooling will generate long chain of consequences. The epoch of the "Cold" Sun that came in about 1990 and appearance of deep minimum TSI in 2043±11 demands continuous monitoring highly precise vibrations Earth's energy balance during the whole period of decline and minimum of TSI.

2. Measurements of the Earth's radiation balance by low-orbit and geostationary satellites.

Determination of homogeneous values of the Earth's Bond albedo based on direct measurements of the reflected and scattered planet's fraction of the energy of solar radiation into space in all directions and the power of the Earth's own thermal radiation by radiometers on orbiting weather satellites [9-13], as well as on geostationary orbit [14] or in any other orbital experiment cannot be performed reliably and with the necessary accuracy (≈0.1%) within a long time; since they are capable of only direct measurements of the energy of radiation emanating only from individual parts of the planet (not from its entire surface) to the upper hemisphere and practically only at a certain angle. Even 14 or any larger number of satellites in geostationary orbit is not able to provide coverage of the Earth's surface at polar latitudes of ±78°-90°. Consequently, they will not allow determining the magnitude of the Earth's energy balance with high accuracy (measurement errors of global climate parameters are estimated either qualitatively or at about 1%) due to the inability to survey and measure the energies of radiation emanating from the entire Earth's surface into space in all directions. On-board DSCOVR radiometers at the Lagrange point L1 of the Sun – Earth system [15] will also be able to provide direct measurements of the energy of reflected and scattered by the Earth solar radiation in almost only one direction, i.e. values of the geometrical albedo of the Earth only in periods when the Moon does not cover the Earth's disk. Moreover, all orbiting spacecraft have a relatively short service life due to the limited energy capabilities and the inability to create a long-term most important fundamental database of homogeneous data on the average annual radiation balance of the planet.

If TSI does not change ($\Delta S_\odot = 0$), then the increment of the Earth's effective temperature is the result of increments of the Bond albedo (ΔA_{BE}) and transmission of atmospheric transparency windows [3]

$$\frac{\Delta T_e}{T_e} = -\frac{\Delta A_{BE} S_\odot}{16\sigma T_e^4}. \qquad (4)$$

An increase in only Bond albedo by $\Delta A_{BE} = +0.0003$ (0.1%) with the same TSI ($\Delta S_\odot = 0$) will lead to a decrease in temperature by $\Delta T_e \approx -0.027$ K. Therefore, it is extremely important to measure the average annual energy balance of the planet with an error of the order of 0.1%

3. Lunar Observatory

To achieve reliable precision measurements of the complex of global climatic parameters of the Earth, the only and most promising place in space is the visible surface of the Moon, which has ideal conditions in lunar night time. The solid surface of the Moon and its slow slight fluctuations when moving around the Earth make the lunar surface an ideal stable platform in space for installing optical telescopes-robots for observing the Earth in perfect conditions of its night time. The installation of optical telescopes on a stable platform - the solid surface of the Moon, has several other unique advantages compared to their installations on board orbiting spacecraft. Such devices are inherent in mechanical vibrations and jitter of the image, regular correction of orbits, accurate tracking of the Earth and thermal stabilization during the operating mode, as well as a relatively short service life due to the limited energy capabilities are necessary. Consequently, it is thus impossible to obtain sufficiently long homogeneous series of high-precision measurements.

Therefore, the most reliable high-precision measurements of the total energy leaving the Earth into space can be carried out only from the visible surface of the Moon. As a result, to solve these vital climate problems, a new space project "Lunar Observatory" was developed to monitor the radiation balance of the Earth and climate (LO). The goal of the project is to create the first stationary astronomical observatory on the lunar surface for long-term research and to be able to predict the most reliable fluctuations in the Earth's global climate which depend on cyclical changes in the incoming, absorbed and consumed by its solar radiation energy on the outer layers of the atmosphere, as well as for the most reliable determination of physical patterns that control the Earth's climate [5,8,16-18]. This is a system of two identical special optical telescopes-robots, functioning sequentially replacing each other in automatic mode only at night during for more than 94% of the lunar day. This will create reliable methods for predicting global climate change.

Two identical special optical telescopes-robots with a main mirror diameter of 300 mm with a retractable visor – SOTR-300V, which make up LO, is a single system of telescopes operating sequentially in automatic mode according to the general scientific program of Earth observation, protected by four patents Russian Federation No. 155044 (2015), 164303 (2016), 2591263 (2016) and 2613048 (2017). Telescopes are installed along the equatorial zone at opposite limbs of the Moon, 80.9 ± 0.1° away from its visible center (figure 1). Such an optical laboratory on the Moon will make it possible to ensure long-term homogeneous series of measurements of variations in the complex of global climatic parameters, namely: the specific power of the reflected and scattered by planet TSI going into space in all directions from the entire atmosphere in the spectral range $\Delta\lambda$ = 0.2-4 µm and outgoing intrinsic thermal radiation in the ranges $\Delta\lambda$ = 4-50 µm and the main window of transparency of the atmosphere $\Delta\lambda$ = 8-13 µm in ideal conditions of lunar nighttime apparently for more than 94% of the time of the lunar day. Measurements are also provided in 10 narrow spectral bands of the range $\Delta\lambda$ = 0.3–3 µm for the global study of cloud and snow-ice cover, vegetation, concentrations of aerosols and ozone, as well as other physical and optical characteristics of the atmosphere and the Earth's surface around the globe. At the same time, monitoring the variations of the energy leaving for space in the spectral range of the window of atmospheric transparency $\Delta\lambda$ = 8–13 µm will allow us to record simultaneously and control the temporal changes in the transmission of thermal radiation of the surface to space by the atmosphere, as well as determine the relative contribution of variations in the concentration of greenhouse gases in the atmosphere to climate changes, regardless of the reasons for their changes. Such possibilities are determined by the fact that on the night side of the Moon its surface temperature decreases to 100 K, which provides a solution to the problem of thermoregulation of the optical system of telescopes and high-precision measurements. Complex measurements of the global climatic parameters of the Earth are carried out sequentially in automatic mode by a system of two identical optical telescopes-robots that are not illuminated by the Sun, replacing each other (figure 2).

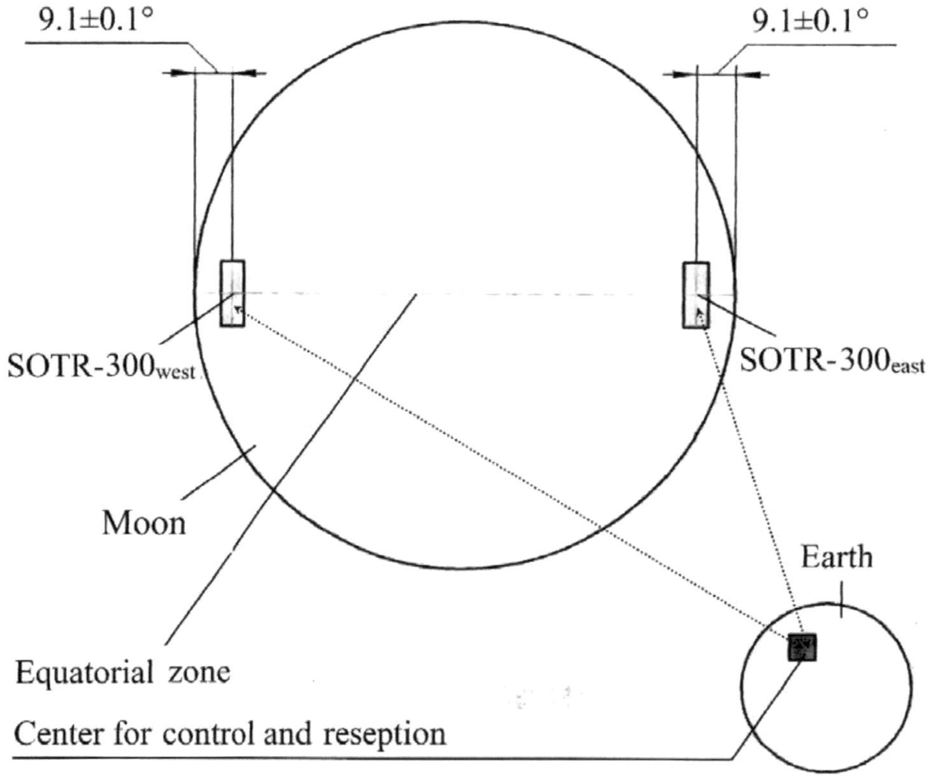

Figure 1. Installation diagram of a system of two optical telescopes-robots on the lunar surface.

SOTR-300V with a working field of view of 132 angular minutes in diameter will be installed in a special protective bowl to protect the telescope mirrors from charged particles of Moon dust. The insignificant remoteness of the Moon provides sufficient energy for radiation emanating from the Earth, allowing the use in a focal plane SOTR-300V one a high-resolution (~2048 pixels x 2048 pixels) micro bolometric matrix with a dynamic range of four orders of magnitude, that records the energy of the radiation flux, the sensitive elements of which are coated with gold black. This matrix uses a system of different filters to measure the distribution of the energy, flux of radiation emanating from the Earth sequentially in the entire working spectral range of wavelengths $\Delta\lambda$ = 0.2–50 μm and in its individual broad and narrow bands. The monthly calibration of the SOTR-300V telescopes with each other is carried out according to the results of their simultaneous parallel measurements of the radiation fluxes emanating from the Earth for 30 hours during each new Moon (figure 2). Telescopes are fully automated and do not require the participation of astronauts during their installation and maintenance in the future. Earth tracking by the telescope will be carried out not continuously, but stepwise with an interval of 6 – 16 hours. Less than 6%, of breaks in the observations of the Earth during the visibility of its night and near-night sides has a negligible fraction of energy, and it will be reliably determined by interpolation and will have virtually no effect on the accuracy of the subsequent determination of the Bond albedo and the Earth's own thermal radiation (figure 3).

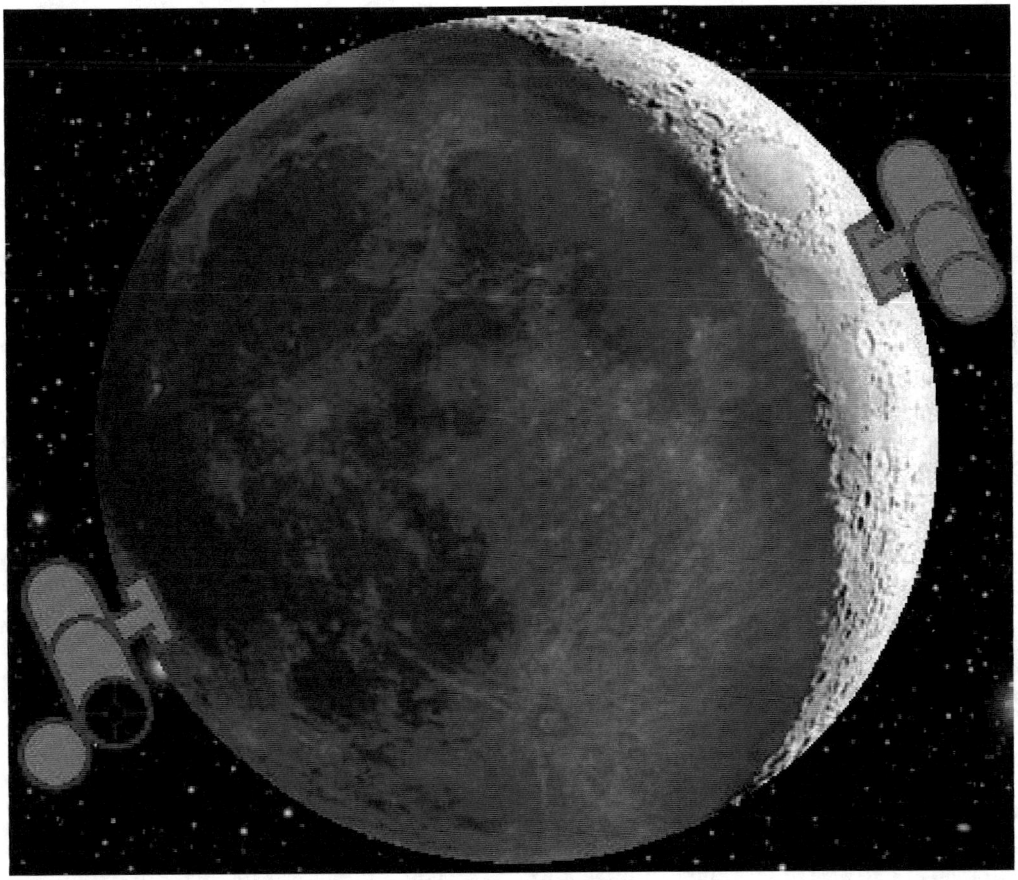

Figure 2. The functioning of the Lunar Observatory, consisting of a system of two identical optical telescopes-robots SOTR-300V.

As a result, for the period of the synodic lunar month, a continuous dependence on the angular coordinate of the measured radiation power over the entire surface of the Earth in the entire working spectrum of the radiation will be obtained. This will allow experimentally reliably determining the phase integral connecting the geometric albedo and the Bond albedo of the planet. The surface power density of the radiation coming from the Earth to a single area of the surface of the Moon [19]

$$I_{\mathrm{E}} = \frac{S_{\odot}}{R_{\mathrm{ES}}^2} A_{\mathrm{gE}} f_E(\alpha) \frac{R_{\mathrm{E}}^2}{R_{\mathrm{EM}}^2}, \tag{5}$$

where $f_E(\alpha)$is the phase function at an angle α, RE is the radius of the Earth, RES, REM are the distances between the Earth and the Sun, Earth and the Moon in astronomical units (AU), respectively. The result of measurements over the entire range of phase angles will be the value of the phase function of the Earth multiplied by the geometric albedo

$$A_{gE} f_E(\alpha) = I_E \frac{R_{ES}^2}{S_\odot} \frac{R_{EM}^2}{R_E^2}. \tag{6}$$

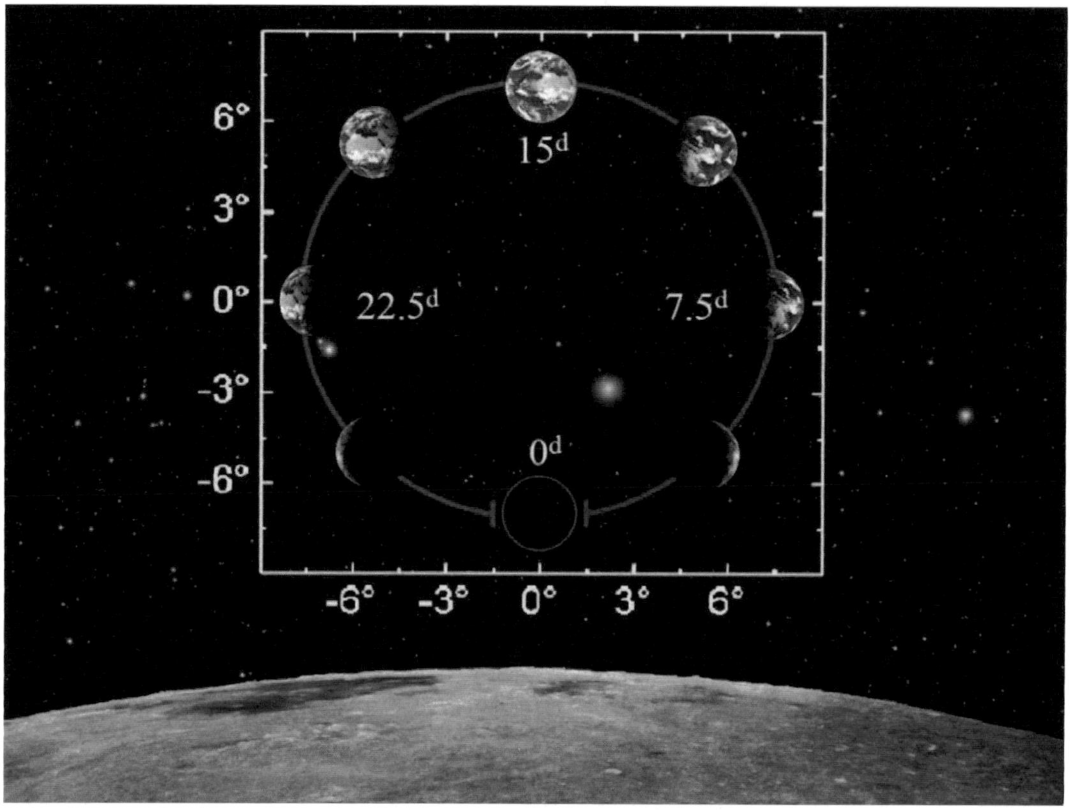

Figure 3. Successive viewing of almost all aspects of the terrestrial surface by a system of two optical robotic telescopes in the ranges of phase angles α from –170.9° to 0° and from 0° to +168.8° during the Moon orbiting the Earth for the lunar synodic month.

In equations (5) and (6), only I_E and $S_\odot$ are sources of error. Distance values can be measured with accuracy well above 0.07%. The values of I_E, R_{ES}, R_{EM}, $S_\odot$ are not directly dependent on the phase angle, however, they change in time, and for each value of the phase angle α they must be measured independently. The measured values of the geometric albedo product and the phase function will allow us to calculate the Earth's Bond albedo with an error of the order of 0.1%. The mass of the complex with one telescope with lightweight mirrors without solar panels and batteries is ~100 kg, and the dimensions are less than 1500x1500x1500 mm^3 (figure 4). Reception and transmission of information from each telescope is carried out in the lunar daytime. The scientific equipment of the SOTR-300V does not require maintenance and tuning, the required electrical power is about 100 Watts. Additionally, LO will simultaneously conduct a continuous search for exoplanets suitable for life only throughout the lunar day using method the transit in the spectral range Δλ = 0.2 – 4 μm. LO telescopes will follow Sun-like and, especially, brighter stars of different spectral classes, since about their planets little is known. Telescopes will observe these planets "directly", tracking how they pass through the disk of the star.

IOP Conf. Series: Earth and Environmental Science 1040 (2022) 012001 doi:10.1088/1755-1315/1040/1/012001

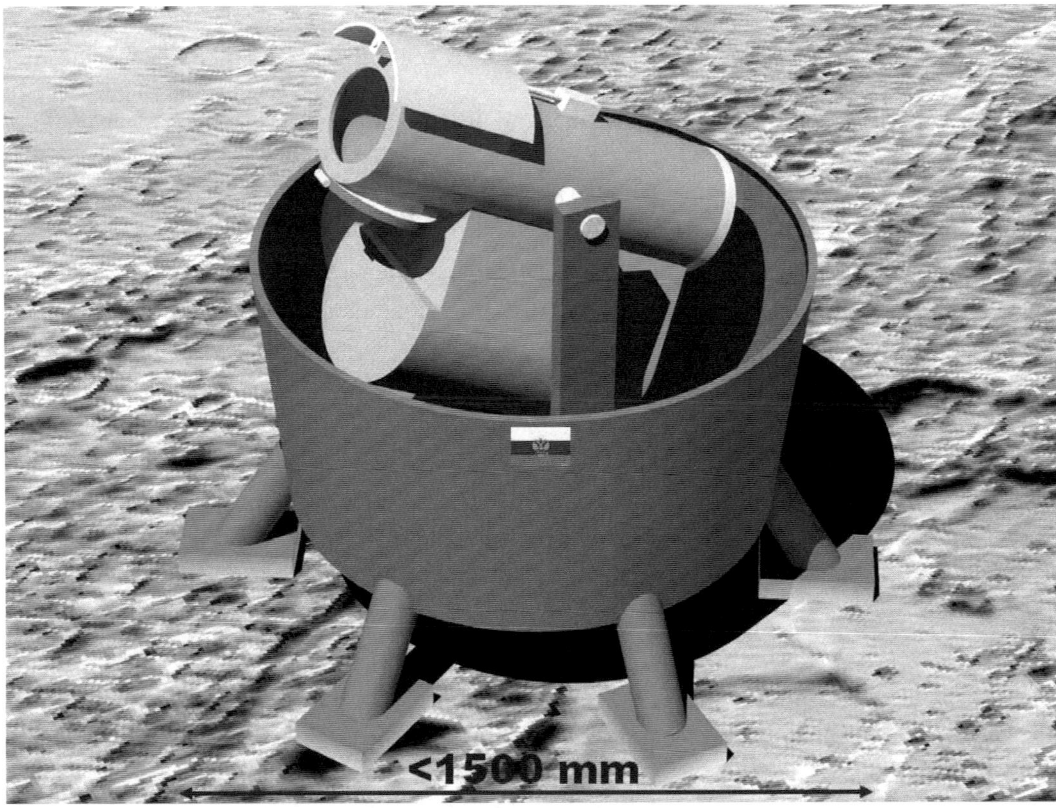

Figure 4. General view of the SOTR-300V complex on the lunar surface (solar panels not indicated).

4. Conclusions

Monitoring and research of long-term variations in global climatic parameters, i.e. high-precision values of the TSI, Bond albedo and the power of the Earth's own thermal radiation, and on this basis the calculation of the planet's energy balance is a task of fundamental importance. A system of two LO telescopes-robots will sequentially survey the entire Earth's surface at almost all phase angles continuously for more than 94% of the time of a lunar day only at lunar night time. The average annual values of the Bond albedo and the Earth's own thermal radiation as a planet, determined by LO, and on their basis (the exact TSI value will be obtained from measurements of other space experiments), the calculation of the average annual energy balance of the planet will be almost an order of magnitude more accurate than that determined by any orbital spacecraft.

LO – for about two 11-year solar cycles will provide for the first time the most important homogeneous missing, extremely important, most reliable global climatic data on the change in the average annual energy balance of the Earth and its excess or deficit of energy accumulated annually in the World Ocean. This will make it possible to determine their dependence on cyclic variations TSI. The resulting dependence will be a reliable indicator for reconstructing changes in the balance of the average annual energy budget of the Earth for the entire period of high-precision TSI measurements since 1978, which will determine the physical mechanisms of formation, causes and patterns of climate changes in our planet and develop the most reliable methods for predicting it. Only when placing telescopes on the visible side of the Moon, the system of two telescopes-robots for the first time will provide the creation of a new fundamental database of the average annual energy balance of the planet, and will also reliably determine the planet's energy condition, the total amount of annual excess of energy received by the Ocean or its deficit. For the first time, LO will provide research on the most important missing, extremely necessary precision global climatic data that cannot be obtained by any other space-based methods for their direct measurements. This is a significant scientific breakthrough and applied area for researching the laws of climate change and the physical mechanisms and reasons that cause these changes.

The first LO to be created is aimed at implementing the Climate Doctrine of the Russian Federation dated December 17, 2009 No. 861-rp, in particular: "... to *ensuring the climate studies of the Russian Federation in accordance with the world level, recognizing the results of Russian studies by the international scientific community, and using them as arguments in the interstate climate policy dialogue* ... ". LO is in line with the strategic plans for the long-term development of the Russian Federation and the interests of its national security. The establishment of the physical causes of global climate change and the timely creation of the most reliable methods for predicting its future profound changes will prevent and mitigate the consequences of global environmental and climate deterioration, social, economic and climate-related changes in the geopolitical situation, which are dangerous to national interests.

Acknowledgments

This study was partly supported by the Project No. KP19-270 of the Praesidium of the Russian Academy of Sciences for Fundamental Scientific Research ("Issues of the Origin and Evolution of the Universe Using Methods of Ground-Based Observations and Space Research").

References

[1] Abdussamatov H I 2015 *Thermal Sci.* **19** S279-S288.
[2] Abdussamatov H I 2012 *Kinematics and Phys. Celest. Bodies* **28** 62-68.
[3] Abdussamatov H I 2016 *Evidence-Based Climate Science* Easterbrook D J (ed) Oxford: Elsevier 307-28.
[4] Exoplanets 2010 Ed. S. Seager. Univ. Arizona Press 425–526.
[5] Abdussamatov H I 2017 *Lunnaya observatoriya dlya issledovanii klimata Zemli v epokhu glubokogo pokholodaniya* (Lunar Observatory for Earth climate studies in the deep Ice Age), St. Petersburg: Nauka.
[6] Trenberth K E, Fasullo J T and Kiehl J 2009 *Bull. Am. Meteorol. Soc.* **90** 311–23.
[7] Stephens G L, Li J, Wild M, Clayson C A, Loeb N, Kato S, L'Ecuyer T, Jr P W S, Lebsock M and Andrews T *Nat. Geosci.* **5**. 691.
[8] Abdussamatov H I 2018 *Izv., Atmos. Ocean. Phys.* **54** 1341-52.
[9] Wielicki B A, Barkstrom B R, Harrison E F, Lee III R B, Smith G L and Cooper J E 1996 *Bull. Amer. Meteorol. Soc.* **77** 853–68.
[10] Thomas S, Priestley K J, Hess P C, Wilson R S, Avery M A, Walikainen D R et al. 2009 Eds. James J. Butler, Xiaoxiong Xiong, Xingfa Gu. *Proc. of SPIE V.* **7452** 74520I.
[11] Loeb N G, Parol F, Buriez J C, Vanbauce C et al. 2009 *J. Clim.* **22** 748–66.
[12] Leckey J 2015 **XL-7/W3** 36th Int. Symp. Rem. Sens. Env. Berlin, Germany. 213–17.
[13] CLARREO 2017 https://eospso.gsfc.nasa.gov/missions/climate-absolute-radiance-and-refractivity-observatory
[14] Harries J.E., Russell J.E., Hanafin J.A., Brindley H., Futyan J. et al. 2005 *Amer. Meteorol. Soc.* **86** 945–60.
[15] DSCOVR 2017 https://www.nasa.gov/feature/goddard/from-a-million-miles-awaynasa-camera-shows-moon-crossing-face-of-earth
[16] Abdussamatov H I 2016 *Issled. Zemli iz Kosmosa* no. **5** 79–88.
[17] Abdussamatov H I 2019 http://www.gaoran.ru/english/lunar%20observatory.pdf
[18] Abdussamatov H I 2018 *J. Remote Sens. GIS* **7** 25.
[19] Qiu J, Goode P R, Palle E, Yurchyshyn V et al. 2003 *J. Geophys. Res.* **108** 4709.
[20] Abdussamatov H I 2019 Abstracts of All-Russian conference "Climate change …" Moscow Russia – M.: Fizmatkniga 2019 97 – ISBN 978-5-89155-324-8.

IOP Conf. Series: Earth and Environmental Science 1040 (2022) 012002 doi:10.1088/1755-1315/1040/1/012002

The studies of the Svalbard glacial surfaces albedo by an unmanned aerial vehicle.

B V Ivanov[1,2], D M Zhuravskiy[1], U V Prokhorova[1], A M Bezgreshnov[1], A V Terekhov[3,1], M V Kurapov[1], A S Paramzin[1] and V S Kashkova[1]

1 Arctic and Antarctic Research Institute, Saint-Petersburg, Russia;
2 Saint-Petersburg State University, Russia;
3 State Hydrological Institute, Saint-Petersburg, Russia.
Corresponded author: b_ivanov@aari.ru

Abstract. Experiments related to the use of unmanned aerial vehicle (UAV) for assessing the albedo of Svalbard glaciers is described. Study area - Esmark Glacier (Isfjord Bay) and Aldegonda Glacier (Greenfjord Bay). The main purpose of the experiments is to estimate the surface albedo in the zone of the edge cracks of the outlet glacier (Esmark), where standard ground-based observations are impossible due to safety conditions, as well as to obtain spatial albedo estimates (Aldegonda) when satellite data cannot be used (overcast). The UAV (DJI Phantom 4 Pro) was retrofitted with a sensor that measures reflected solar radiation. The data on the incoming solar radiation at the surface level, measured by a similar sensor, were used to calculate the albedo. The albedo measurements were carried out along several profiles across the Aldegonda Glacier and along one profile above the Esmark Glacier, which was laid from a flat plateau (ice dome) through a zone of cracks to the open water surface of the fiord. For the first time, estimates of the surface albedo of the outlet glacier in the zone of edge cracks were obtained. Ground-based verification observations carried out on the Aldegonda glacier confirmed the results obtained by the UAV.

1. Introduction

Thermal balance observations carried out in natural conditions allow to evaluate the processes of interaction in the "*underlying surface - surface layer of the atmosphere*" system. Albedo is one of the most important elements of the surface heat balance, especially in the Arctic regions, where mechanisms of positive feedbacks cause "dramatic" changes in the conditions of the snow-ice surfaces (e.g. sea ice, glaciers). As a general rule, albedo values obtained at a point or a limited area of the surface cannot be extrapolated to the entire area of interest. That is why, special space ("*ground true*") surveys are carried out. Data on the albedo of snow-glaciers covers are also necessary for correct calculations of the melted layer of snow or ice during the ablation period. Together with the data of direct measurements (e.g. glaciological monitoring), these data make it possible to verify and improve the existing thermodynamic models of glaciers. What is even more important, it allows to evaluate the processes in the marginal zone of glaciers (zone of cracks), where any "*ground true*" observations are impossible. These zones absorb and reflect the incoming solar radiation with intensity different from the flat areas of a glacier. However, these processes are not taken into account in the classical thermodynamic models of glaciers sufficiently.

The UAV practice demonstrates their high efficiency in the research of hard-to-reach snow-ice surfaces areas, where standard "*ground true*" surveys are impossible for safety reasons. Such studies make it possible to obtain objective estimates of the spatial distribution of the albedo in the hard-to-reach areas and perform detailed mapping of similar areas mentioned above [1, 2, 3]. Previously, this could be done (and even then, only partially) during the long and laborious walking routes - profile or space surveys [4]. The studies of the albedo of the antarctic land-fast ice and the zone of cracks in the margin zone of the glacier dome in the area of the Russian "Progress" station were carried out within the framework of the 63rd Russian Antarctic Expedition [5]. Such study was done for the first time in domestic practice. The research was based on the use of an original portable measuring device (an analog-to-digital converter paired with a PaspberryPi microprocessor device) mounted on a UAV (DJI Mavic Pro). Finally, within the framework of the international expedition TransArktica-2019 on the RV "Akademik Treshnikov" (Russia), the high level spatial resolution estimates of the albedo and surface temperatures of arctic hummocks were obtained using an UAV, for the first time. Also, there were evaluated vertical turbulent fluxes of sensible heat from the windward and the leeward sides of the hummock ridge [6]. The main goal of this article is to demonstrate the original results related to obtaining albedo estimates for hard-to-reach areas of Svalbard snow-glacial surfaces.

2. Research locations

The main objects of our research were the glaciers of the Svalbard, located near the Russian Scientific Center in the settlement of Barentsburg. The first of them is the Aldegonda mountain-valley glacier, located on the western shore of the Greenfjord Bay, West Spitsbergen Island (see Figure 1a). The surface of the glacier is rather uniform, but there are several water streams extending from its top to the foot, as well as cracks located in its upper part. The second object, the Esmark glacier, is also a mountain-valley type, but its marginal part goes into the eponymous bay, meaning the glacier is an outlet (see Figure 1b). The marginal zone of the glacier is covered with numerous deep cracks and is absolutely inaccessible for standard "*ground true*" observations.

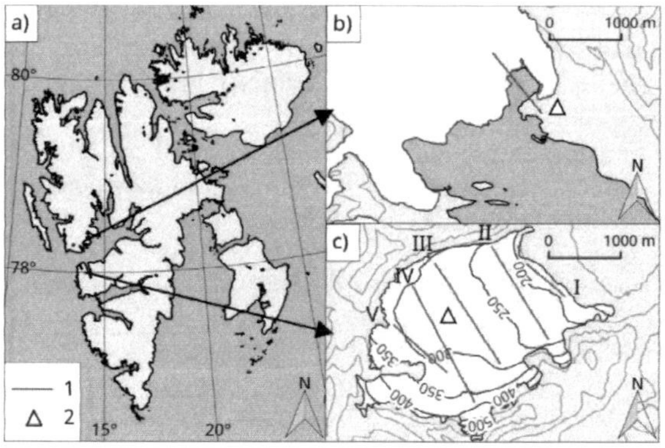

Figure 1 Location of the Esmark (b) and Aldegonda (c) glaciers on Svalbard (a)
1 - routes (profiles) of the UAV albedo surveys; 2 - position of ground observation points.

3. The methodology

An original portable device based on a PaspberryPi microprocessor (see Figure 2), a PiCamV2.0 camera and an LI-190SA photometer was designed and manufactured at the AARI for measurements using a UAV. The choice of these elements was associated with the flexibility of the Pi microprocessor architecture, the coincidence of the spectral coverage of the camera and the sensor (400-700 nm), as well as the typicality of the camera, which is a fairly common example of a good quality photographic recording device.

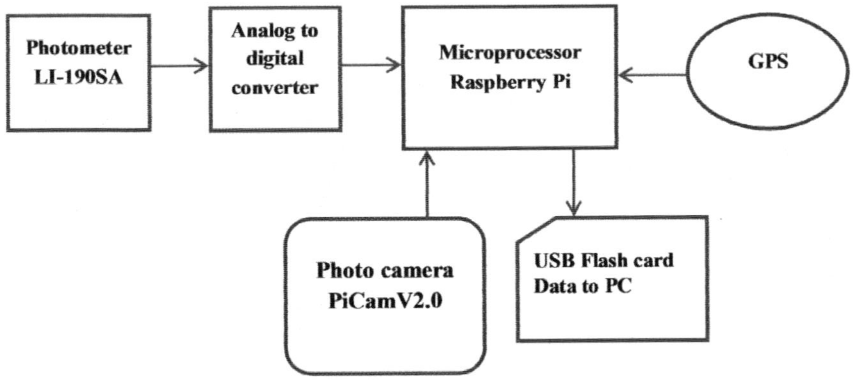

Figure 2 - Schematic diagram of the measuring unit based on the RaspberryPi microprocessor device

The device was made in a compact lightweight body and was intended for use on UAVs (model DJI Phantom 4 Pro). Autonomy was provided by a built-in rechargeable battery up to several hours of continuous operation. The built-in GPS / GLONASS receiver provided data on the spatial position of the UAV. Piloting was carried out in manual control mode, so that the UAV was constantly in the operator's line of sight. Data logging was performed in accordance with a specified time interval from 2 to 10 seconds. The device performed several actions simultaneously during working flights:
- registration of photometer readings (μmol/m²/sec);
- photographing the underlying surface with saving the image in the internal memory of the device;
- recording of parametric information (time, coordinates, image and exposure data).

Thus, "at the output" we received synchronous data on the reflective and exposure characteristics of the underlying surface. Photographic materials made it possible to exclude unrepresentative materials and provide a qualitative assessment of the morphometric characteristics of the filmed surface. The collected information was transmitted to a computer using a Wi-Fi connection. The analysis of the data obtained and the calculation of the surface albedo were carried out using licensed software.

The ground observation point with a special heat balance mast (HBM) was organized in the central part of the Aldegonda glacier to verify the data obtained using the UAV (model DJI Phantom 4 Pro). The HBM was equipped with radiation sensors to assess both incoming and reflected solar radiation. This was necessary in order to obtain representative estimates of the albedo at the surface level and compare it with the data obtained by the UAV. A similar ground-based measuring complex (HBM) was installed on the moraine in the vicinity of the Esmark glacier (see Figure. 1). We used LI-192SA low-response pyranometers (USA, *LICOR*), which record solar radiation in the range of 400-700 nm. A similar sensor (LI-190SA) was used in the UAV measuring device. Both sensors (LI-190SA and LI-192SA) recorded incoming and reflected solar radiation in the same spectral range and with the same resolution as the photo sensor of the measuring device camera. During the flights over glaciers, the UAV sometimes deviates from the horizontal position (roll, pitch, and yaw) due to changing wind loads (wind speed and direction). Unlike the built-in automatically stabilized UAV camera, the measuring device (LI-190SA sensor) is rigidly attached to the UAV body and this must be taken into account when taking measurements and data analysis. During each flight, a special file was created with complete flight telemetry - speed, altitude, tilt angles, data from onboard sensors, etc. When data processing we selected only the data collected during the horizontal position of the UAV and attachments. The error in calculating albedo values based on measurements of the incoming (HBM) and reflected solar radiation (UAV) did not exceed 5% in comparison with the data obtained at the ground observation point (HBM). On the Aldegonda glacier, the flight routes were planned along profiles oriented across the glacier from northwest to southeast, where the surface elevations were approximately the same above sea level (see Figure 1). Observations were made at a height of 20 m above the glacier surface. The flights were carried out at a speed of 5 m/s and a constant measurement interval of 5 seconds. Thus, the resolution along the flight profile (discreteness of the calculated albedo values) was 25 m. The albedo values calculated using the UAV reflected radiation data were interpolated to the nodes of a regular grid using the "*natural neighbor*" algorithm included in the standard set the "*gridding*" package of the Golden Software Surfer [7]. The algorithm uses a subset of the input data closest to the measurement point and applies a weighting method to them: each point in the subset has its own weight depending on the proximity of its location to the desired one. The key feature of this algorithm is that the interpolated values will always be within the range of the used primary data set which rules out the formation of false peaks or dips in the interpolation results. Thus, we managed to obtain the spatial distribution of albedo for a limited period of time under constant illumination conditions (sun height, cloudiness, horizontal visibility range, and other atmospheric conditions). The main area of the research at the Esmark glacier is the edge zone (zone of cracks) of the outlet glacier (see photo in Figure 3). The flights were carried out only along the profile from the dome area (a relatively flat area of the glacier surface) to the open water surface of the fjord.

Figure 3 - Area of work on the Esmark glacier.

Taking into account the changing height of the glacier surface above sea level (from 30 m in the outlet part to 150 m in the plateau area), the UAV flights were performed at different altitude depending on the glacier section, but not less than 50 m from its surface. In both cases (objects of research), the interval of data acquisition from the HBM was synchronized with the interval of data logging and image by the UAV.

4. Results

The marginal zone is a characteristic feature of the Svalbard outlet glaciers. There are discontinuities of the glacier cover, the formations of deep faults oriented mostly across the direction of the glacier movement along the valley, (i.e. parallel to its front) are observed in such areas. Deep cracks should determine a different nature of the reflect solar radiation from the glacier surface, which should affect the albedo values of such surface areas, from our point of view. For example, we recorded a decrease in the surface albedo in the zone of deep cracks formed by the snow-ice cover failures over subsurface water streams in the area of the antarctic "Progress" station by the UAV [5]. We expected to obtain similar results at the Esmark glacier. Figure 4 shows the profile of the albedo values, calculated using the measured reflected solar radiation data (UAV) and the incoming solar radiation measured at the ground observation point near the glacier (HBM).

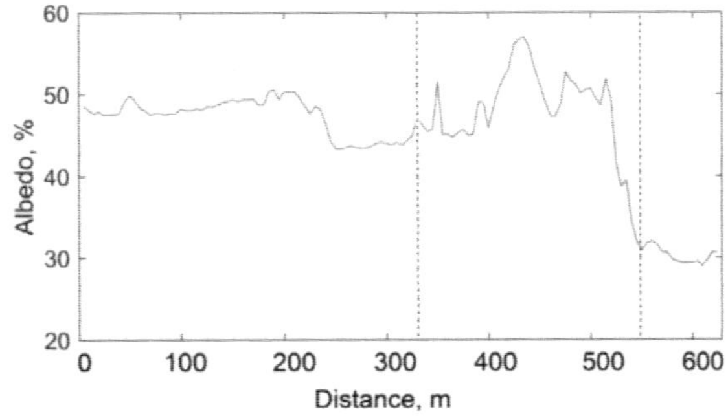

Figure 4 Albedo values along the UAV flight profile over the marginal zone.

The actual zone of cracks is limited by dotted lines (see photo in Figure 3). As follows from the figure, the albedo of the sea surface (adjacent to the ice barrier of the fjord water) is approximately 30%, which significantly exceeds the average albedo of sea waters. This can be explained by the large amount of fine and grated ice on the surface of the fjord waters, which is formed as a result of the constant spalling (collapse) of fragments of the ice barrier (see photo in Figure 5a). The albedo values are characterized by significant variability directly in the edge zone of the glacier (zone of cracks). The average albedo values in the area immediately adjacent to the flat surface of the dome (see photo in Figure 5b) do not exceed 50%. Directly in the zone of cracks (see photo in Figure 5c), the albedo varies from 30 to 58%. This circumstance may be due to the increasing slope of the glacier surface directly in the edge zone, in comparison with the relatively horizontal sections adjacent to the dome. Taking into account the height of the sun, the angles of incidence of sunlight (direct solar radiation) can reach 90 degrees on such areas of the glacier surface. In such cases, the reflection is minimal and the albedo decreases. On the other hand, the incidence of sunlight on the vertical walls of deep cracks can lead to multiple re-reflection of the incident ray, including into the depth of the crack. This should lead to a decrease in the reflected radiation and be the reason for a decrease in the albedo values. Also, the marginal zone is characterized by the alternation of sunlit and shaded areas of the glacier surface. This can also explain the decrease in the surface albedo, all other things being equal.

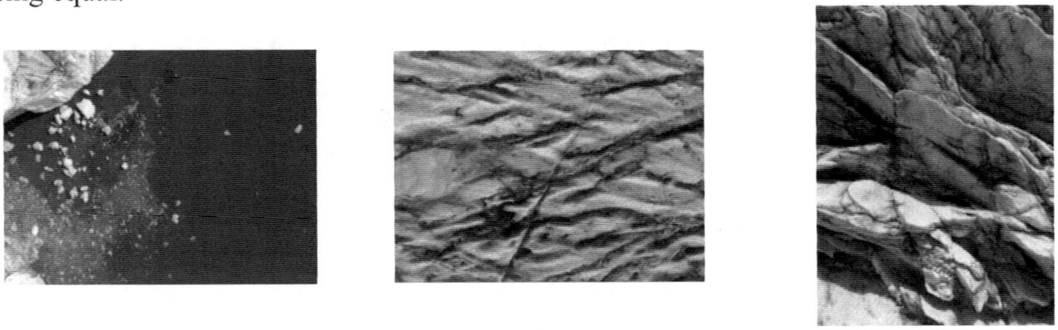

(a) (b) (c)

Figure 5 Typical surface types of the Esmark glacier:
(a) – see water surface; (b) – flat area (dome); (c) – crack zone.

The use of the UAV made it possible to estimate the albedo of a significant part of the surface of the Aldegonda glacier, spending no more than one hour on flights along the profiles (see Figure 1). In the framework of standard ground-based observations, this requires at least a day, which is associated with limited routes for safe movement along the glacier [4]. The spatial distribution of albedo was obtained for various weather conditions and the state of the glacier surface. One UAV flight was carried out in the complete absence of snow on the surface (the snow melted), the second - after a short snowfall. In both cases, cloudy conditions did not allow using the capabilities of high spatial resolution satellites in the visible range (Landsat or Sentinel). The results are shown in figure 6.

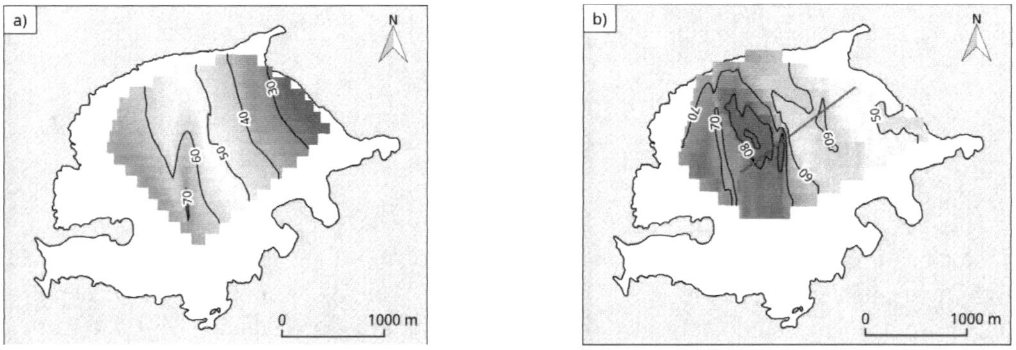

Figure 6 Distribution of the albedo before (a) and after snowfall (b), red line – the route of ground observations.

We compared ground observations made along the profile ("*ground true*" measurements) with the data obtained by UAV for the data verification. The results are shown in Figure 7.

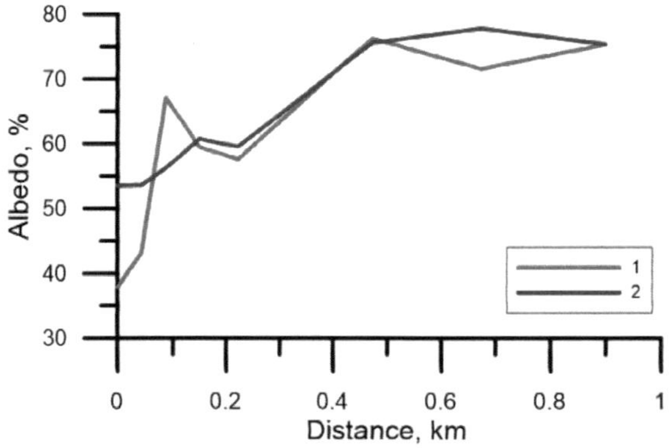

Figure 7 The albedo of the glacier surface obtained by "*ground true*" measurements (1) and UAV (2).

As follows from the figure, the profile obtained by UAV is smoother than the profile drawn from the ground-based observations. This phenomenon can be explained by the difference of measuring technique. When measuring the reflected radiation from a certain altitude (UAV), the signal is formed by a larger area comparing with the one during the ground measurements. This leads to the well-known "*smoothing*" of the morphometric characteristics and color features of the glacier surface during this period of the year. The surface itself can have numerous features such as firn, fresh snow, melt water on the surface, pollution by rock particles, etc. Another possible reason is the use of interpolated values to build a profile by UAV. The main trend is an increase in albedo with height. It is clearly visible in both approaches, while the absolute values coincide, especially in the central part of the profile.

5. Conclusions

Field experiments carried out over the Svalbard glaciers have shown the high efficiency of using UAV for assessing the albedo of the underlying surface. For the first time, it was possible to estimate the albedo in the marginal zone of the cracks of the outlet type glacier (Esmark Glacier, Isfjord Bay). Such measurements cannot be performed by standard methods and means. The estimates obtained will make it possible to more correctly calculate the heat balance of the glacier surface in such a dynamic zone, which will undoubtedly contribute to an objective assessment of the mass balance of the glacier as a whole. Areal surveys of the Aldegonda Glacier (Greenfjord Bay) made it possible to obtain estimates of the albedo of its surface for various snow and surface conditions. The use of UAVs makes it possible to estimate short-term changes in the albedo of the underlying surface, which is impossible when using satellite images due to the discreteness of flights and cloud conditions.

6. Acknowledgments

This work was carried out in accordance with the Roshydromet project "Monitoring of the state and pollution of the environment, including the cryosphere, in the Arctic basin and in the areas of the research base "Ice Base Cape Baranova", Tiksi Hydrometeorological Observatory and the Russian Science Center at Svalbard Archipelago" and the Polar Geophysical Institute (Apatity) project "Carrying out stationary observations of magnetic and optical phenomena on the Spitsbergen archipelago, study of geophysical processes in the high-latitude atmosphere of the Earth" . We are extremely grateful to the Norwegian Meteorological Institute because these studies were supported by Russian–Norwegian "Agreement in area of analysis of climate and sea ice data in the North part of Barents Sea" (2019-2020), financed by the Ministry of Climate and Environment in Norway (RUS-19/0001).

7. References

[1] Hann R, Hodson A, Jonassen M Glacier mapping and wind estimation with UAVs on Svalbard https://forskningsradet.pameldingssystem.no/svalbard-science-conference-2019)

[2] Thomson E, Spiegal M, Malhi Y, Macias-Fauria M Mapping plant functional groups, traits and degradation in the High Arctic using multispectral UAV imagery https://forskningsradet.pameldingssystem.no/svalbard-science-conference-2019)

[3] Burkhart J F, Pedersen C, Bogren A, Gerland S, Storvald R, Kohler J, Kokhanovsky A, Ström J Variability of Albedo Using an Unmanned Aerial Vehicle (VAUUAV). American Geophysical Union. Fall Meeting 2009 2009AGUFM.C31E0469B

[4] Ivanov B V, Svyashchennikov P N 2015 Albedo of the snow-glacier surface of Svalbard *Atmosphere and Oceanic Physics, Issue of RAS* **51** No 9 pp. 943-8.

[5] Zhuravskiy D M, Prokhorova U V, Ivanov B V, Janjura A S, Kuprikov N M, Kurapov M V 2020 Field Tests of the Procedure for a Photogrammetric Estimation of Snow-Glacial Surface Albedo *Atmospheric and Oceanic Physics, Izvestiya* **56** No 9, pp. 927–35

[6] Ivanov B V, Urazgildeeva A V, Paramzin A S, Surovetkin S S, Drabenko D V 2020 Features of turbulent heat transfer near hummocky areas of sea ice *Problems of arctic and Antarctic* **66**(3) pp 364–80

[7] Sibson R 1981 A brief description of natural neighbor interpolation. In V. Barnett, editor, *Interpreting Multivariate Data* pp. 21–36. John Wiley, Chichester

IOP Conf. Series: Earth and Environmental Science 1040 (2022) 012003 doi:10.1088/1755-1315/1040/1/012003

Ionospheric disturbance statistics according to the Swarm satellite mission in 2014

L A Voloboev[1*] and V I Zakharov[1,2]

[1] Faculty of Physics, M.V.Lomonosov Moscow State University, Leninskie Gory, 119991, Moscow, Russia
[2] Schmidt Institute of Physics of the Earth of the RAS, B. Gruzinskaya str., 10, build. 1, 123242, Moscow, Russia

[*] l.voloboev@mail.ru

Abstract. In the present work, based on the database of magnetic field indices Kp, AE, and Dst, we consider the relationship between geomagnetic activity and ionospheric scintillation indices S4 and RODI based on in-situ satellite measurements. 37 events with quiet and disturbed geomagnetic conditions during 2014 were selected and analyzed. The correlation coefficients of the scintillation indices with high-latitude AE and low-latitude Dst are obtained. Thus, in periods of disturbed geomagnetic conditions, the RODI10s and S4 indices are, in principle, equally related to the high-latitude AE index (with a small, up to 4% in the number of realizations, the advantage of the RODI index) In periods with quiet geomagnetic conditions, the situation changes - in 17% of cases the S4 index is better reflects a relationship with mid-latitude Dst than the RODI10s index. The indicated relationship means the presence of a linear correlation of more than 0.3 between the processes.

1. Introduction

The paper deals with the description of the intensity of fluctuations of the electron density in the Earth's ionosphere under various geo-heliophysical conditions according to the data of satellite in-situ measurements in 2014 using the S4 and RODI scintillation indices.

1.1. About mission

The experimental data in the study is an array of high-frequency observations from the Swarm ESA satellite mission, designed to study the Earth's magnetic field. The mission was launched in 2013 and consists of three satellites in different polar orbits with an altitude of 460 km to 530 km. Each spacecraft, in addition to the main magnetometric equipment, is equipped with Langmuir probes, which measure electron density with a sampling frequency of 2 Hz.

1.2. Scintillation indices

The scintillation indices are related to the calculated average power of the fluctuation noise - an index similar to the amplitude index of scintillations or radio scintillations S4 at a certain time interval. The index is calculated in equation (1), where Ne is the measured in-situ electron density. Filtration windows are selected that correspond to the required spatial resolution (so a window t of 20 samples corresponds to 10 s or 74 km).

$$S_{4_t} = \left(\frac{\langle N_e^4 \rangle_t - \langle N_e^2 \rangle_t^2}{\langle N_e^2 \rangle_t^2} \right)^{1/2} \quad (1)$$

The RODI index (Rate of Change of Density Index) [1] shows the rate of local deviation of electron density variations from the current average value, determined at a certain time interval - the filtering window. In both cases, the window is selected so as to provide the required spatial resolution. For example, a 10 sec window corresponds to a spatial latitude resolution of about 0.8 deg. The RODI10s is calculated, which is the standard deviation with a window 10s of the time derivative of the electron density ROD, calculated as in equation (2), where $\delta t = 0.5$ seconds.

$$\mathrm{ROD}\,(t) = \frac{N_e\,(t + \delta t) - N_e\,(t)}{\delta t} \quad (2)$$

2. Data processing

Based on the database of indices Kp, AE and Dst of the magnetic field, the influence of geomagnetic activity on the fluctuations of the electron density in 2014 is considered. [2]. For the study, we selected days with disturbed (Kp > 4) and quiet (Kp = 1-2) geomagnetic conditions. In each of the selected cases, the indices S4 and RODI10s were calculated. The analysis results are categorized by high, mid, and low latitudes of current Swarm flybys.

3. Results and conclusions

The correlation coefficients of the S4 and RODI10s indices with the high-latitude index AE and low-latitude Dst were calculated for 37 selected cases during 2014, based on the data from the Swarm A, B, and C satellites. Examples of the obtained and calculated data in various geophysical conditions are presented in figures 1 and 2.

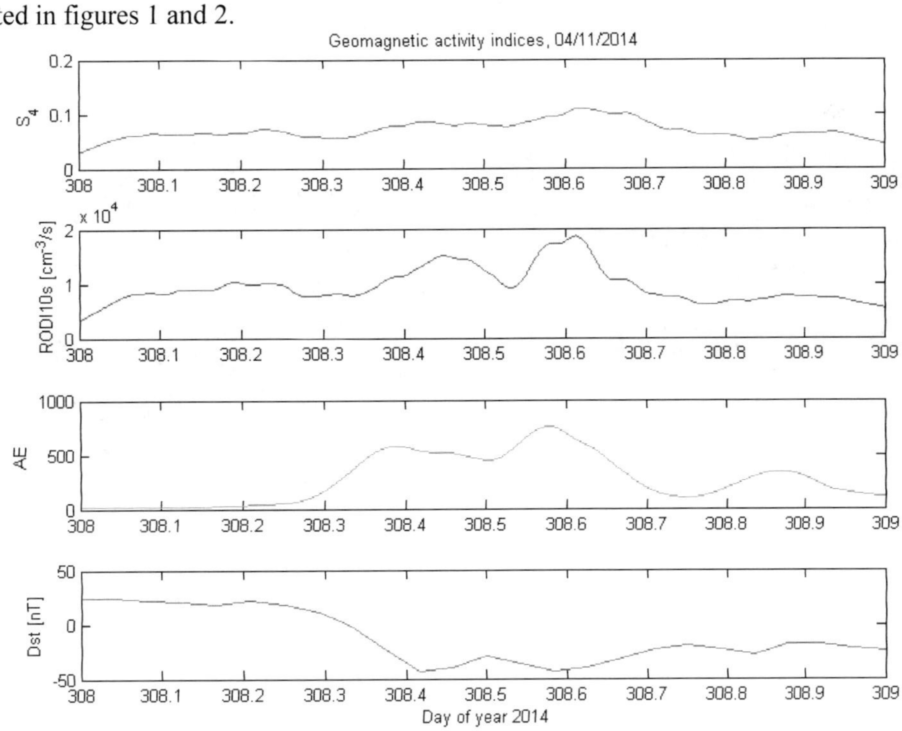

Figure 1. Indices S4, RODI10s, AE, Dst in terms of disturbed geomagnetic conditions (November 4, 2014, Kp = 4).

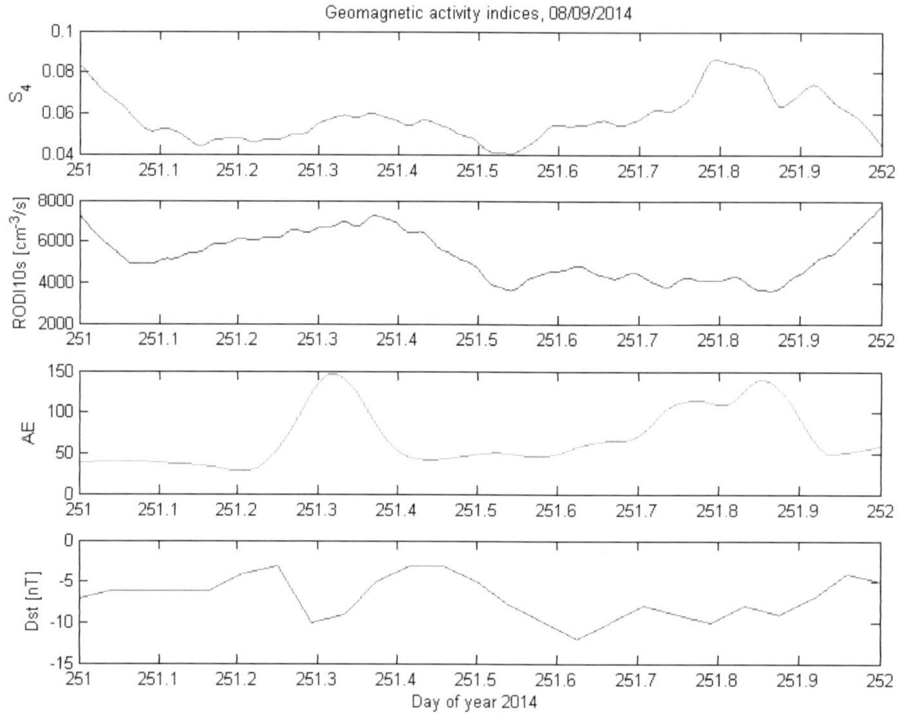

Figure 2. Indices S4, RODI10s, AE, Dst under quiet geomagnetic conditions (September 8, 2014, Kp = 2).

Polar maps of geophysical indices were calculated, shown in figure 3, based on weekly measurements from the Swarm C satellite.

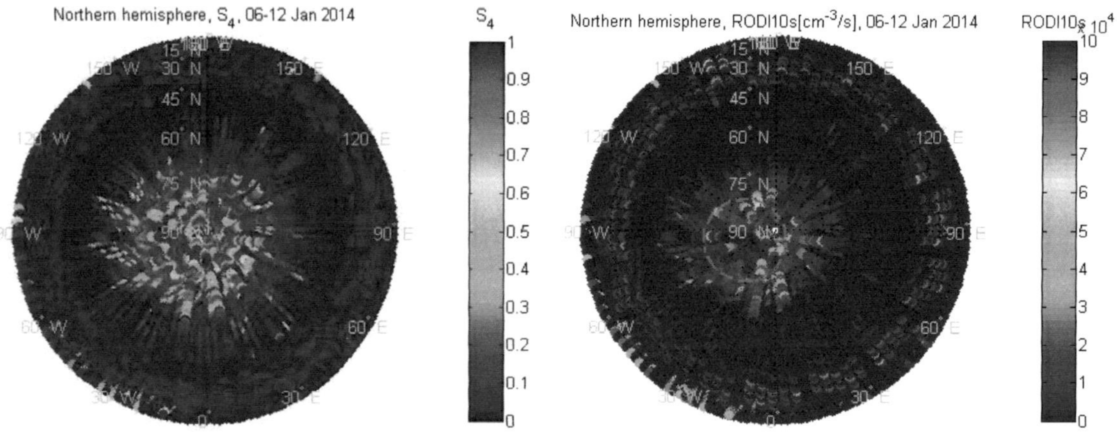

Figure 3. Polar maps of S_4, RODI10s indices according to weekly measurements (from January 6 to 12, 2014) from Swarm C.

So, for 2014, in periods of disturbed geomagnetic conditions, the RODI10s and S4 indices are, in principle, equally related to the high-latitude AE index (with a small, up to 4% in the number of realizations, the advantage of the RODI index) In periods with quiet geomagnetic conditions, the situation changes - in 17% of cases the S4 index better reflects the relationship with the mid-latitude Dst than the RODI10s index. The indicated relationship means the presence of a linear correlation of more than 0.3 between the processes.

IOP Conf. Series: Earth and Environmental Science 1040 (2022) 012003 doi:10.1088/1755-1315/1040/1/012003

Preliminary studies have shown that the amplitude scintillation index S4 is also sensitive to global variations in the AE and Dst indices, which makes it possible to use the proposed index for studying ionospheric disturbances.

References

[1] Jin Y, Spicher A, Xiong C, Clausen L B N, Kervalishvili G, Stolle C and Miloch W J 2019 Ionospheric plasma irregularities characterized by the Swarm satellites: Statistics at high latitudes *J. Geophys. Res. Space Physics* **124** 1262–82

[2] Nose M, Iyemori T, Sugiura M and Kamei T 2015 Geomagnetic Dst index *World Data Center for Geomagnetism* (Kyoto)

IOP Conf. Series: Earth and Environmental Science 1040 (2022) 012004 doi:10.1088/1755-1315/1040/1/012004

Submicron aerosol and nitrogen dioxide in the atmospheric near-surface layer at the Zvenigorod Scientific Station of the A. M. Obukhov Institute of Atmospheric Physics RAS: Thirty years of measurements

A N Gruzdev, A A Isakov, A S Elokhov and P P Anikin

A. M. Obukhov Institute of Atmospheric Physics, Pyzhevsky per. 3, 119017 Moscow, Russia

E-mail: a.n.gruzdev@mail.ru

Abstract. The paper presents results of measurements and trend analysis of the mass concentration of submicron aerosol and the NO_2 content in the atmospheric surface layer at the Zvenigorod Scientific Station of the A. M. Obukhov Institute of Atmospheric Physics, Russian Academy of Sciences. The station is located in a rural area of the western Moscow region. Aerosol concentration is determined by directional light scattering in a flow nephelometer. The NO_2 content is measured spectrometrically by zenith-scattered solar radiation. Three decades of aerosol and NO_2 measurements give a possibility to analyze long-term trends in near-surface aerosol and NO_2. Seasonally dependent estimates of the aerosol trends are presented for three periods differing in character of aerosol variability: 1991–2002, 2003–2012, and 2013–2020. The common feature of the trends is a general decrease in the aerosol concentration. There are however significant interdecadal differences in the strength of the trends and in their seasonal dependence. The NO_2 content in the near-surface layer undergoes a significant positive trend. The NO_2 increase is likely associated with an increase in the population and human activity resulting, in particular, in the increase in motor traffic in the region. Effects of wind direction on aerosol and NO_2 anomalies are also studied.

1. Introduction

Aerosol and nitrogen oxides are typical gaseous and particulate species of the atmospheric surface layer (ASL). Their concentrations are determined by local (regional) sources and by atmospheric transport from remote sources. Both can also be generated in the atmosphere through photochemical processes.

Measurements of the near-surface concentration of submicron aerosol started at the Zvenigorod Scientific Station (ZSS) of the A. M. Obukhov Institute of Atmospheric Physics in 1991. One year before, in 1990, spectrometric measurements of nitrogen dioxide, a major representative of the nitrogen oxides family - were begun. Three decades of aerosol and NO_2 measurements at the same site give unique possibility to analyze and compare their long-term variability. It is especially interesting since the two species are affected by regional pollution and long range transport to varying degrees.

This paper reviews shortly the results of measurements of the aerosol concentration and the NO_2 content in the ASL at the ZSS and presents analysis of their long-term trends.

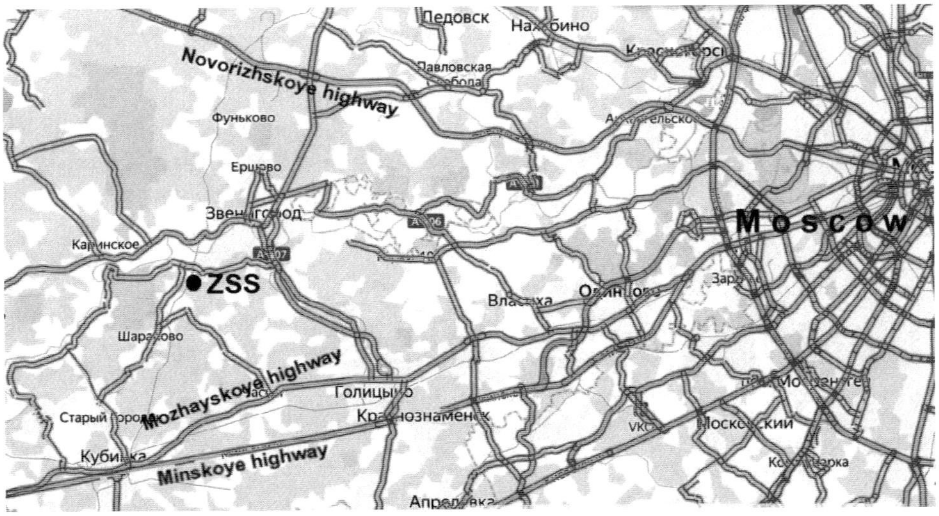

Figure 1. Map of the near western Moscow region.

2. Site and methods of measurements

Zvenigorod station is located in a rural area in the west Moscow region at ~55 km distance from the Moscow center and 40 km distance from the Moscow ring highway (Figure 1). At distances of 10 km and 13 km to the south of the ZSS are Mozhaiskoe and Minskoye highways, and 17 km to the north is Novorizhskoe highway, which are major highways with heavy traffic.

Near-surface aerosol measurements at the ZNS in 1991–2003 were performed around the clock with a flow nephelometer by V. N. Sidorov [1]. In 2001–2012, measurements were carried out with a spectropolarimeter [2]. Since 2012, measurements were done with a serial FAN nephelometer. Spectropolarimeter and FAN measurements were performed in morning around 9 a.m. local time. Comparison and mutual binding of data obtained with different instruments was done for the overlap intervals of 2001–2003 and 2012.

Values of the mass concentration of aerosol are obtained by multiplying the directional light scattering coefficients measured at an angle of 45° by a factor of 2000. The factor was estimated by comparing our data with the results of $PM_{2.5}$ measurements at the Mosecomonitoring station located at the ZNS [3]. An uncertainty of application of this coefficient is about 20%.

The NO_2 content is measured by the spectrometric method registering zenith-scattered solar radiation during morning and evening twilight [4]. The measurement method allows retrieving the vertical distribution of NO_2. The bottom value of the vertical profile characterizes the NO_2 content in the atmospheric boundary layer. Under typical conditions, the main contribution to it is made by the ASL [5]. An important feature of the NO_2 contents in the ASL determined in our measurements is that it is not a local concentration but the content in the ASL column. This feature makes the ASL NO_2 content insensitive to local pollution, for example, due to individual cars.

3. Measurement data

Figure 2a shows a series of daily aerosol concentrations. The period from 2003 to 2012 is characterized by frequent and strong anomalies of large concentration values. Their cause has not yet been fully established, although analysis of the anomalies and their relation to atmospheric transport was done [6].

Figure 2b shows decimal logarithm of the aerosol concentration. It is the logarithm, and not the concentration, that is preferable to use for the trend analysis [3].

IOP Conf. Series: Earth and Environmental Science 1040 (2022) 012004 doi:10.1088/1755-1315/1040/1/012004

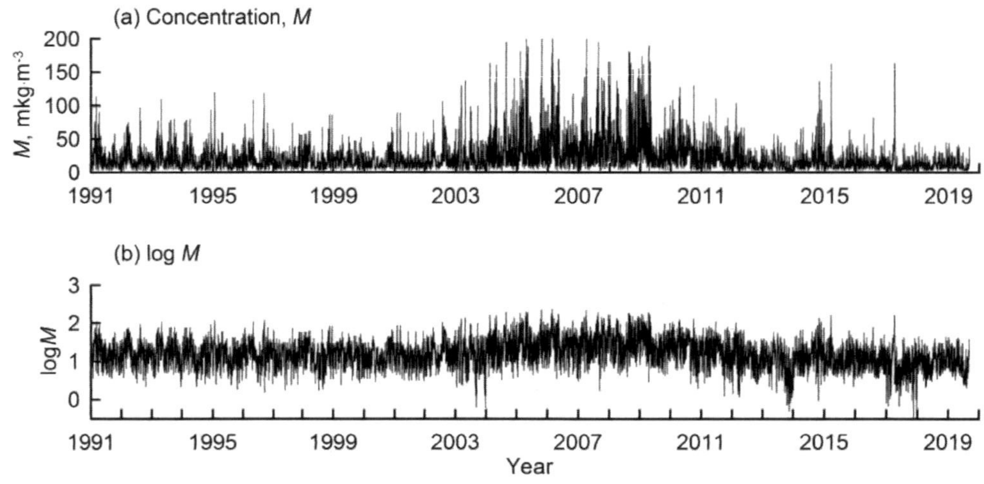

Figure 2. (a) Mass concentration of near-surface aerosol and (b) its decimal logarithm.

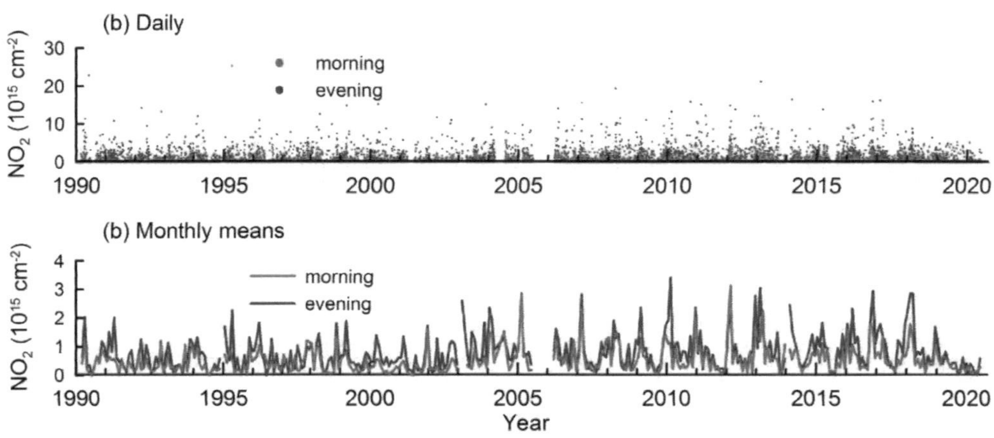

Figure 3. (a) Daily and (b) monthly mean values of the morning (red) and evening (blue) NO_2 content in the atmospheric surface layer.

According to figure 2 three periods can be distinguished: before 2003, after 2012 with "normal" level of aerosol concentration and variability, and the "abnormal" period of 2003–2012. The three periods will be considered separately.

The NO_2 abundance in the ASL according to morning and evening measurements, as well as the corresponding monthly mean values are shown in figure 3. Evening values are generally larger than morning values. This is due to the additional formation of NO_2 as a result of the photolysis of N_2O_5 and more intensive motor traffic during daytime. Monthly averages indicate the presence of the annual variation of NO_2 with the winter maximum and the summer minimum. It is due to the fact that the most frequent and strong episodes of ASL pollution by nitrogen oxides are observed in winter. The proportion of days with close to zero NO_2 content in the ASL in summer is 30-40%, while it is less than 10% in winter.

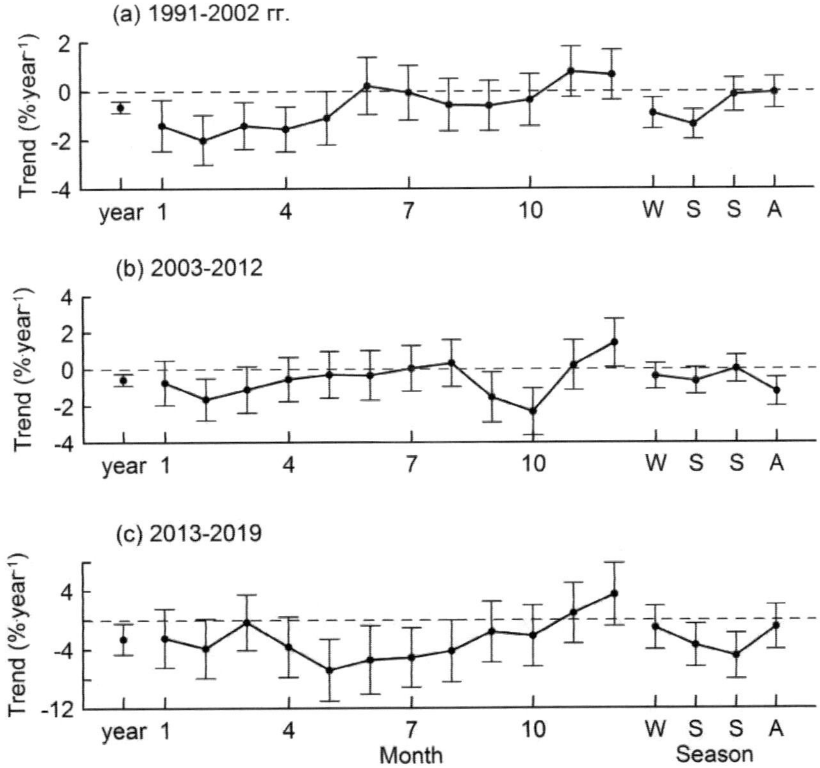

Figure 4. Annual (dots on the left side), monthly (curves in the middle), and seasonal (curves on the right side) estimates and associated 95% confidence intervals of linear trends in logarithm of aerosol concentration in (a) 1991–2002, (b) 2003–2012, and (c) 2013–2019.

4. Method of trend analysis

Trend analysis is done with a multivariate regression method. Analysis of NO_2 in this paper is a part of the analysis of the long-term trends and interannual variability in the NO_2 vertical distribution reported in a paper published in this issue [7]. In this case the regression model includes as predictors the following proxies of long term and interannual variability: the free member, the linear term (linear trend), the quasi-biennial oscillation, the sea surface anomaly index for the equatorial east Pacific in the Nino 3.4 region, the solar activity index $F10.7$, the North Atlantic Oscillation index, and the aerosol optical depth. References to these data can be found in [7]. The regression model is applied to monthly mean NO_2 contents. Here we keep all these features of the regression model unchanged to be consistent with [7].

In case of aerosol, the regression model includes as predictors a free term, a time-linear term (to describe a linear trend), a quadratic term (to account for a quadratic trend), and variables describing changes in wind direction. The quadratic term is symmetric with respect to the midpoint of the analyzed period. Thus, it does not affect the value of the linear trend. Its purpose is to approximate the domed shape of aerosol course in 2003–2012. The quadratic trend is also kept in the regression model when analyzing data for the other periods (also with centering on the middle of the corresponding period).

The proxies of the wind direction in the regression model are sine and cosine of the angle corresponding to the wind direction. The wind direction is determined from velocities of zonal and meridional winds at 975 hPa level derived from the ERA-Interim reanalysis data (https://www.ecmwf.int/en/forecasts/datasets/reanalysis-datasets/era-interim).

Technique of estimating statistical significance in the regression problem is mostly applicable when a residual series (error) obeys the Gaussian distribution [8]. This is not the case if the regression analysis is applied to aerosol concentration. We use for the trend analysis the logarithm of aerosol concentration since the residual series in this case follows the Gaussian low [3].

All regression coefficients in the regression model are expanded into Fourier pairs corresponding to the annual and higher harmonics in order to account for seasonal dependence of trends in aerosol and NO_2. In case of NO_2, two harmonics, the annual and semiannual ones, are used. In case of aerosol, four harmonics are used.

To solve the system of the regression equations, the method is used taking into account the serial correlation (autocorrelation) of the residual on long time scales [9]. The serial correlation is due to memory in a system and affects confidence intervals of regression estimates [8].

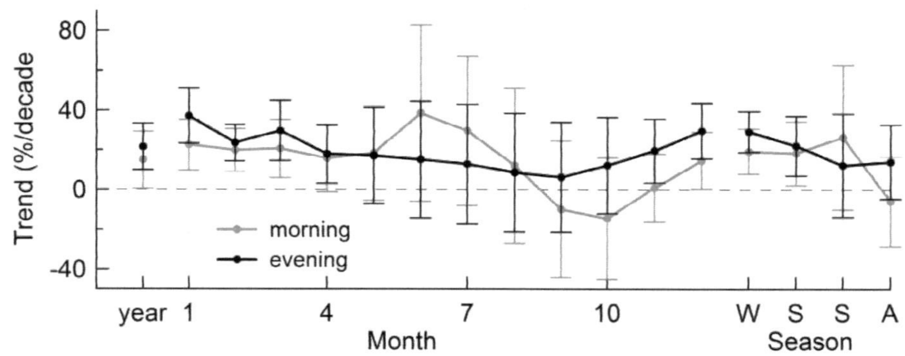

Figure 5. Annual (dots on the left side), monthly (curves in the middle), and seasonal (curves on the right side) estimates and associated 95% confidence intervals of linear trends in the morning (gray) and evening (black) NO_2 contents in the ASL.

5. Results of trend analysis

Figure 4 shows the annual, monthly, and seasonal estimates of the linear trends in the aerosol concentration for the three periods. The annual trends are negative for all the periods. The annual trend, linear in logarithm of concentration, is about –0.6% per year in 1991–2002 and 2003–20012 and –2.6% per year in 2013–2020. Seasonal dependences of monthly trend estimates in the first two periods have common features. These are the negative trend in the winter–spring season, the absence of trend in summer, the negative trend in autumn (statistically insignificant for the first period), and the change in the trend sign in late autumn to positive value in early winter. Some fragments of this dependence are also characteristic of the third period.

The main feature of the third period is the negative aerosol trend in late spring and summer (Figure 4c). It was discussed in [3] and is presumably associated with a decrease in the summer surface temperature in the region. However, a special analysis is needed to quantify the contribution of temperature to the aerosol trends.

Figure 5 shows the NO_2 trends derived from data of morning and evening measurements. The trends are positive and statistically significant in winter and spring, and the evening trend for this period is larger than the morning trend. The NO_2 trends in summer are also positive but statistically insignificant. However we would like to note a larger value of the morning trend in this season. The lower evening value of the summer NO_2 trend is likely associated with turbulent (convective) transport of near-surface NO_2 during daytime to higher tropospheric heights (see trends in the NO_2 vertical profiles in [7]).

The annual estimates of the NO_2 trends are about 15% per decade in morning and 29% per decade in evening. The winter trends approach 20% per decade in morning and almost 40% per decade in evening.

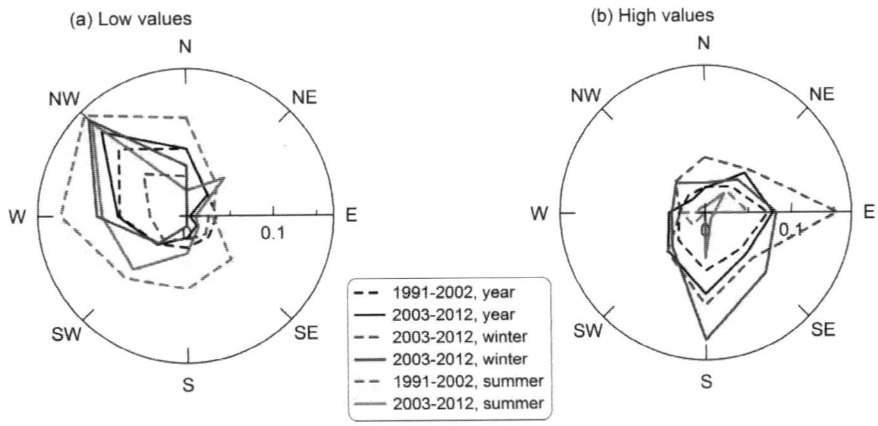

Figure 6. Probability of (a) low and (b) high values of the aerosol concentration as function of wind direction in 1991–2002 (dashed curves) and 2003–2012 (solid curves) in winter (blue) and summer (red) and for entire year (black).

6. Effects of atmospheric transport

Distribution of aerosol and its sources in the ASL is non-uniform [10–13]. Aerosol abundance in the ASL at an observation site is determined by local conditions and atmospheric transport. Changes in transport patterns bring air from various regions with different aerosol content and composition.

Concentration of nitrogen oxides in the ASL near industrial centers and big cities is affected significantly by anthropogenic activity [14]. Changes in wind direction can result in significant fluctuations of the NO_2 content at a measurement site.

Figure 6 shows the probability of low and high aerosol concentrations at the ZSS depending on the wind direction for the first and second periods of measurements in winter, summer, and for the entire year. Low and high values are defined as the concentrations for which their logarithm is out of 1.6 standard deviation (SD) of the logarithm below or over its mean value, respectively. Since the logarithm of the aerosol concentration obeys the Gaussian distribution (see [3]), 90% of the logarithm values falls within 1.6 SD of the mean, and the low and high values contain each 5% of the distribution.

According to the diagrams in figure 6a, low aerosol concentrations are usually observed under winds from the western sectors. At north-westerly winds, low concentrations are most probable. For example, probability of low concentrations under the north-westerly wind in summer is about 0.15 (see red diagrams). The probability of low aerosol concentrations in an easterly wind is small, and during the second period is close to zero.

High concentrations of aerosol at the ZSS in winter are most probable at winds with the easterly, southerly, and south-easterly directions (Figure 6b). The probability diagrams for the winter seasons of the 1st and 2nd periods differ significantly. The highest probability, of about 0.15, of high aerosol anomalies was associated with southerly winds in 2003–2012 and with easterly winds in 1991–2002.

IOP Conf. Series: Earth and Environmental Science 1040 (2022) 012004 doi:10.1088/1755-1315/1040/1/012004

The southerly wind in 1991–2002 also often brought high anomalies (with 0.1 probability). In addition, we note that the probability of high anomalies under south-westerly winds in winters of the two periods is not small (about 0.05), which, in combination with the wind rose extended in this direction, provides a significant number of high aerosol concentration values.

The angular diagrams of summer probabilities of high aerosol anomalies for the two periods have the same main qualitative features as the diagrams of winter probabilities (Figure 6b). The summer diagram extends in the eastern direction in 1991–2002 and in the southern direction in 2003–2012. Along with this we note that the winter diagrams for high anomalies are wider than the summer diagrams, and the probability of abnormally high aerosol concentrations in winter is significantly larger than in summer.

Figure 6 shows that high and low aerosol concentrations at the ZSS are usually associated with winds of generally opposite directions. In winter conditions, these directions are oriented toward remote regions with relatively strong and weak sources of aerosol. In summer conditions with slower winds, the role of local conditions and regional transport increases.

The NO_2 content in the ASL at the ZSS changes in a very wide range from zero to values, which in order of magnitude are larger than the column NO_2 content above the ASL [5, 7]. Zero values mean that the NO_2 content in the ASL is less than the detection threshold of the method used. They correspond to background conditions.

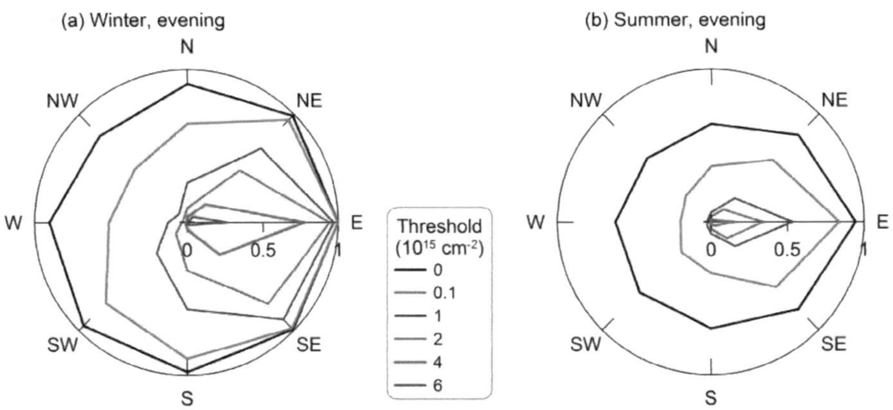

Figure 7. Probability of exceeding the specified threshold value of the NO_2 ASL content at ZSS as function of wind direction in (a) winter and (b) summer, for evening.

Figure 7 shows probabilities of evening NO_2 contents in the ASL that are larger than prescribed threshold contents, as function of wind direction, in winter and summer. The threshold values are specified in the legend. We note for information that the threshold value of $2 \cdot 10^{15}$ molecules/cm^2 is approximately equal to the typical winter value of the column NO_2 content above the ASL (see, for example, Figure 3 in [7]).

The probability of the ASL pollution by NO_2 at the station decreases with increasing threshold under all wind directions (Figure 7). Pollution is possible at any wind (see diagrams for zero threshold). In winter, easterly and south-easterly winds always bring pollution (probability at zero threshold is equal to 1). Probability of pollution is also very high in winter under winds of north-east, south, and south-west directions.

The strongest pollution is mainly related to easterly winds blowing from Moscow. Sector of winds associated with strong pollution in winter is wider and include also orientations to the north-east, south-east, and south (Figure 7a), i.e. toward the major highways and Moscow suburb (see Figure 1).

Note that strong pollution is not observed in westerly and north-westerly winds. The north-westerly wind direction is the direction of the least pollution probability at all thresholds.

Weak NO_2 pollution, which can be characterized by differences between diagrams for the zero and the closest to it thresholds in figure 7, is observed in winds not blowing from a major highway or large locality. This level of pollution is a feature of "background" parts of the region itself, which, however is also affected by motor traffic and other human activity in the region.

7. Discussion and conclusions

The long-term trends in the near-surface aerosol concentration and in the ASL NO_2 content at the ZSS are, in general, of opposite signs. The NO_2 content in the ASL is most sensitive to human activity and strongly affected by transport from polluted zones. Due to relatively short lifetime of nitrogen oxides, especially in summer, the NO_2 content is primarily affected by transport from a surrounding area. Proximity to Moscow and major highways has defining value to strong and moderate pollution of the ASL by nitrogen oxides. The increase in the NO_2 content reported in this study can be attributed to an increase in population of the megapolis and associated increase in anthropogenic activity, especially to the increase in motor traffic.

Unlike near-surface NO_2, the mass concentration of near-surface aerosol at the ZSS undergoes a long-term decrease that is not uniform on the interdecadal scale and depends on season but determines most general long-term evolution of aerosol. The negative aerosol trends at the ZNS are consistent with a general decrease in the aerosol mass concentration in West Europe and some other regions of the world [15, 16] (see also [3] for more references).

Small aerosol concentrations and small NO_2 contents at the ZSS are associated with winds directions of the north-west and neighbor sectors. In the NO_2 case, this direction is most clean in the sense of absence of significant sources of nitrogen oxides at a distance of up to a few hundred km from the ZSS. In the aerosol case, it orients toward the Baltic region and, farther toward the northern Atlantic since aerosol can be transported over larger distance than chemically active species like nitrogen oxides.

The observed large NO_2 concentrations are related to pollution episodes at the ZSS caused by air transport from the sides of Moscow with its suburb and major highways. The directions associated with high aerosol anomalies are approximately the same. However we do not believe that the location of aerosol sources should be the same. The south-east direction is the orientation toward steppes, which are sources of soil aerosol. In summer, forest fires in Siberia could be also responsible for high aerosol concentrations under appropriate advection conditions. However, high aerosol concentrations are most probable in winter.

It is worth to note a general decrease in the aerosol optical thickness (AOT) over the ZSS reported in [17] for the warm periods of 2007–2017, which is less pronounced, however, than the negative AOT trend over Moscow. The difference is attributed in [17] to opposite long-term tendencies of pollutant emissions at the two sites: an emission decrease in Moscow and an emission increase in the ZSS area. Our results show a very distinctive negative aerosol trend for the spring-summer periods after 2013. We can therefore deduce that local urban emissions do not affect significantly this trend,

Acknowledgments

ECMWF ERA-Interim data used in this study have been obtained from the ECMWF data server. The study was supported by the Russian Foundation for Basic Research, project No, 20-05-00274.

The authors are grateful to the anonymous reviewer for useful comments.

References

[1] Sidorov V N 1999 Seasonal and diurnal variations of the dry fraction content of submicron aerosol in the surface layer *Proceed. Conf. "Physics of Atmospheric Aerosol"* (Moscow) 356–367 (in Russian)

[2] Isakov A A and Gruzdev A N 2009 Long-period variations in the optical and microphysical

parameters of surface aerosol in a Moscow suburb *Izvestiya, Atmos. Oceanic Phys.* **45** 233–241

[3] Gruzdev A N, Isakov A A, and Anikin P P 2020 Multiyear trends in the mass concentration of near-surface aerosol at Zvenigorod Research Station, A. M. Obukhov Institute of Atmospheric Physics, Russian Academy of Sciences *Atmos. Oceanic Optics* **33** 274–281

[4] Elokhov A S and Gruzdev A N 2000 Nitrogen dioxide column content and vertical profile measurements at the Zvenogorod Research Station *Izvestiya, Atmos. Oceanic Phys.* **36** 763–777

[5] Gruzdev A N and Elokhov A S 2011 Variability of stratospheric and tropospheric nitrogen dioxide observed by visible spectrophotometer at Zvenigorod, Russia *Int. J. Remote Sens.* **32** 3115–3127

[6] Gruzdev A N and Isakov A A 2021 Anomalies and trends in the mass concentration of near-surface aerosol in the western Moscow region *Izvestiya, Atmos. Oceanic Phys.* submitted

[7] Gruzdev A N and Elokhov A S 2021 Three decades of remote sensing of NO_2 vertical distribution and column content at the A. M. Obukhov Institute of Atmospheric Physics *IOP Conf. Ser. Earth Environ. Sci.* This issue

[8] Draper N R and Smith H 1998 *Applied Regression Analysis* (New York: John Wiley & Sons)

[9] Gruzdev A N 2019 Accounting for autocorrelation in the linear regression problem by an example of analysis of the atmospheric column NO_2 content *Izvestiya, Atmos. Oceanic Phys,* **55** 65–72

[10] Hatzianastassiou N, Gkikas A, Mihalopoulos N, Torres O, and Katsoulis B D 2009 Natural versus anthropogenic aerosols in the eastern Mediterranean basin derived from multiyear TOMS and MODIS satellite data *J. Geophys. Res.* **114** D24202

[11] Chazette P, Raut J-C, Dulac F, Berthier S, Kim S-W, Royer P, Sanak J, Loaëc S, and Grigaut-Desbrosses H 2010 Simultaneous observations of lower tropospheric continental aerosols with a ground-based, an airborne, and the spaceborne CALIOP lidar system *J. Geophys. Res.* **115**. D00H31

[12] Taylor M, Kazadzis S, Amiridis V, abd Kahn R A 2015 Global aerosol mixtures and their multiyear and seasonal characteristics *Atmos. Environ.* **116** 112–129

[13] Liu Y, Zhu Q, Wang R, Xiao K, and Cha P 2019 Distribution, source and transport of the aerosols over Central Asia *Atmos. Environ.* **210** 120–131

[14] Finlayson-Pitts B J and Pitts J N Jr. 2000 *Chemistry of the Upper and Lower Atmosphere: Theory, Experiment, and Application* (San Diego: Academic Press))

[15] Barmpadimos I, Keller J, Oderbolz D, Hueglin C, and Prévôt A S H 2012 One decade of parallel fine ($PM_{2.5}$) and coarse (PM_{10}–$PM_{2.5}$) particulate matter measurements in Europe: trends and variability *Atmos. Chem. Phys.* **12** 3189–3203

[16] Wang K C, Dickinson R E, Su L, and Trenberth K E 2012 Contrasting trends of mass and optical properties of aerosols over the Northern Hemisphere from 1992 to 2011 *Atmos. Chem. Phys.* **12** 9387–9398

[17] Zhdanova E Y, Chubarova N Y, and Lyapustin A I 2020 Assessment of urban aerosol pollution over the Moscow megacity by the MAIAC aerosol product *Atmos. Meas. Tech.* **13** 877–891

Transport of biomass burning products from Siberian wildfires into the Arctic

S A Sitnov[1] and I I Mokhov[1, 2]

[1] Obukhov Institute of Atmospheric Physics, Russian Academy of Sciences, Moscow, Russia
[2] Lomonosov Moscow State University, Moscow, Russia

E-mail: [1]sitnov@ifaran.ru

Abstract. The study of the long-range transport of biomass burning products from Siberian wildfires into the Arctic atmosphere during the period of 2000-2019 is presented. An analysis of the characteristics of forest fires over the past 20 years revealed an increase in radiation power of an average Siberian wildfire, which is characterized by a statistically significant linear trend of 1.7 ± 1.0% / year. A joint analysis of fire activity in Siberian forests, as well as the contents of the black carbon (BC) and carbon monoxide (CO) contents in the Arctic atmosphere, indicates that extreme fire events force the development of regional anomalies in BC and CO. Correlation between the anomalies of BC (CO) over the Russian segment of the Arctic and the number of Siberian wildfires is found to be statistically significant at the $\alpha = 0.05$ level and reach the value $r = 0.77$ (0.48) during the summer months. Using a linear regression model, an estimate of the sensitivity of changes in the total BC content and in the volume mixing ratio of CO at the 700-hPa level in the $1.9 \cdot 10^{-8}$ kg·m^{-2} per 1000 fires and 0.4 ppbv per 1000 fires, respectively. The results of a detailed analysis of the long-range BC transport into the Arctic during catastrophic Siberian wildfires in the summer of 2019 are presented. It is shown that the considered episode was conditioned by the features of the large-scale atmospheric circulation characteristic for the atmospheric blocking event.

1. Introduction

With the ongoing global warming, climatic changes in the Arctic are attracting special attention [1,2]. The increase in the average annual near-surface temperature in the Arctic latitudes during last decades is several times higher than the average temperature rise over the planet (so called Arctic amplification) [3,4]. An additional factor contributing to the heating of the Arctic system under global warming is the BC content in the air, which is a component of the submicron aerosol of the atmosphere, which effectively absorbs solar radiation [5]. When deposited on snow-covered and ice surfaces, BC reduces the albedo and increases the absorption of solar radiation, contributing to additional heating of the underlying surface [6].

Among the sources of BC in the atmosphere are anthropogenic and natural factors, associated with incomplete fuel combustion [7]. On the territory of Russia, the main source of BC during the warm period of the year is massive wildfires, which are responsible for more than 80% of the total BC emission [8], moreover the wildfires in Western Siberia and Eastern Siberia account for 23% and 50% of the total number of forest fires in Northern Eurasia [9]. In recent years, there has been a positive trend in the radiation power of an "average" Russian fire of 0.48 W m-2 per year in magnitude [9]. According to model estimates, with the continuation of global warming in the XXI century, in the

Asian part of Russia one should expect an increase in the frequency of wildfires [10] and an increase in the duration of the fire hazardous period.

Due to the small number of local sources, the main regulator of the BC content in the Arctic atmosphere is the processes of long-range transport of BC [11. The transport of pollution from middle latitudes to the Arctic is often associated with atmospheric blocking events. It was shown that seasonal variations in the concentration of anthropogenic aerosol over Arctic Alaska are associated with seasonal variations in the frequency and position of mid-latitude blocking anticyclones [12]. The rapid (7-10 days) transport of aerosols from the south to the north occurs over the western periphery of quasi-stationary anticyclones.

The aim of this work is to analyze the connection of pollution of the Arctic atmosphere by BC and CO with the Siberian forest fires, as well as to analyze possible physical mechanisms of long-range transport of biomass combustion products from Siberia to the Arctic.

2. Data

The characteristics of Siberian wildfires are analyzed using measurements from MODIS satellite instruments operated on board of the Terra and Aqua platforms from November 2000 to December 2020 with a reliability of fire diagnosis of not less than 80%. MODIS Active Fire Products (C6, L2), processed by standard MCD14ML algorithm are obtained through the FIRMS (Fire Information for Resource Management System, https://earthdata.nasa.gov) system. A detailed description of the fire detection algorithm MODIS C6 and its differences from previous one can be found in [13].

Hourly BC column mass densities (M2T1NXAER v5.12.4) and BC surface mass concentrations (M2T1NXAER v5.12.4) as well as 10-m northward wind (M2T1NXSL v.5.12.4 10M) from the MERRA-2 reanalysis were used. MERRA-2 (Modern-Era Retrospective analysis for Research and Applications, version 2) is a modern reanalysis, created in NASA with an emphasis on assimilation of various satellite measurements since 1980. MERRA-2 uses the new version of the GEOS-5 (Goddard Earth Observing System model, version 5) data assimilation system to synthesize regular time series of gridded data with a spatial resolution 0.5° x 0.625° (latitude x longitude) at 72 pressure levels (from the surface to 0.01 hPa) both instantaneous and time-averaged (hourly, 3-hourly and monthly). Among the advantages of MERRA-2, in comparison with its previous version, is joint assimilation of aerosol and meteorological fields with accounting their interaction through the radiative effects of atmospheric aerosol [14].

The daytime measurements of the CO mole fraction in air (AIRS3STM_006_CO_VMR_A) obtained by high-resolution spectrometer AIRS (Atmospheric InfraRed Sounder, C6) were also used. AIRS operates onboard the Aqua satellite from September 2002. It measures radiation in 2378 IR channels (3.7-15.4 μm) and 4 channels in the visible range (0.4-1 μm) using measurements near a wavelength of 4.6 μm for the CO retrieval . The accuracy of AIRS CO measurements is 15%, the spatial resolution in the nadir is 13.5 km, while the vertical resolution is 1 km [15]. BC and CO data were obtained through the Giovanni Internet server https://giovanni.gsfc.nasa.gov/giovanni developed and maintained by the NASA Goddard Earth Science Information Technology Information Center [16.

In addition, the geopotential heights at 500-hPa level (H500) taken from the NCEP/NCAR reanalysis [17] were used. Daily mean data of H500 with a spatial resolution of 2.5° × 2.5° were obtained from the https://www.esrl.noaa.gov/psd/data/gridded/data.ncep.reanalysis.html server.

IOP Conf. Series: Earth and Environmental Science 1040 (2022) 012005 doi:10.1088/1755-1315/1040/1/012005

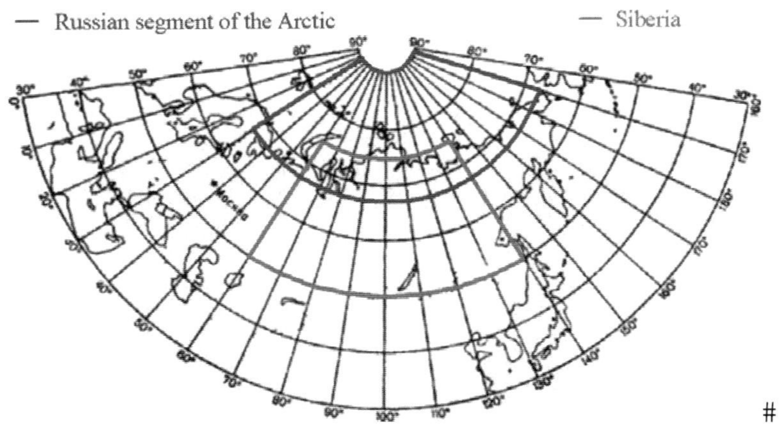

Figure 1. Territories of Siberia and the Russian segment of the Arctic considered in this study.

3. Method

The paper analyzes the relationship between the content of BC and CO in the atmosphere over the territory restricted by coordinates 66°-90°N, 30°-180°E (hereinafter the Russian segment of the Arctic) with wildfires detected in the area bounded by coordinates 50°-75°N, 60°-140°E (hereinafter Siberia) (figure 1). The regions under consideration are territories close in location. The area of the Russian segment of the Arctic is 9.2 million km^2, the area of the considered territory of Siberia is 11.3 million km^2.

4. Results and discussion

4.1. Siberian forest fires

The results of the analysis of hotspots detected by MODIS instruments indicate a regular manifestation of fire activity in Siberian forests with a strong interannual and intra-annual variability of forest fires. Formally, the regional amount of wildfires in Siberian forests (N) and the total fire radiation power (FRP) are characterized by weak positive linear trends. However, the calculations of the coefficients of variation N and FRP ($C_V = \sigma / \mu$) show that C_V exceeds 100%, which indicates on inhomogeneity of the data and makes it impossible to obtain statistically valid characteristics from the analyzed samples, including estimates of trends in N and FRP.

On the other hand, monthly variations of the ratio of the total FRP to the total fire counts are characterized by a much lower value of C_V (19%), which allows a statistical analysis of the data (figure 2a). The ratio is an estimate of the radiation power of the average Siberian fire (FRP_{avg}). Analysis of FRP_{avg} variations during the summer periods of 2003 - 2019 shows that the radiation power of the average Siberian fire is characterized by a noticeable increase, which can be approximated by a positive linear trend of 1.25 ± 0.82% / year, which is statistically significant at the α= 5% level (trend is calculated in % of the average value of FRP_{avg} in 2003-2019). The results obtained confirm the trend of an increase in the radiation power of the Siberian wildfire, obtained earlier on the basis of the analysis of shorter time series [9].

IOP Conf. Series: Earth and Environmental Science 1040 (2022) 012005 doi:10.1088/1755-1315/1040/1/012005

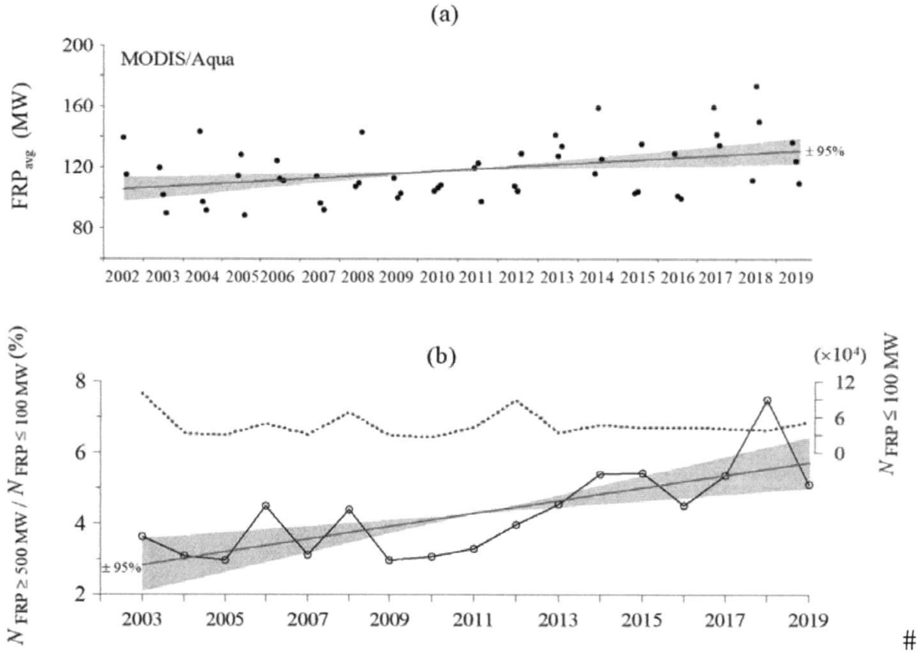

#

Figure 2. (a) - Yearly variations of FRP_{avg} of the average Siberian wildfire in June-August during the period of 2002-2019 and its linear trend (red line) with a 95% confidence interval (shaded area); (b) - the ratio of the number of intense fires (FRP $\geq$ 500 MW) to the number of weak fires (FRP $\leq$ 100 MW) (black line) and its linear trend (red line) with a 95% confidence interval (shaded area).Variations in the number of weak fires (blue dashed line, right scale) are also shown. MODIS/Aqua data.

The total mass of forest fires is dominated by hotspots characterized by a low FRP [9]. At the same time in [9] it was shown that from west to east in Russia the frequency of weak wildfires decreases while the frequency of intense forest fires increases. It is of interest to analyze the tendencies of long-term changes in Siberian wildfires of various intensities. figure 2b shows the annual variations in the ratio of the number of summer wildfires in Siberia with an intensity of more than 500 MW and those less than 100 MW. The results obtained indicate that in the last two decades, in the territory of Siberia, there has been a predominant increase in the number of intense forest fires, characterized by a statistically significant linear trend of 1.8 ± 0.9% / 10 years. It is characteristic that long-term changes in the number of summer Siberian fires with a FRP of less than 100 MW does not show any pronounced tendencies (figure 2b).

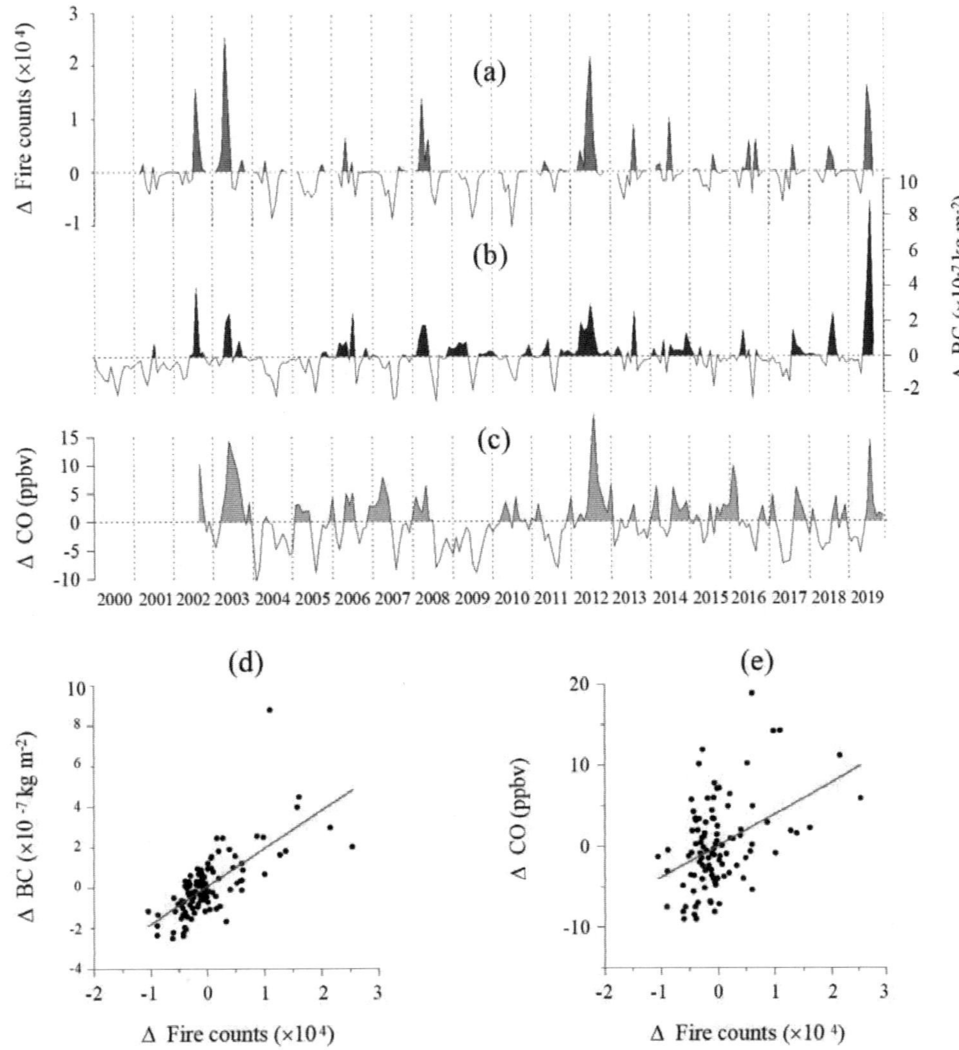

#

#

Figure 3. Monthly variations in the number of Siberian wildfires (a) and also monthly variations in the BC column density (b) and in CO VMR at 700 hPa (c) in the Arctic atmosphere. Scatter plots between Siberian wildfires and BC column density in the Arctic (d) and between Siberian wildfires and CO VMR at 700 hPa in the Arctic (e). The data sets in (a-c) were deseasonalized by subtraction of the long-term monthly means from the time-series. The scatter plots show relations between the anomalies in the fire hazardous periods (Apr-Sept).

4.2. Relationship of the BC and CO anomalies in the Arctic atmosphere with Siberian wildfires

figure 3a shows the differences between the current monthly number of Siberian wildfires and corresponding long-term monthly mean values, which characterize the anomalies of fire activity in the forests of Siberia. Calculated in a similar way the monthly mean BC and CO anomalies in the atmosphere over the Russian segment of the Arctic are shown in figure 3b and figure 3c, respectively. Comparison of figure 3a and figure 3b indicates that the increased fire activity in the forests of Siberia, noted in August 2002, May 2003, April 2008, June-July 2012 and in July-August 2019 (figure 3a), was accompanied by positive anomalies of the BC content in the Arctic atmosphere (figure 3b). The correlation of the monthly BC anomalies in the Arctic with the number Siberian wildfires in the period 2001-2019 reaches r = 0.70 (95% confidence interval: 0.62, 0.77), and moreover increases in the summer period to r = 0.77 (0.64, 0.86). Because the contextual algorithm of MODIS detects fires in ground pixels with a size of 1 km × 1 km, the total number of fires is at the same time an lower estimate of their total area. The high correlation between BC ond wildfires is explained by a relationship between the emission of biomass burning products with the total area of wildfires [18]. The direct relationship between the Arctic BC anomalies and Siberian wildfires is confirmed by the scatter diagram (figure 3d). Figure 3d shows that the increase of BC is associated with an increase in wildfire activity in Siberian forests. Obtained on the basis of a linear regression model the estimate of the sensitivity of changes in the BC column mass density over the Russian segment of the Arctic to changes in the number of wildfires in Siberia, is 1.9 10^{-8} $kg \cdot m^{-2}$/1000 wildfires.

Comparison of figure 3a and figure 3c indicates that increased fire activity in Siberian forests in 2003, 2012, and 2019 was also accompanied by positive CO anomalies in the Arctic troposphere (figure 3c). The correlation of the monthly mean values of CO anomalies in the Arctic with the number of fires in Siberia in the period 2002-2019 is r = 0.39 (0.26, 0.51), increasing in the summer period to r = 0.48 (0.24, 0.67). A weaker correlation of CO with wildfires (in comparison with BC) can be associated with a greater variety of CO sources in the Russian sector of the Arctic. The results indicate, in particular, the manifestation of positive CO anomalies in winter, as well as in early spring and late autumn, i.e. outside the fire hazardous season (see figure 3c). This fact can be associated with anthropogenic influence. The longer lifetime of CO in the atmosphere (from several weeks to several months) increases the contribution of long-range CO transport from urban territoriest in the formation of regional CO anomalies. The weaker relationship between CO in the Arctic atmosphere and forest fires in Siberia is also confirmed by the scatter diagram (figure 3f). The sensitivity of CO changes in the troposphere of the Arctic to changes in the number of fires in Siberia, obtained on the basis of a linear regression model, was estimated to be $4.0 \cdot 10^{-4}$ ppbv / 1000 wildfires.

4.3. Mechanism of the long-range transport of biomass burning products from Siberian wildfires into the Arctic atmosphere

When considering the Russian territory as a whole, the relation of massive forest fires with atmospheric blockings was found in a number of papers [19-21]. An analysis of the topography of geopotential heights of the 500-hPa pressure level (H500) showed that episodes of extreme fire activity in Siberian forests (figure 3a) were associated with large-scale circulation characteristic for atmospheric blockings (not shown due to lack of space). Atmospheric blockings contribute to the formation of regional anomalies of atmospheric impurities, including CO, aerosol, ozone (with the formation of mini-ozone holes) [19-25].

The results of a detailed analysis of the long-range transport of BC into the Arctic during severe forest fires in Siberia in the first ten days of August 2019 are presented in figures 4.The episode under consideration was accompanied by the formation of the most strong positive BC anomaly over the Russian segment of the Arctic during the period of 2001-2019 (comp. figure 3a and figure 3b).

An analysis of longitudinal-temporal variations in the blocking index (figure 4b), calculated according to [26], indicate the manifestation of atmospheric blocking in the longitudinal sector 95°-125°E in the first ten days of August 2019. The spatial distribution of H500 (figure 4a) in this period was characterized by a stationary high-pressure system situated over the central part of Siberia (60°N, 110°E), and two stable low-pressure systems located over the Northern Urals and the Far East. The noted features of H500 are characteristic for the large-scale atmospheric circulation during atmospheric blocking event. An analysis of fire activity indicates that atmospheric blocking in the first ten days of August 2019 was accompanied by the development of massive fires in Siberian forests (figure 4e). A joint analysis of wildfires and H500 reveal that the bulk of the sources of air pollution by biomass burning products were concentrated in the center of the blocking area (figure 4a).

The mechanism of long-range transport of pyrogenic pollutants in August 2019 is closely related to atmospheric blocking. Figure 4a indicates that when approaching the blocking anticyclone, the western zonal flow deviated far to the north, and when bending around it, to the south, dragging the air contaminated with combustion products with it. Since the spatial gradient of H500 to the west of the anticyclone exceeded the gradient of H500 to the east (figure 4a), the wind speed over the western periphery of the anticyclone was higher than the wind speed over its eastern periphery. Therefore, in a whole, the blocking area was dominated by the flow of combustion products from the south to the north, i.e. from the territory of forest fires to the Arctic.

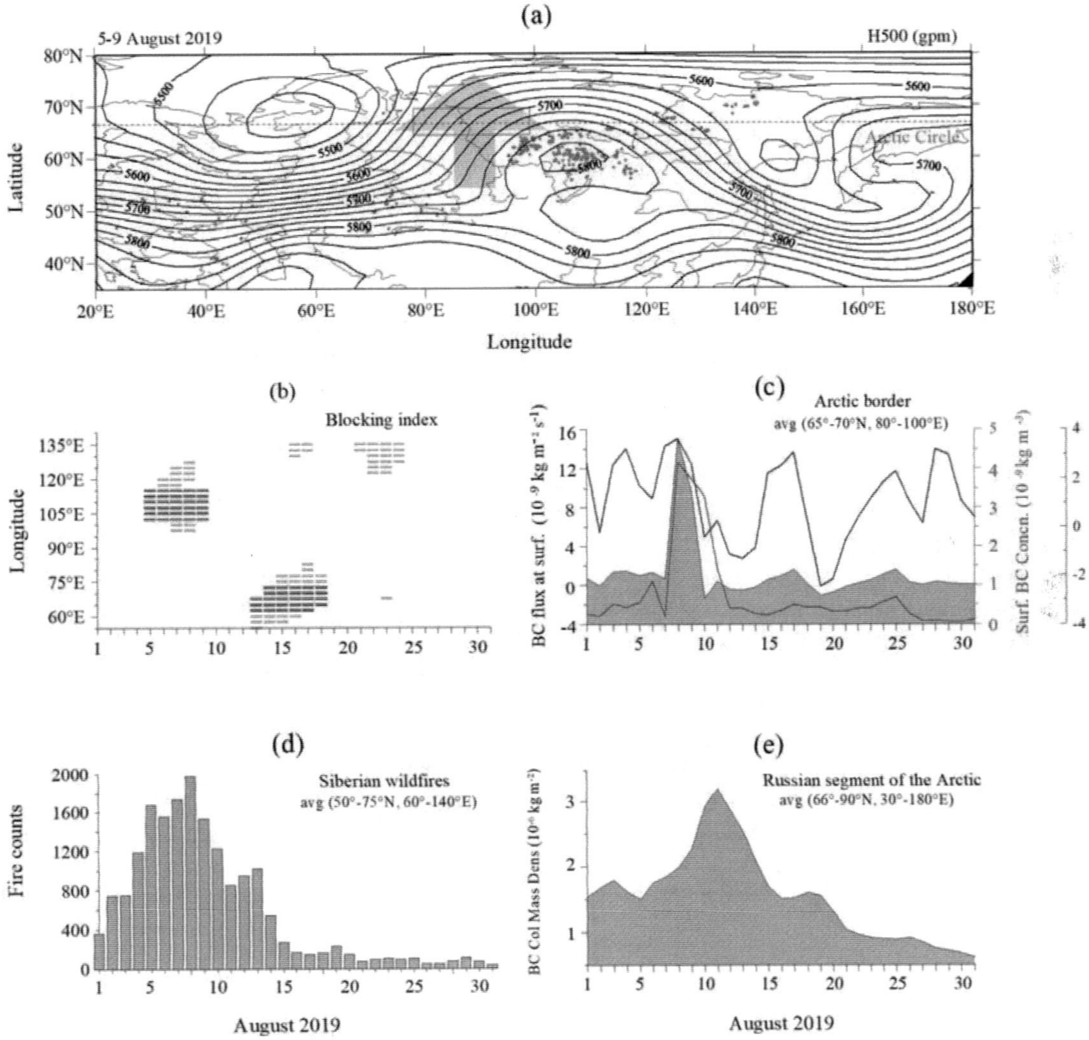

#

Figure 4. (a) - Spatial distributions of H500 (black isolines) and wildfires (red circles) during 5-9 August 2019, daily variations of: (b) - blocking index (red symbols stand for blocking conditions lasting at least 5 days), (c) - meridional BC flux in the surface layer through the Arctic Circle (shaded); BC mass concentration (red line) and meridional wind at the 10-m height (blue line) in the surface layer close to the Arctic Circle (see also (a)), (d) , the number of Siberian wildfires, (e) - BC total column over the Russian segment of the Arctic. In (a) blue rectangle denotes territory considered in (c), thick arrow shows the northward flux of burning products.

As part of the work, an estimate of the BC fluxes from wildfires in Siberia into the Arctic atmosphere was obtained. For calculations, the formula was used:

$$F_{\mathrm{BC}} = C_{\mathrm{BC}} \cdot V,$$

where F_{BC} - is the BC flux through the Arctic Circle (AC) [kg m^{-2} s^{-1}], C_{BC} - is the surface BC mass concentration [kg m^{-3}], V - is the meridional wind at the height of 10-m [m s^{-1}].

Figure 4c shows the day-to-day variations in the meridional flux of BC, regional BC mass concentration and meridional wind in the atmospheric surface layer on August 2019. All data are averaged over the spatial region (66°-69°; 80°-100°E) centered on the Arctic Circle (AC) (see also figure 4a). In early August, due to not enough developed wildfires, even despite the strengthening of the southerly wind, a weak BC flux to the Arctic was noted near the AC. It can be seen from figure 4c that at the end of the first ten days of August, forest fires in Siberia reached their maximum development, that has saturate subarctic atmosphere by combustion products. The followed strengthening in the southerly wind over the southern boundary of the Arctic, led to a sharp increase in the BC flux to the Arctic, which reached its maximum value of $1.5 \cdot 10^{-8}$ kg m^{-2} on 8 August 2019. At the beginning of the second decade of August, a decrease in the number of fires and a change in wind from south to north led to a weakening and even reversal of the BC flux, contributing to a decrease in the regional BC concentration in Arctic (figure 4e). The subsequent strengthening of the southerly wind led to a slight increase in the BC flux to the Arctic due to a significant decrease in the number of forest fires in the Siberian forests (figure 4d).

Changes in the total BC content in the atmosphere over the Russian sector of the Arctic (figure 4e) indicate an increase in the regional atmospheric BC abundance in the first ten days of August 2019, which has reached the maximum of $3.21 \cdot 10^{-8}$ kg m^{-2} on 11 August 2019. The three-day delay exists between the maximum of BC content in the Arctic for and the maximum of BC flux over AC, which can be explained by the time required for the BC transport from forest fires in Siberia into the Arctic.

5. Conclusion

The conducted analysis of the causes and mechanisms of the formation of large-scale anomalies in BC and CO contents in the atmosphere of the Russian sector of the Arctic revealed a direct relationship of the regional BC and CO anomalies with forest fires in Siberia. The estimates of the sensitivity of changes in the regional content of aircraft and CO in the atmosphere of the Russian Arctic to the corresponding changes in the number of Siberian fires were obtained: $1.9 \cdot 10^{-8}$ kg m^{-2} / 1000 fires and $4.0 \cdot 10^{-4}$ ppbv / 1000 fires, respectively.

The analysis of fire activity in Siberian forests showed that in the last two decades, there has been a predominant increase in the number of intense forest fires, characterized by a statistically significant linear trend of 1.8 ± 0.9% / 10 years (figure 2a), while long-term changes in the number of weak forest fires reveal no somehow trend.

A detailed analysis of extreme BC pollution of the high-latitude atmosphere during catastrophic forest fires in Siberia in the summer of 2019 reveals that regional pequliarities of large-scale atmospheric circulation asociated with atmospheric blocking event contributed to the long-range transport of BC from the fires of the Siberian forests to the Arctic. Estimates of BC flux from Siberian forest fires through the Arctic circle into the Arctic have been were obtained. The maximum BC flux in August 2019 reached the value of $1.5 \cdot 10^{-8}$ kg m^{-2} s^{-1}.

Acknowledgements
This work was carried out with the financial support of a project of the Russian Federation represented by the Ministry of Science and Higher Education of the Russian Federation (agreement No. 075-15-

2020-776) using results of regional analysis of blocking effects, obtained within the framework of project 19-17-00240 of the Russian Science Foundation.

References

[1] *Climate Change 2013: The Physical Science Basis*. Contribution of Working Group I to the Fifth Assessment Report of the Intergovernmental Panel on Climate Change ed T Stocker et al (New York: Cambridge University Press) pp 659–740

[2] Serreze M C, Barrett A P, Stroeve J C, Kindig D N and Holland M M 2009 The emergence of surface-based Arctic amplification *Cryosphere* **3** 11-19

[3] Screen J A, Deser C and Simmonds 2012 I Local and remote controls on observed Arctic warming *Geophys. Res. Lett.* **39** L10709

[4] Mokhov I I 2020 Russian climate research in 2015–2018 *Izvestiya, Atmos Oceanic Phys.* **56** (4) 325-343

[5] Haywood J M and Boucher O 2000 Estimates of the direct and indirect radiative forcing due to tropospheric aerosols: A review *Rev. Geophys.* **38** 513–43

[6] Hansen J and Nazarenko L 2003 Soot climate forcing via snow and ice albedos *Proc. Natl. Acad. Soc. USA.* **101** 423-28

[7] Bond T C, Doherty S J, Fahey D W, Forster P M, Berntsen T, DeAngelo B J, Flanner M G, Ghan S,. Kärcher B, Koch D et al 2013 Bounding the role of black carbon in the climate system: A scientific assessment *J. Geophys. Res. Atmos.* **118** 5380–552

[8] Evans M, Kholod N, Kuklinski T, Denysenko A, Smith S J, Staniszewski A, Hao W M, Liu L and Bond T C 2017 Black carbon emissions in Russia: A critical review *Atmos. Environ.* **163** 9-21

[9] Sitnov S A and Mokhov I I 2018 A comparative analysis of the caracteristics of active fires in the boreal forests of Eurasia and North America based on satellite data *Izv., Atmos. Oceanic Phys.* **54** 966–78

[10] Mokhov I I and Chernokulsky A V 2010 Regional model assessments of the risk of forest fires in the Asian part of Russia under climate changes *Geogr. Prir. Resur.* **2** 120–126

[11] Vinogradova A A, Smirnov N S, Korotkov V N and Romanovskaya A A 2015 Forest fires in Siberia and the Far East: Emissions and atmospherict transport of black carbon to the Arctic *Atmos. Oceanic Optics* **28** . 566-74

[12] Iversen T and Joranger E 1985 Arctic air pollution and large scale atmospheric flows *Atmos. Environ.*, **19**, 2099–108

[13] Giglio L, Schroeder W and Justice C O 2016 The collection 6 MODIS active fire detection algorithm and fire products *Rem. Sens. Environ.* **178** 31-41

[14] Gelaro R, McCarty W J, Suarez M J, Todling R, Molod A, Takacs L, Randles C A, Darmenov A, Bosilovich M G, Reichle R et al 2017 The modern-era retrospective analysis for research and applications, Version 2 (MERRA-2) *J. Clim.* **30** 5419–54

[15] Aumann H H, Chahine M T, Gautier C, Goldberg M D, Kalnay E, McMillin L M, Revercomb H, Rosenkranz P W, Smith W L et al 2003 AIRS/AMSU/HSB on the Aqua mission: Design, science objectives, data products, and processing systems *IEEE Trans. Geosci. Remote Sens.* **41** 253-264

[16] Acker J C and Leptoukh G 2007 Online analysis enhances use of NASA Earth science data *Eos Trans. AGU* **88** 14-17

[17] Kistler R, Collins W, Saha S, White G, Woollen J, Kalnay E, Chelliah M, Ebisuzaki W, Kanamitsu M, Kousky V et al 2001 The NCEP-NCAR 50-year reanalysis: Monthly means CD–ROM and documentation *Bull. Amer. Meteor. Soc.* **82** 247-268

[18] Bondur V G, Gordo K A and Kladov V L 2017 Spacetime distributions of wildfire areas and emissions of carbon-containing gases and aerosols in northern Eurasia according to satellite-monitoring data *Izv. Atmos. Ocean. Phys.* **53** 859-874

[19] Bondur V G, Mokhov I I, Voronova O S and Sitnov S A 2020 Satellite monitoring of Siberian wildfires and their effects: Features of 2019 anomalies and trends of 20-year changes *Dokl. Earth Sci.* **492** 370–375

[20] Mokhov I I, Bondur V G, Sitnov S A and Voronova O S 2020 Satellite monitoring of wildfires and emissions into the atmosphere of combustion products in Russia: Relation to atmospheric blockings *Dokl. Earth Sci.* **495** 921–24

[21] Mokhov I I, Sitnov S A, Tsidilina M N and Voronova O S 2021 NO_2 emission into the atmosphere from forest fires in Russia in relation with atmospheric blocking event *Atmos. Oceanic Optics* **34** (in press)

[22] Peters D, Egger J, Entzian G 1995 Dynamical aspects of ozone mini-hole formation *Meteorol. Atoms. Phys.* **55** 205-214

[23] James P M 1998 A climatology of ozone miniholes over the Northern Hemisphere *Int. J. Climatol.* **18** 1287–1303

[24] Barriopedro D, Anto M, Garcia J A 2010 Atmospheric blocking signatures in total ozone and ozone miniholes *J. Climate* **23** 3967-3983

[25] Sitnov S A, Mokhov I I 2015 Ozone mini-hole formation under prolonged blocking anticyclone conditions in the atmosphere over European Russia in summer 2010 *Dokl. Earth Sci.* **460** (1) 41-45

[26] Tibaldi S and Molteni F 1990 On the operational predictability of blocking *Tellus* **42A** 343-365

IOP Conf. Series: Earth and Environmental Science 1040 (2022) 012006 doi:10.1088/1755-1315/1040/1/012006

Meteorological multi-rotor unmanned aerial complex and its application for monitoring of the atmosphere

Yu B Popov[1], **P A Karpushin**[1], **N P Krasnenko**[1,2], **S A Kurakov**[2], **K Yu Popova**[1] **and A S Rakov**[2]

[1] Radio Engineering Faculty, Tomsk State University of Control Systems and Radioelectronics, Tomsk 634050, Russia

[2] Department of Geophysical Research, Institute of Monitoring Climatic and Ecological Systems SB RAS, Tomsk Russia

E-mail: popovyub@mail.ru

Abstract. A design of a meteorological complex based on multi-rotor aeromobile platforms of unmanned aerial vehicles and its possible application for measuring vertical profiles of the main meteorological parameters of the atmosphere, including the air temperature, humidity, and pressure are considered. An example of a comparison of multicopter measurements with the data obtained using an MPT-5 meteorological temperature profiler is given.

1. Introduction

Recently manufacture and use of unmanned areal vehicles (UAVs) for various applications [1–13] has become one of the fast developing scientific and technical directions. An explosive growth of the market of the unmanned aerial vehicles is observed all over the world. In addition to solving traditional problems of video registration and photographing of various localities [1, 2], the UAVs (or in other words, the drones) can also solve geophysical [3, 4, 7] and meteorological problems [4-13], in particular, they can measure the parameters of the atmospheric state [5–13]. The aircraft-type UAVs have a higher flight resource stock and, as a rule, a higher load lifting capacity for the same masses and sizes. However, it is sufficiently difficult to automate the measurement process for the aircraft-type UAVs. The existing accessible autopiloting systems with automatic takeoff and control of the UAV flight route do not allow the aircraft-type UAVs to land automatically. As a rule, a parachute is used to land these UAVs, which is obviously inconvenient for their use in measurements of the vertical structure of the atmosphere. The aircraft-type UAVs [4, 10] are well suited for measurements along horizontal paths, but they do not provide strictly vertical ascent over the start point. The dynamic characteristics of these UAVs do not allow high-sensitive low-inertia sensors to be placed onboard, which inevitably limits their measurement accuracy.

For the UAVs with vertical ascent and descent, it is possible to automate completely the measurement process. Takeoff, rising, flight along a selected trajectory, and landing of the UAV can be performed according to a preset program. In this case, the operator only gives the command to ascent. The flight, measurements, and descent can be performed in completely automatic mode. Such systems are best suited for monitoring of the vertical structure of the atmosphere [4–13]. Therefore, in our opinion, such multi-rotor UAVs based on aeromobile platforms (MUAVs or multi-copters) are most promising for measuring the vertical profiles of the meteorological parameters in the atmospheric boundary layer (ABL). At present, the multicopters are widespread. The UAVs of these types are universal bearing platforms with a wide range of applications. Their high stability in flight, stability against wind loads, and availability of programmable onboard navigating devices allow the MUAVs to be used in a wide variety of weather conditions, round the clock, in manual control mode, and in independent flight. The MUAP equipped with meteorological sensors allows contact measurements of vertical profiles of the main meteorological parameters to be performed in the ABL during each ascend and descent, which is important for the study of the ABL and verification of ground-based systems of remote measurements of the atmospheric parameters. The maximum altitude of measuring the vertical profiles of the meteorological parameters can reach several kilometers and depends on the capabilities of the employed device and weather conditions. The special features of the design and application of the MUAVs allow one to increase the volume of reliable measurements of meteorological parameters and to verify in real time remote sounding data on new qualitative level.

In contemporary scientific publications [1–13], immediate prospects for UAV application in solving meteorological problems and their advantages and disadvantages are widely discussed. Technologies of meteorological measurements with the use of multicopters are developed by a number of scientific groups [3, 5-13]. They are used not only to measure the profiles of meteorological parameters, but also to verify the meteorological parameters measured remotely by other instruments. This is caused by the fact that the data of remote sounding of the atmosphere are gradually more and more involved in weather forecasting. Moreover, they are used by weather forecasters in their work and as input data for prognostic models. It is obvious that as any meteorological data, they need testing and calibration. From this point of view, the UAV is a universal and very mobile system; it can be used for acquisition of data on the state of the atmosphere in the gaps between measurements with aerologic or any other meteorological stations and in hard-to-reach regions. An additional equipment of the UAVs with meteorological sensors of various designations will allow one to carry out multipurpose measurements and to reorient fast the platform for solving specific problems. The development of sounding techniques using meteorological UAVs allows one not only to obtain a large volume of data, but also to solve problems of fundamental study of the atmosphere on qualitatively new level.

The present work generalizes results of our work on the development of unmanned instruments of measuring the meteorological parameters in the ABL based on multicopters and their application for verification of data measured with the MTP-5 temperature profiler. Design of the meteorological complexes based on multicopters of different types is considered. Requirements are given to meteorological sensors, design of the measuring complex, and aerial platform. The developed models and their possibilities are described. Results of their outdoor testing and verification of the data measured with the MTP-5 temperature profiler are given. The verification of remote instruments intended for measuring the meteorological parameters, especially passive ones, of the MTP-5 type, is an urgent problem, because it allows one to calibrate measurements and to improve data processing algorithms.

2. Design of the meteorological complex based on a multi-copter

Technologies of manufacture of modern UAVs and their electronics make it possible to carry out contact measurements of the meteorological parameters in the ABL at altitudes from 10 to 2000 m and higher with high temporal and spatial resolutions depending on the UAV type. The multicopter UAVs are best suited for solving this problem. The multicopter strictly follows a preset trajectory, hangs over preset places, and ascents and descends strictly vertically with a chosen velocity. Reusability of these UAVs reduces expenses on their operation and allows one to increase significantly the frequency of their launch, whereas their controllability makes it possible to change the flight program or to stop the flight depending on circumstances.

Thus, the meteorological UAVs should involve a meteorological complex with a number of meteorological sensors (a portable electronic meteorological station for onboard measurements of the meteorological parameters) capable of measuring the air temperature, pressure, and humidity in the atmosphere and to calculate the wind speed and direction.

The requirements to and the principles of designing the onboard measuring complex (OMC) intended for measuring the meteorological parameters, placed onboard the rotor-type unmanned aerial vehicle (UAV), and capable of recording spatial distributions of the fields of the main meteorological parameters, involving the air temperature, humidity, and pressure of the atmosphere (their vertical or horizontal profiles) are considered below.

2.1 Requirements to meteosensors intended for measuring the air temperature, humidity, and pressure
The characteristics and the design of the individual OMC sensors placed onboard the UAV should meet the requirements to the stationary meteorological stations. This concerns protection against solar radiation. Additional requirements are imposed on meteosensors and conditions of their placing onboard UAVs. Thus, hot airflows from the UAV propellers and engines are sources of errors.

2.1.1 Temperature sensor. To eliminate measurement errors due to direct and reflected solar and IR radiation, the temperature sensor should be specially protected. In addition, the temperature sensor should be operational in the presence of vibrations in UAV flights with stable characteristics to maintain calibration during measurements. For the meteosensors placed onboard the UAVs, one of the main requirements is the speed of response (the time constant) that should provide reliable measurements of the meteorological parameters during UAV ascent and descent with a given speed. Considering the foregoing, the air temperature sensor in radiation protection is better to arrange either outside of the UAV clamped on a long horizontal bar or on a special vertical holder above the UAV case. Such arrangement of the sensor provides its natural ventilation and influences insignificantly the aerodynamic stability of the complex. The temperature was measured with a 701-101BAA-B00 platinum sensor [14].

2.1.2. Humidity sensor. Successful functioning of the meteorological sensor of relative air humidity placed onboard the UAV depends on the rate of exchange by water molecules between the sensor and the atmosphere. When the temperature decreases, the sensor operation requires good external ventilation because the free exchange by water molecules between the sensor and the atmosphere decreases; otherwise, considerable measurement errors occur. Protection against atmospheric precipitations is also needed. The sensor should be well ventilated, and the requirements to its arrangement are the same as to the temperature sensor: it should be placed outside of the AUV and clamped on an external bar. A HIH-5031-001 sensor was used to measure the air humidity [15].

2.1.3. Atmospheric pressure sensor.
The air pressure profile is used in modern numerical weather forecast models. It is also used to calculate (from the barometric formula) the UAV flight altitude along with its determination from signals of global satellite positioning systems. The pressure sensor should not be influenced by external factors (wind, air temperature, and vibration); it should be calibrated or tested regularly. Digital membrane piezoresistive or capacitive pressure sensors with temperature compensation and miniature metal or semiconductor aneroid boxes used as sensitive elements one of the sides of which is connected to a piezoresistor or is a capacitor plate are well suited for atmospheric pressure sensors. The systematic error of these sensors is determined by the dependence of indications on the internal sensor temperature and the hysteresis and elasticity of the aneroid box. To measure the atmospheric pressure, an MPL3115A2 digital pressure sensor [16] was used.

2.2. Requirements to the OMC design
Based on the requirements to the meteosensors considered above, we now can formulate the main requirements to the design of the OMC placed onboard the UAV. It should provide:

- Reliable operation at air temperatures, pressures, and humidities in the required altitude range year round,
- Protection of sensors against solar and IR radiation,

- High rate of heat exchange between the temperature sensor and the surrounding medium,
- Inadmissibility of ingress of hot air flows from UAV engines on meteosensors,
- Protection of sensors and electronics against precipitations and moisture,
- Writing of measurement data to non-volatile memory with periods from 1 till 10 s,
- Radio transmission of the measured meteorological parameters to a ground-based station in real time,
- Vibration resistance,
- High maintainability and ease of installation on and removal from the copter carrier,
- Simple replacement and charging of power supply units,
- Compactness and small weight.

2.3. Requirements to the copter carrier platform

The multicopter can flight in different ways: it can hang over the earth at a preset altitude, rotate round an axis, and move in any direction including strictly upwards. It comprises the frame with 4, 6, 8, or 12 brushless motors clamped on the frame, propellers, electronics, and lithium battery.

We consider the principle of multicopter flight on an example of a quadcopter (platform with 4 motors). Pairs of its motors rotate in opposite directions. When hanging, all propellers rotate with the same speeds, the forces acting between them are equalized thereby allowing the quadcopter to soar steadily in air without rotating about its axis. To rotate about the axis, one opposite pair of motors starts to rotate faster, and another pair rotates more slowly to keep the overall balance of the thrust and altitude. To rotate in another direction, the first pair of motors rotates more slowly, and the second pair rotates faster. A flight in a preset direction is provided by simple tilting of the copter in the corresponding direction and increase in the rotation speed of the motors on the opposite sides. From the viewpoint of aerodynamics, the multicopter is an unstable design and it flies accurately due to the controller that controls motors and sensors of position, acceleration, flight direction, air pressure, temperature, and satellite navigation. Nevertheless, the embodiment in the comparatively small complex of many of the latest technical achievements has allowed us to develop a sufficiently reliable complex for solving a wide class of problems. To measure the vertical profiles of the air temperature, humidity, and pressure, the multicopter platform with the following characteristics is required:

- Weight of the OMC placed onboard the copter platform no less than 500 g (this weight is sufficient to measure the vertical distributions of the main meteorological parameters using modern electronic units),
- The maximum flight altitude no less than 2 000 m (as a rule, this altitude allows the meteorological parameters to be measured in the ABL);
- Maximum vertical speed no less than 5 m/s (to obtain data on transient atmospheric processes, it is required to scan the ABL during a limited time and to use the data processing procedures well-developed for radiosondes ascending with speeds from 4 to 6 m/s),
- Maximum horizontal speed no less than 20 m/s (to operate under conditions of strong winds),
- Flight time no less than 20 min (sufficient for ascending to altitudes no less than 2 km),
- Simple control and operation,
- Automatic or semi-automatic flight regime,
- Compactness and small weight.

2.4. UAV models intended for measuring the meteorological parameters

At the first stage, we searched for optimal location of meteosensors and optimal OMC design and measurement procedure with allowance for perturbations of the surrounding medium caused by air flows from the working motors. Two design variants had been developed and tested. In the first variant (figure 1a), the sensors were arranged at the edges of the horizontal bars at a distance of 70 cm from the motors. In most cases, this provided minimal influence of air jets from the motors on the sensors, but increased the overall dimensions and windage of the entire system and worsened the flight conditions. It was experimentally established that thermal flows from the running UAV motors propagated at distances of several tens of meters. This prevents placing of sensors under the carrier and also decreases the accuracy of temperature and humidity measurements during descents when the sensors are in air heated by running engines. In the second variant (figure 1b and c), the sensors were arranged on the vertical rod above the UAV at a distance of 40–50 cm. This arrangement of the meteorological sensors was more acceptable.

In the developed models, the temperature and humidity sensors are enclosed in a radiation shield (Figure 1c, d, and e). On the printed-circuit board of the measuring unit, a pressure sensor and a radio modem with an antenna were located. The ground-based part of the meteorological complex included an external directed antenna, a telecommunication unit, and a computer. The telecommunication unit matches the USB interface with the radio modem module. As a result, several operating OMC models had been developed and realized at the IMCES SB RAS (figure 1) in the following sequence: onboard an MK Quadro XL quadcopter, DJI (model S900) hexacopter [8], and Walkera Voyager 3 quadcopter.

The meteorological parameters measured by OMCs and their main characteristics are given in tables 1 and 2.

Table 1. Measurable meteorological parameters.

Measurable parameter	Measurement range	Error
Air temperature, °C	-70 … +55	±0.2 at $T \leq 30$°C ±0.3 at $T > 30$°C
Relative air humidity, %	0 … 100	±2.5 at $T > 0$°C ±5 at $T < 0$°C
Atmospheric pressure, GPa	500 … 1100	±0.5 Resolution of 0.0025

Table 2. Main OMC characteristics.

Characteristic	Parameters
Range of working temperatures of the controller, °C	-50 … +50
Time constant of temperature measurement, s	2.5
Time constant of pressure measurement, s	1
Time constant of humidity measurement, s	3
Weight, g	450
Supply voltage, V	3,6 … 10
Period of continuous operation, h	4
Software, remote server for data storage, user attachment for controller configuration, data read-out from the controller and the server, visualization, export of data in the csv text format	Available
Adjustable measurement period, s	1 … 3000
Volume of non-volatile memory	9 000 measurements
Data transmission interface	USB, SDcard, WI FI, GSM
Degree of protection against external actions	IP 66

IOP Conf. Series: Earth and Environmental Science 1040 (2022) 012006 doi:10.1088/1755-1315/1040/1/012006

a) b)

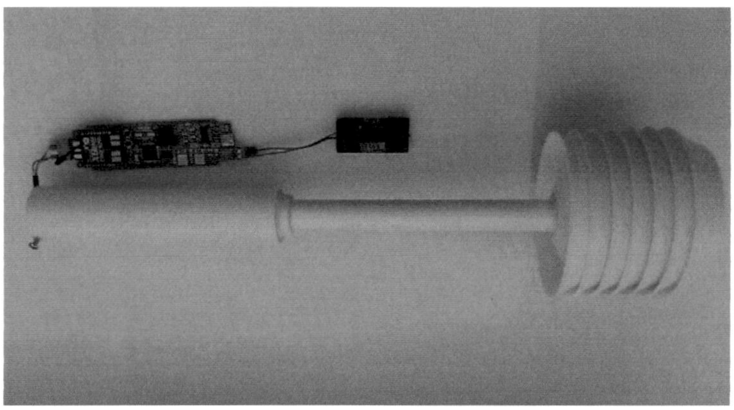

c)

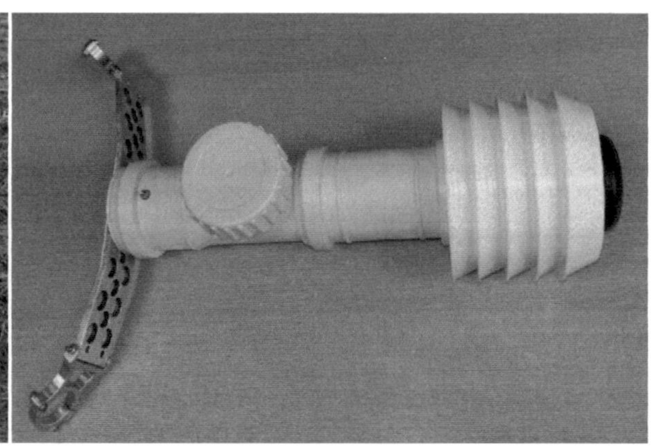

d) e)

Figure 1. OMC placed onboard the MK Quadro XL quadcopter and clamped on the bars 0.7 m long (a) and OMC placed onboard the DJI (model S900) hexacopter and clamped on the rod 0.5 m long (b); the rod with the measuring unit and power supply elements extracted from it (c); OMC placed vertically above the Walkera Voyager 3 quadcopter and clamped on the rod 40 cm long (d); and the rod with clamping element (e).

3. Outdoor tests of the meteorological complex

The developed OMC with UAVs of the above-indicated types were tested outdoors. The time of their flight was limited by the capacity of the storage battery; it was 20…40 min for the complexes of the above-indicated types; this was sufficient to ascent the UAVs at altitudes up to 2…3 km and to descent them. Several storage batteries can be used to perform long-term measurements. Our tests showed that for example, unlike the weather balloon the trajectory of which depends on the wind, strictly vertical profiles of the meteorological parameters can be measured, and horizontal flight along a preset trajectory over a given territory can be performed. The vertical resolution depended on the ascend speed and the time constant of the employed sensor. During experimental flights, the vertical resolution was 10–15 m. The data obtained allowed us to estimate the achievable measurement altitude. The time of ascend to an altitude of 3 km was up to 15 min, and the descent time was up to 8 min. The multicopter could be launched several times along a preset trajectory, which allowed us to study the dynamics of the profiles of the measurable meteorological parameters. To reduce the time of recording transient processes in the atmosphere, we propose to use simultaneously several multicopters arranged as an aerial shelf.

Our tests allowed us to consider such measuring complex as an effective tool for real-time measurements of vertical profiles of the atmospheric meteorological parameters, their verification, and application of remote methods of atmospheric sounding.

4. Verification of measurements of the vertical temperature profile in the atmosphere

The meteorological UAVs perform contact measurements of vertical profiles of the meteorological parameters with high spatial resolution using calibrated meteosensors. Therefore, it is an ideal means for verification of systems of remote measurements of the atmospheric parameters. Only high meteorological masts or radiosonde measurements are alternative to them, but all of them have lower spatial resolution. The meteorological complexes placed onboard the UAVs allow one to measure and to study a thin structure of the ABL.

All systems of remote sensing of the atmosphere require testing of procedures of indirect measurements of the atmospheric parameters and periodic verification – testing of the correspondence between their measurements and actual values of the meteorological parameters. Recently the UAVs are increasingly employed for this purpose.

Figure 2. Walker Voyager 3 quadcopter in flight.

In the last few years, in particular, the MTP-5 temperature profiler is widely used to solve this problem [17]. It is based on the passive method of recording thermal radio emission of the atmosphere at the frequency of oxygen absorption oncoming from different elevation directions. The vertical profile of the air temperature is determined at altitudes up to 1 km by processing the measured signals using a special program. The MTP-5 profiler is approved as a measuring device with a limiting error of temperature measurement of 1.2 °C and sampling temperature discrete no more than 25 m at altitudes from 0 to 100 m for temperatures within ± 50 °C and no more than 50 m at altitudes from 100 to 1000 m. The profiler data have already been compared with radiosonde and sodar data and measurements on high meteorological masts and showed fairly good results.

We also measured the vertical profiles of the temperature and air humidity with the meteorological complex placed onboard the Walkera Voyager 3 quadcopter (figure 2) in the lower atmosphere and compared them with data of the MTP-5 temperature profiler for different atmospheric stratifications. The most interesting results of comparison were obtained for the stable atmospheric stratification in the presence of elevated temperature inversions. They are presented below.

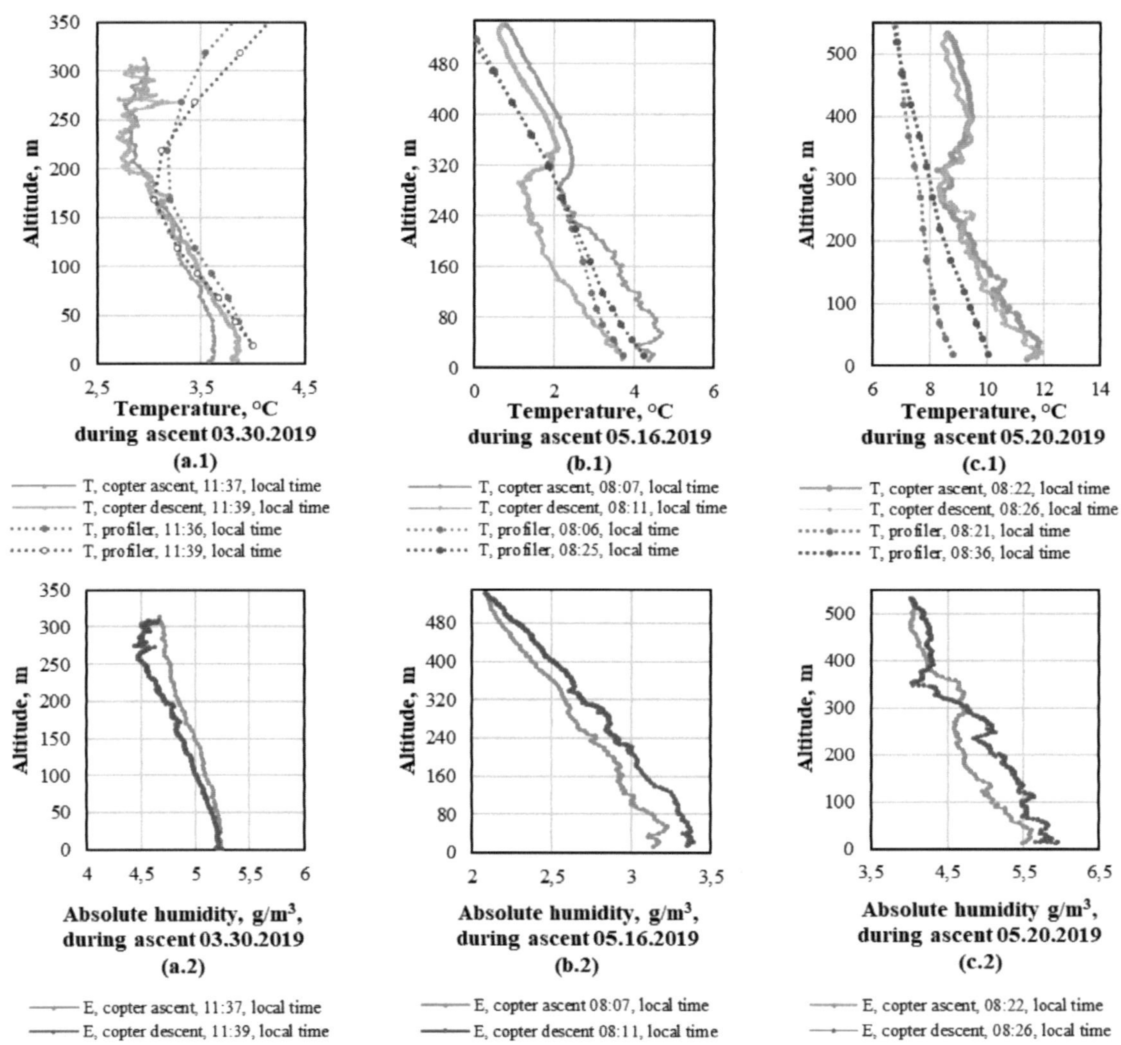

Figure 3. Vertical profiles of the temperature (T, °C) and absolute humidity (E, g/m3) obtained during the UAV ascent and descent and data of temperature measurements with the MTP-5 temperature profiler.

Figure 3 shows some results of such measurements in the presence of temperature inversions in the lower layer of the atmosphere. From figure 3 it can be seen that the curves of the MTP-5 temperature profiler are smoother, which is caused by lower spatial and temporal resolutions of the profiler. Each temperature profile was recorded with the profiler with a discrete period of 5 min. In figure 3a, the UAV profiles have the weak surface temperature inversion to altitude of ~40 m and the elevated inversion with the lower boundary at 200 m. The profiler also shows the elevated inversion, but its data diverge with increasing altitude. In figure 3b, the UAV shows the presence of a thin layer of elevated temperature inversion at altitudes of 300–350 m, whereas the profiler did not detect it. The same can be seen in figure 3c: the UAV shows the elevated temperature inversion at altitudes of 250–330 m, which is not seen in the temperature profile recorded with the profiler. For other atmospheric stratifications, the curves behave synchronously with high values of the correlation coefficient.

Results of our comparison demonstrate the divergence of the data obtained with the OMC and MTP-5 for the stable atmospheric stratification and indicate the necessity of obtaining additional experimental data for higher altitude range.

5. Conclusions

Our investigations and outdoor tests have shown that the meteorological complexes placed onboard multicopters are effective tools for measuring the vertical profiles of the main meteorological parameters (the air temperature, humidity, and pressure) in the atmospheric boundary layer. The vertical resolution of these measuring systems depends on the speed of ascent and the time constant of the employed meteosensors (~1 s). During our experiments, the vertical resolution was 10–15 m. The flight altitude or the altitude range depends on the type of the employed multicopter. Usually it took us 17 min to ascent to an altitude of about 3 km, and it took us up to 8 min to descent. The multicopter can be launched several times along a preset trajectory in real time, which allows one to trace the dynamics of the measurable vertical profiles of the meteorological parameters. To reduce the recording time of transient processes in the atmosphere, simultaneously several multicopters can be used arranged as an aerial shelf. To measure the wind speed and direction, algorithms of their measurements should be developed and tested outdoors. Results of comparison of the measured vertical temperature profiles with the data of the MTP-5 temperature profiler allow us to consider the meteorological complex placed onboard the UAV as an effective tool for verification of the methods of remote sounding of the atmosphere.

This work was carried out within the framework of the State Assignment to the IMCES SB RAS.

References

[1] Khan A C, Alvi B A, Safi E A and Khan I U 2018 Drones for Good in Smart Cities: A Review *Proc. Int. Conf. on Electrical, Electronics, Computers, Communications, Mechanical and Computing (EECCMC)* (India) 28th & 29th January 2018 7 pp

[2] Floreano D and Wood R J 2015 Science, technology and the future of small autonomous drones *Nature* **521** p 460

[3] Watts A C, Ambrosia V G and Hinkley E A 2012 Unmanned aircraft systems in remote sensing and scientific research: Classification and considerations of use *Remote Sensing* **4** pp 1671-92

[4] Rastorguev I P 2014 Unmanned technologies of monitoring of weather conditions *Geliogeofizicheskie Issledovaniya* 8 pp 51-54

[5] Brosy C, Krampf K, Zeeman M, Schafer K, Emeis S and Kunstmann H 2017 Simultaneous multicopter-based air sampling and sensing of meteorological variables *Atmospheric Measurement Technique* 10 pp 2773-84

[6] Sitnikov N M , Akmulin D V, Borisov Yu A et al. 2013 The Use of Unmanned Aerial Vehicles for the Atmospheric Monitoring *Meteorology and Hydrology* 1 pp 90-99

[7] Chin-Chung Chang, Chin-Yuan Chang, Jia-Lin Wang et al. 2020 An optimized multicopter UAV sounding technique (MUST) for probing comprehensive atmospheric variables *Chemosphere 254 1*26867

[8] Kurakov S A , Zuev V V 2016 Unmanned aerial vehicle for measuring vertical profiles of the meteorological parameters in the atmospheric boundary layer *Atmospheric and Oceanic Optics* **29** 11 pp 994-999 (In Russian)

[9] Bell T., Greene B., Klein P., Carney M., & Chilson P. (2019). Confronting the Boundary Layer Data Gap: Evaluating New and Existing Methodologies of Probing the Lower Atmosphere. Atmospheric Measurement Techniques Discussions, (December), 1–23

[10] Kral S., Reuder J., Vihma T., Suomi I., O'Connor E., Kouznetsov R., Wrenger B., Rautenberg A., Urbancic G., Jonassen M., Båserud L., Maronga B., Mayer S., Lorenz T., Holtslag A., Steeneveld G.-J., Seidl A., Müller M., Lindenberg C., Langohr C., Voss H., Bange J., Hundhausen M., Hilsheimer P., & Schygulla M. (2018). Innovative Strategies for Observations in the Arctic Atmospheric Boundary Layer (ISOBAR)—The Hailuoto 2017 Campaign. Atmosphere, 9(7), 268

[11] Segales A. R., Greene B. R., Bell T. M., Doyle W., Martin J. J., Pillar-Little E. A., & Chilson P. B. (2020). The CopterSonde: an insight into the development of a smart unmanned aircraft system for atmospheric boundary layer research. Atmospheric Measurement Techniques, 13(5), 2833–2848

[12] Varentsov M. I., Yu Artamonov A., Pashkin A. D., & Repina I. A. (2019). Experience in the quadcopter-based meteorological observations in the atmospheric boundary layer. IOP Conference Series: Earth and Environmental Science, 231, 012053

[13] Varentsov M., Stepanenko V., Repina I., Artamonov A., Bogomolov V., Kuksova N., Marchuk E., Pashkin A., & Varentsov A. (2021). Balloons and Quadcopters: Intercomparison of Two Low-Cost Wind Profiling Methods. Atmosphere, 12(3), 380

[14] Honeywell 2020 Sensing and Internet of Things 701-101BAA https://sensing.honeywell.com/701-101BAA-B00-rtd-sensors

[15] Honeywell 2020 Sensing and Internet of Things. HIH-5031-001 http://sensing.honeywell.com/HIH-5031-001-Humidity-Sensors

[16] NXP Semiconductor. MPL3115A2: 20 to 110 kPa, Absolute Digital Pressure Sensor http://www.nxp.com/products/automotive-products/sensors/pressure-sensors-for-automotive/barometric-pressure-15-to-115-kpa/20-to-110kpa-absolute-digital-pressure-sensor:MPL3115A2?fsrch=1&sr=1&pageNum=1.

[17] Kadygrov E N and Kuznetsova I N 2015 *Methodical recommendations for application of remote measurements of temperature profiles in the atmospheric boundary layer with microwave profilers: Theory and practice* (Dolgoprudnyi: Fizmatkniga) 171 pp.

IOP Conf. Series: Earth and Environmental Science 1040 (2022) 012007 doi:10.1088/1755-1315/1040/1/012007

Influence of air temperature on air composition in Moscow

M A Lokoshchenko[1], A Yu Bogdanovich[1] and N F Elansky[2]

[1]Department of Meteorology and Climatology, Faculty of Geography, Lomonosov Moscow State University, 119991, Lengory, Moscow, Russian Federation
E-mail: loko@geogr.msu.su
[2]Obukhov Institute of Atmospheric Physics, Russian Academy of Sciences, 119017, Pyzhevskii lane 3, Moscow, Russian Federation

Abstract. Empirical relations between T and surface concentrations of CO_2 and five minor air gases have been studied on a base of hourly data during 12 years. In wide range -6...+15 °C significant changes of minor gases are absent. Real tendencies are increase of O_3 and, vice versa, fall of NO and NO_2 with increasing T from +15...+17 to +38 °C. Both effects are a consequence of unstable stratification that enhances vertical mixing. In cool weather -7...-18 °C O_3 falls whereas nitrogen oxides grow with decreasing T due to frequent inversions and slowing down the NO oxidation rate. At an even lower temperature up to -30 °C NO and NO_2, vice versa, decrease with decreasing T – probably, due to strong cold advection of clean Arctic air. Unlike minor gases, CO_2 decreases with increasing T up to 25 °C due to photosynthesis intensification from winter to summer. Seeming growth of CO at T>27 °C is fully explained by smoky haze during heat waves in 2010 and 2002. The CO_2 growth in hot weather is also created by heat stress of trees. Thus, except only oxidation rate, any influence of T is indirect as a result of stratification, photosynthesis, smoky haze, advection, etc.

1. Introduction

As is known a ratio between major air gases (N_2, O_2 and Ar) in the atmosphere is constant everywhere up to 80÷100 km heights. Changes in the dry air chemical composition relate to different content of CO_2 and minor air gases (ozone, nitrogen oxides, carbon monoxide, sulfur oxides, methane and others). In fact, carbon dioxide surface concentration in mid latitudes in the Northern hemisphere demonstrates clear annual course and geographical differences between dry land and oceans due to photosynthesis production by forests. Spatial distribution of minor air gases and their temporal changes depend on their main sources which are concentrated in big cities and industrial regions for NO, NO_2, CO; in wetlands and around gas pipelines for CH_4, etc. [1]. Studying the air composition in cities is important because of the strong influence of minor gases on human health. Topical issue is the relation between meteorological parameters and concentrations of minor gases. These relations have both scientific and practical importance, including their possible use for air pollution forecasting in big cities. Some examples of similar empirical relations are presented in the literature, but they are usually calculated on the basis of relatively short data series and for a limited range of considered parameters. Among others, positive formaldehyde function of the air temperature T in Beloyarsky city (Khanty-Mansi autonomous region of the Russian Federation) for one year is presented in [2]. The CO functions of the wind velocity V under the conditions of both the surface inversion and the elevated inversion are demonstrated in [3]. The dependences of the concentrations of CO_2 and minor gases in the air on V are presented by authors in [5].

The aim of this work was to obtain for the first time statistically significant empirical functions of T for various air gases based on long-term hourly data for 12 years. Preliminary results were published earlier in [6].

2. Data and methodology

The special ecological station was created jointly by the Obukhov Institute of Atmospheric Physics (IAP) and Lomonosov Moscow State University (MSU) on the territory of the MSU Meteorological Observatory in the south-west of Moscow at about 8 km from the centre of Moscow Kremlin (city centre). This area, located close to the Botanical Garden of Moscow State University, has low-density urban development (only scientific buildings). No plants or other industrial objects operate in the vicinity of MSU area, and two of the three nearby major roads with heavy traffic are located 500 m from the Observatory. Another road (Michurinsky avenue) is closer (at only 200 m), but it was closed for several years due to the subway construction. Thus, the air quality is comparatively high here and represents the cumulative impact of all urban emissions (the so-called 'urban background' without any prevailing emission source). As a result, the air composition measurements there are representative and show real dynamics of air pollution.

The continuously measured surface concentrations of CO_2 and various trace gases, including O_3, NO, NO_2, CO and SO_2 were taken at this station. Ozone was measured using a Dasibi 1008-RS gas analyzer based on recording radiation attenuation with the 253 ηm wavelength. Carbon monoxide was measured by infrared absorption using a TE48S analyzer. Nitrogen oxides were measured by the chemiluminescence method using a TE42C-TL gas analyzer. Sulfur dioxide was detected by the fluorescence method using a Horiba APSA-360 analyzer. The error in measuring of O_3 amount is ±1 ppb (by volume); in measuring NO and NO_2 is up to ±0.05 ppb; in measuring SO_2 is up to ±0.5 ppb; the errors in measuring CO and CO_2 amount are ±10 ppb and ±1 ppm, respectively. The measuring instruments used at the ecological station fully complied with the state standard and international requirements for measuring systems operating at the network of stations of the Global Atmosphere Watch of the World Meteorological Organization (GAW WMO) [5]. The height of all gas intakes at the station was 4 m above the ground.

The total sampling of all data from 2002 to 2014 for O_3 is 100,412 hours; for NO and NO_2 is 72,740 and 72,738 hours, respectively; for CO and CO_2 is 101,074 and 97,948 hours, respectively; for SO_2 is 79,773 hours (measurements of this gas started later, in 2004).

Hourly measurements of T are carried out at the MSU Meteorological observatory uninterruptedly using the classic thermograph records, the data of which are corrected according to the thermometer readings in the Stevenson box every 3 hours. Both liquid thermometers and a thermograph (supplied with a bimetallic sensor) are regularly tested and calibrated in accordance with [7] in the department of verification of the Russian Hydrometeorological Service. As is known, the accuracy of T measurements with a thermometer in the Stevenson box is ±0.2 °C. Both the thermometer and thermograph are installed at a height of 2 m.

It should be noted that this analysis represents the first step without separately considering the daily and annual courses of T. Evidently, any value of T can be detected at different seasons and at different times of the day (e.g., the value of +15 °C in mid-latitudes is typical both for summer nights and for mid-spring and early autumn midday). In fact, the response of the air composition to a certain value of T can be different depending on other factors (presence or absence of short-wave solar radiation, etc.). However, we do not study here the possible influence of other factors on the air composition.

3. Results of analysis

3.1. Air temperature

Let us discuss the main results of our studying. All continuous stationary data on T during the operation of gas analyzers from February 1st, 2002 to July 21st, 2014, were collected and analyzed. The total sample of T values is 109,287 hours during 4554 days. The histogram of T (close to the distribution function) is presented in figure 1. As is seen, it is bimodal. The main winter mode of 1 °C (i.e. from 0.5 to +1.4 °C) is sharp: 5079 hours. The secondary summer mode is wider and less: 3455 and 3459 hours for 15 and 16 °C, respectively. Evidently, the minimum between these maxima (2530 hours at 7 °C) means a comparatively fast transition of T in the annual course in spring and autumn. Distribution wings extend far in both directions: down to -30 and up to +38 °C, thus a range of the T values for 12.5 years is 68 °C. The coldest day was January 18th, 2006 due to strong cold advection of Arctic air [8], when the minimal T was -30.1 °C: it was 4 hours in the early morning of this day and another hour the next day January 19th with T≤ -29.4 °C. On the other hand, an extremely high air temperature was detected during the intense heat wave in the summer of 2010: a total of 5 hours were detected with $T \geq 37.5$ °C at midday: one hour – on July 26th, one hour – on July 29th, three more hours – on August 2nd. It should be noted that the maximal T value for the entire history of meteorological observations at MSU since 1954 was observed on July 29th, 2010: +38.1 °C [9]. It is also important to note that the range from -25 °C (51 hours) to +35 °C (33 hours) is provided with sufficient sampling for statistical analysis (at least 30 separate values for each gradation).

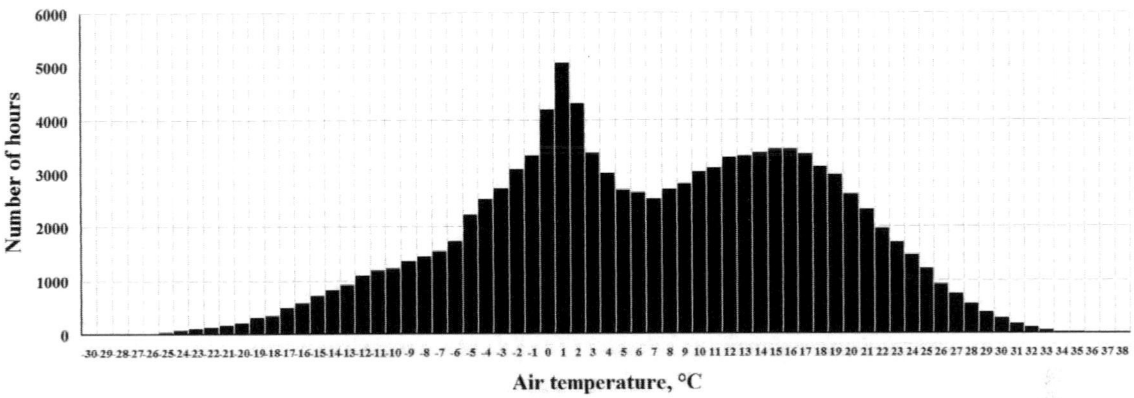

Figure 1. Air temperature T distribution function by hourly data from February, 2002 to August, 2014 in Moscow.

3.2. Ozone

Empirical ozone function of T based on the entire measurement period 2002÷2014 is presented in figure 2. As one can see, there are no significant changes in the wide T range from -7 to +15 °C. For higher air temperature, there is a clear tendency to an increase in O_3 with an increase in T. Evidently, such high T values are mostly observed in mid latitudes under unstable stratification in midday in summer or late spring. Thus, the increase in O_3 is explained by both its chemical source (photochemical dissociation of NO_2 molecules in the presence of ultraviolet solar radiation with a wavelength $\lambda<410$ ηm), and the downward O_3 transport from upper air layers under conditions of intense thermal convection and vertical mixing. It should be noted that extremely high ozone values in strong heat ($T > 30$ °C) are mainly associated with two strong heat wave cases in the summer of 2010 and 2002. In both cases, there was a dense smoky haze from forest and peat fires around the Russian capital above the city. Therefore, in these conditions, additional ozone generation took place in these fire plumes [4]. Thus, at $T = 37$ °C the average O_3 amount is extremely high: 84.7 ppb (data sampling for this T value is 13 hours).

On the other hand, at low air temperature ($T < -7$ °C) there is a tendency for a gradual decrease in O_3 with a decrease in T to at least -21 °C; at even lower T this drop goes to nothing, so the O_3 level is stabilized with an extremely low amount (3÷4 ppb only at T from -25 to -29 °C). Probably, the decrease in the ozone content is due to the prevailing stable stratification, including frequent surface inversions, which prevent vertical mixing and downward O_3 transport from the upper air layers to the surface. Inside the inversion layer, the ozone content decreases due to the dry deposition of this gas on the surface. In addition, in late autumn and in winter, the incoming ultraviolet radiation is extremely low due to the low altitude of the Sun and frequent dense low clouds. Thus, the photochemical generation of O_3 in the lower atmosphere is almost absent at these seasons. As a result, on average the lower T, the less the surface ozone concentration. However, during severe strong frosts at $T < -21$ °C, this tendency is not clear.

3.3. Nitrogen oxide and nitrogen dioxide

Nitrogen oxide and dioxide, unlike other minor air gases, demonstrate significant changes in the middle of the T range (from about -7 to 20 °C for NO and from -7 to 16 °C for NO_2), but these changes are small and non-directional. At higher T, a clear and statistically significant tendency towards a decrease in NO and NO_2 with an increase in T is seen from figure 3.

Evidently, this is due to the predominance of unstable stratification at midday in summer and late spring. Thermal convection leads to the upward dispersion of both gases which are mostly produced by low sources of fuel burning (moving cars). In contrast, at lower T (from -7 to -18 °C), both pollutants grow rapidly with decreasing T. Probably this is a result of both the prevailing stable stratification in winter, preventing dispersion, and the slowing down of NO oxidation rate to NO_2 under negative T values. Therefore, it is not surprising that at $T < -10$ °C the NO/NO_2 ratio becomes >1 whereas at any higher T it is less than 1.

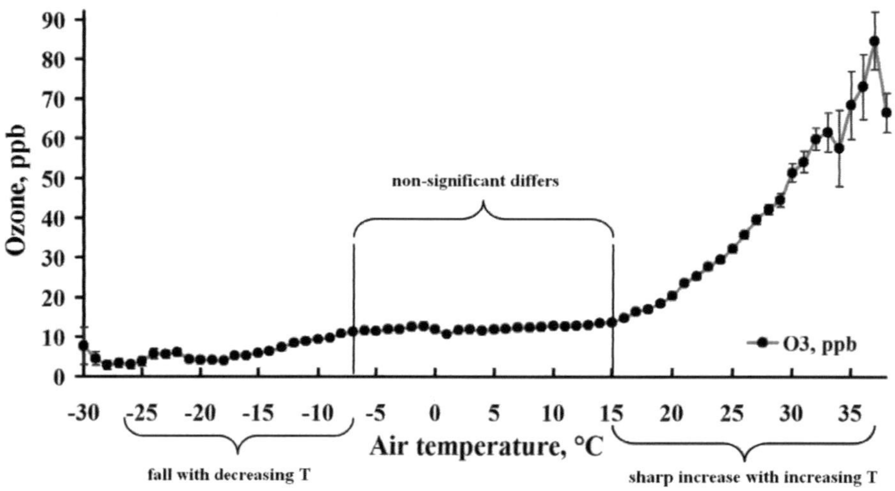

Figure 2. Dependence of the surface ozone concentration on air temperature T by the data from 2002÷2014 in Moscow. Confidence intervals are calculated with a 0.95 confidence probability.

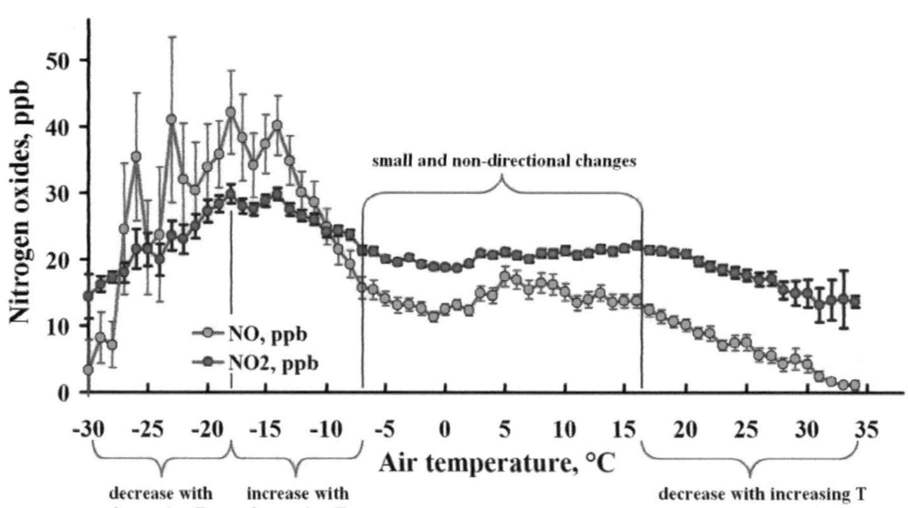

Figure 3. Dependence of NO and NO_2 surface concentrations on air temperature T by the data from 2002÷2014 in Moscow. Confidence intervals are calculated with a 0.95 confidence probability.

Finally, at extremely low T < -18 °C the opposite tendency towards a decrease in NO and NO_2 with a decrease in T is evident. This is probably the result of strong cold advection of Arctic air with low background levels of NO and NO_2.

3.4. Carbon oxide

The CO function of T is presented in figure 4. As with ozone, there are no significant changes in CO amount over the wide range of T from -6 to 27 °C. The sharp increase in CO with increasing T at an even higher air temperature, especially at T > 32 °C, requires an explanation. For this, we separately considered all cases of haze detected at the MSU Meteorological observatory from February 2002 to July 2014. As is known, haze is a meteorological phenomenon of strictly limited horizontal visibility (< 10 km) not due to the presence of small water droplets (like mist), but because of dry particles in the air: smoke, dust, soot particles, etc. Both mist and haze lead to limited visibility, but haze, unlike mist, is accompanied by grey or brown sky color and as a rule by low relative humidity [7]. Thus, in total during the operation of the station, there were 67 separate cases (total – 849 hours) of haze; mainly in extremely hot summers of 2010 and 2002, when forest and peat fires often occurred around Moscow. The longest episode of strong smoky haze (plumes from multiple fires) lasted 112 hours 15 min continuously from August 6th to August 11th, 2010. During these days, horizontal visibility dropped to only 100 m on August 7th and extremely high air pollution was detected.

Figure 4 b) shows a separate calculation of CO average surface concentrations during haze and at all other times. As one can see, the sharp CO growth under extremely high T, especially at T=34 °C, is fully explained only by the presence of haze. In the absence of haze, the CO function of T is smoothed to its highest 37 °C value (all 5 hours of CO measurements with T = 38 °C were detected during a strong smoky haze). Thus, the only significant tendencies of CO levels depending on T is their gradual increase with decreasing T from -6 to -26 °C and, vice versa, a fall under extremely low air temperature below -26 °C. The first tendency is the probable result of the prevailing stable stratification including frequent strong surface inversions which contribute to the accumulation of CO, as well as other fuel combustion products from low sources (mostly emitted by urban transport). In addition, the rate of CO oxidation as well as NO slows down under low T. The opposite tendency for CO to fall with decreasing T for T< -26 °C seems surprising but is apparently explained by the strong cold advection of continental Arctic air from the Arctic region. It should be noted that all 54 hours with T ranging from -27 to -30 °C took place in four days during just two synoptic periods: January 18th, 19th, 20th and February 06th, 2006. Clean Arctic air has a low background CO level and, due to its quick invasion into Moscow region, it did not have enough time to saturate the fuel combustion products above the city. In addition, thermal stratification during intense cold advection is unstable, which also leads to air cleaning due to upward dispersion of pollutants.

Thus, the influence of air temperature on surface CO concentrations, except for only a slowdown in the oxidation rate with decreasing T, is indirect through stratification, advection, smoky haze and other possible factors.

3.5. Carbon dioxide

Besides minor air gases, the surface CO_2 concentration also was measured at the station. As is known, it has been continuously growing globally at almost the same rate, at least since the middle of the 20th century according to the well-known so-called 'Keeling curve' at the Mauna Loa Observatory – e.g., [10]. For the period 2002÷2014, CO_2 there has increased from 373 to 399 ppm. In European Russia (in Obninsk town in Kaluga region, 96 km south-west of the centre of Moscow, according to RPA 'Typhoon' measurements), the CO_2 content is higher than in the Pacific Ocean and ranged from 380 ppm in 2002 to 400 ppm in 2014 [11]. In such a large city as Moscow, it is still slightly higher according to the measurements at IAP and MSU ecological station: from 386 ppm in 2002 (by the data of 11 out of 12 months) to 406 ppm in 2013.

Unlike minor gases, the CO_2 function of T (see figure 5) demonstrates a steady decrease with increasing T over a wide range from -18 to +25 °C. The evident reason for this is the annual course of carbon dioxide, which is especially clear in mid latitudes in the Northern hemisphere due to changes of photosynthesis intensity in different seasons. According to [11], the CO_2 annual course in Obninsk is characterized by minimal values in summer from June to August and maximal values in November and December. In Moscow, according to the data of IAP and MSU ecological station, the annual CO_2 course is similar: the minimum is the same from June to August (382÷385 ppm), but the maximum is shifted forward in time to January and February (404÷405 ppm). Thus, a decreasing function in figure 5 demonstrates a total transition from high winter CO_2 values to low summer ones with increasing T.

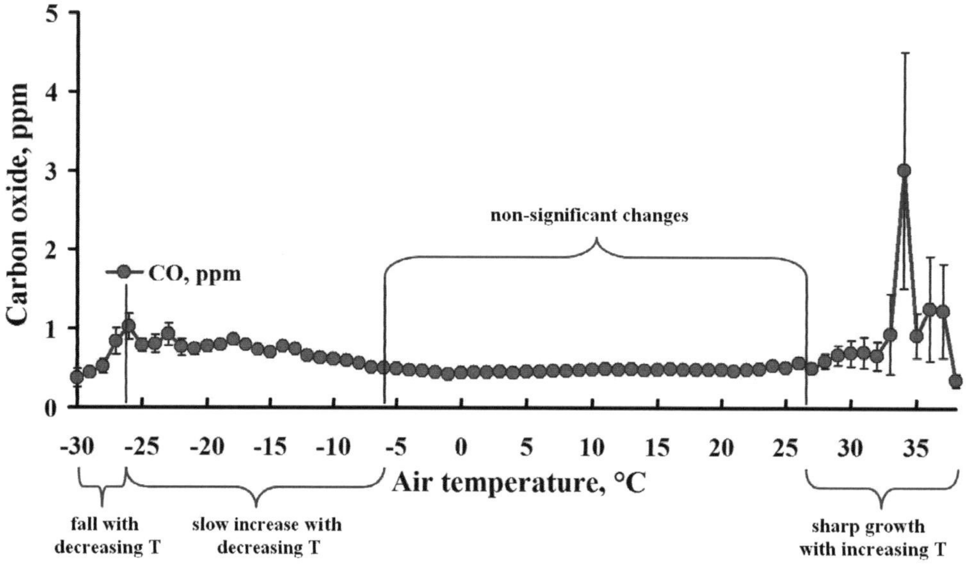

a) Whole data sampling;

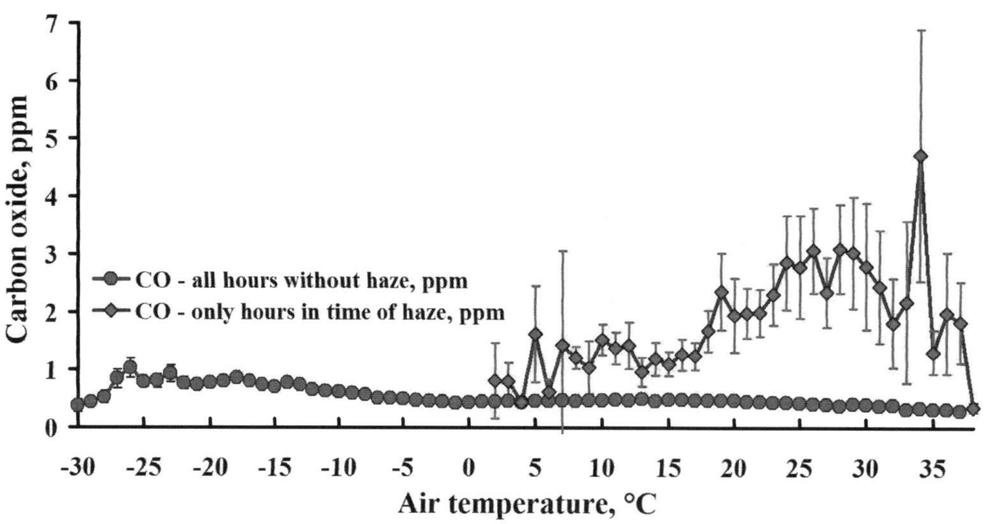

b) Data separation based on the presence of haze.

Figure 4. Dependence of the CO surface concentrations on air temperature T in Moscow according to 2002÷2014 data. Confidence intervals are calculated with the 0.95 confidence probability.

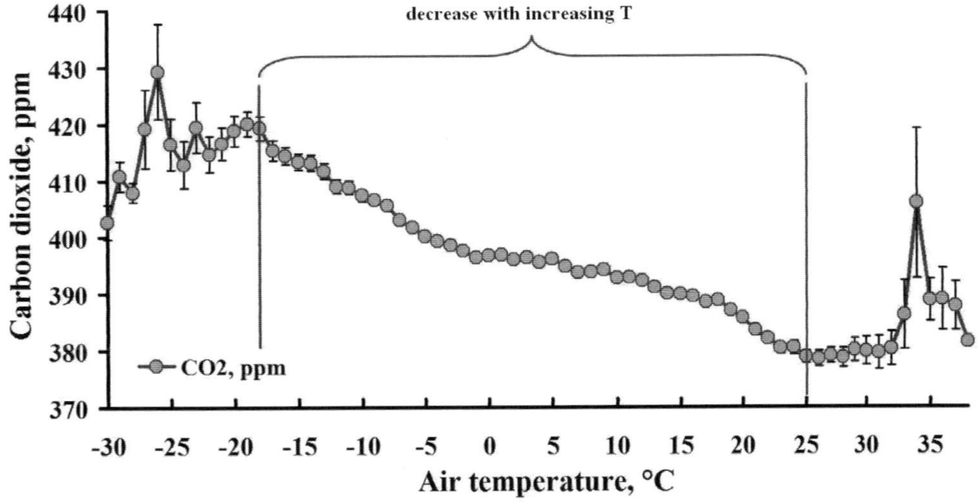

a) Whole data sampling;

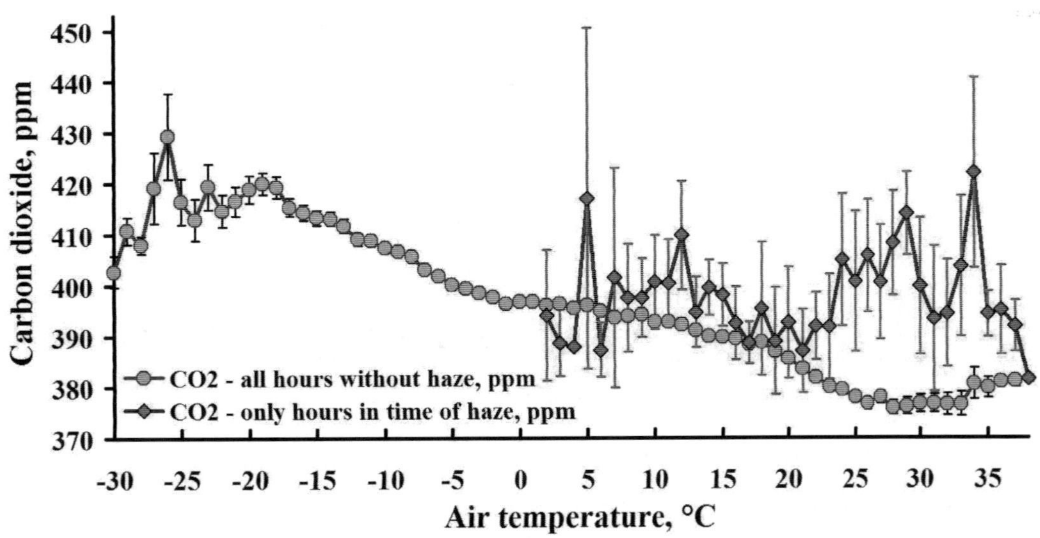

b) Data separation based on the presence of haze.

Figure 5. Dependence of surface CO_2 concentrations on air temperature T in Moscow according to 2002÷2014 data. Confidence intervals are calculated with the 0.95 confidence probability.

A sharp growth of CO_2 concentration for T from +33 to +37 °C relates to strong smoky haze, as well as for CO. Separate recalculation of all values, both during haze and in the absence of this phenomenon (figure 5 b), demonstrates that a sharp maximum of 406 ppm at T = +34 °C is fully explained by the increase of CO_2 as a product of combustion in fire plumes. Haze was detected during 22 out of 36 hours of CO_2 measurements at +34 °C, and the average value for this partial sampling is 422 ppm, whereas the average value for the rest 14 hours without haze at +34 °C is only 380 ppm.

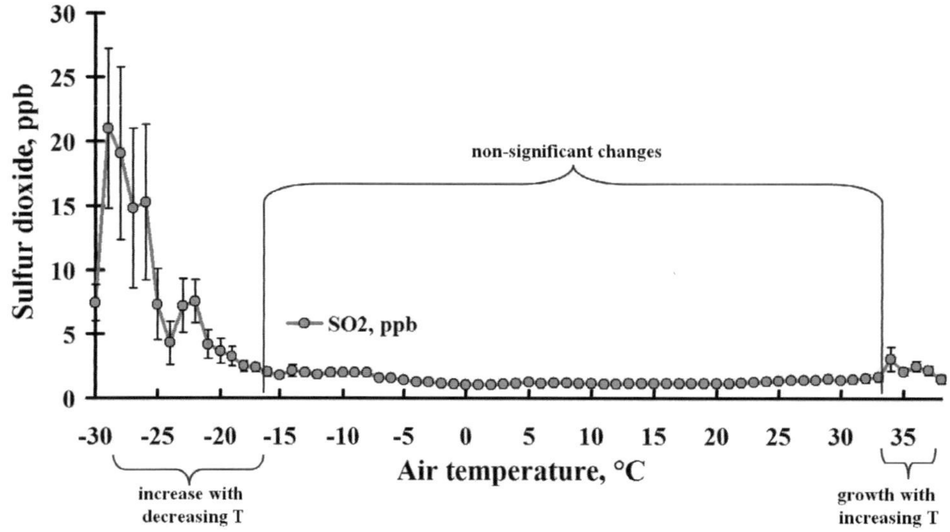

a) Whole data sampling;

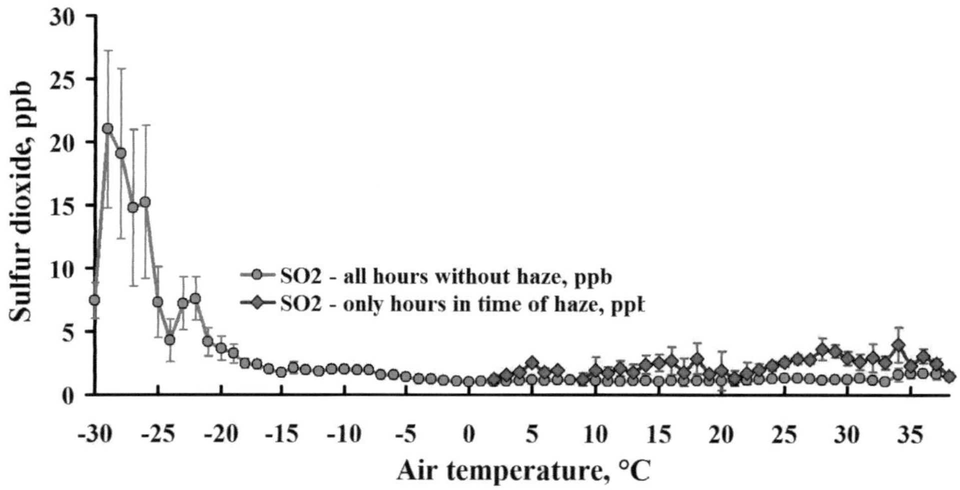

b) Data separation based on the presence of haze.

Figure 6. Dependence of surface SO_2 concentrations on air temperature T in Moscow according to 2004÷2014 data. Confidence intervals are calculated with the 0.95 confidence probability.

However, it should be noted that in the conditions of extremely hot weather, even in the absence of haze, the CO_2 levels are a bit higher than at lower air temperature. As one can see in figure 5 b), the average concentration at T from +28 to +33 °C is 376÷377 ppm, whereas at T from +34 to +37 °C it is 380÷381 ppm (at the highest air temperature of +38 °C, as well as for CO, all 5 hours of measurements were received during haze). The difference between the last two average values at +36 and +37 °C and any value between +28 to +33 °C is statistically significant with the 0.95 confidence probability as seen from the confidence intervals. The reason for this effect is the oppression of the activity of trees during strong heat wave, which leads to the slowing of photosynthesis rate [12]. Thus, the CO_2 growth in extremely hot weather is the result of at least two factors: both oppression of trees and a high concentration of this gas in plumes of fires during smoky haze. At low values T < -18 °C, the CO_2

IOP Conf. Series: Earth and Environmental Science 1040 (2022) 012007 doi:10.1088/1755-1315/1040/1/012007

function of T is not monotonous. In general, this gas decreases with decreasing T during severe frosts, which may be the result of low background values of CO_2 in the Arctic air.

3.6. Sulfur dioxide

One more minor air gas we are studying is sulfur dioxide. As can be seen in figure 6, it has the longest range of non-significant changes depending on T among other gases: from -16 to 33 °C. At higher T, the growth of SO_2 is weaker than that of CO and CO_2. Nevertheless, the effect of smoky haze is also detected for SO_2: a separate calculation without hours of haze demonstrates a smoothed SO_2 function. Thus, the plumes of fires during the catastrophic heat wave in summer 2010 contained sulfur dioxide along with other combustion products. A sharp increase with decreasing air temperature at T < -16 °C is explained not only by stable stratification, but also by an increase in urban heating – especially during severe frosts. As is known, in the extremely cold winter of 2006, reserve fuel (fuel oil and coal) with an increased sulfur content was used in Moscow [5,8]. Thus, unlike NO, NO_2 and CO, the SO_2 concentration in Moscow remained anomalously high even at extremely low T values.

Thus, almost any influence of air temperature on air composition, as a rule (except for only changes in the oxidation rate) is indirect and manifests itself in thermal stratification, smoky haze, photosynthesis activity, advection, the intensity of urban heating emissions, and other factors.

4. Conclusions

1. The influence of air temperature T on air content is usually relatively weak and, as a rule, indirect. In a wide range of T from -6÷-7 °C for O_3, NO, NO_2 and CO to 15÷16 °C for O_3, NO_2, 20 °C for NO and 27 °C for CO, no significant changes or any direct tendencies were found. For SO_2 this range is even wider: from -16 to 33 °C. Unlike minor air gases, carbon dioxide demonstrates a gradual decrease with increasing T from winter to summer up to 25 °C due to its annual course connected with intensive photosynthesis during the summer growing season.
2. At high T values (>15 °C for O_3, >16 °C for NO_2, and >20 °C for NO), clear tendencies towards the growth of ozone and, vice versa, a decrease in nitrogen oxides are revealed. Both tendencies seem to be the consequence of unstable stratification, i.e. thermal convection, which enhances vertical mixing during daytime in summer and in late spring. More intense mixing leads to upward dispersion of NO and NO_2 and, on the contrary, to downward transport of O_3 to the surface from upper air layers. In addition, the incoming solar radiation also leads to an additional ozone increase due to its chemical generation.
3. At comparatively low T values (< -7 °C for O_3, from -7 to -18 °C for NO and NO_2, from -7 to -26 °C for CO and < -16 °C for SO_2) a fall of ozone and, vice versa, an increase in NO, NO_2, CO and SO_2 with decreasing T were detected due to prevailing stable stratification, including frequent surface inversions, which prevent vertical mixing of air dispersion of minor gases. This leads to O_3 decrease due to its dry deposition on the surface and to the accumulation of other minor air gases near the surface. Besides, slowing down the oxidation rate of NO to NO_2 also leads to increase of nitrogen oxide. The SO_2 growth is also the result of increased urban heating.
4. At extremely low T during severe frosts (< -18 °C for NO, NO_2 and < -26 °C for CO), an opposite tendency is revealed for a decrease in NO, NO_2 and CO with a decrease in T is detected as a result of strong and rapid cold advection of clean Arctic air with low a background of these gases to Moscow region. Unlike these gases, the SO_2 levels remain high during severe frosts due to the use of reserve oil fuel for urban heating.
5. The seeming growth of CO at high T (> 27 °C) is fully explained only by the influence of smoky haze from forest and peat forests as a result of strong heat waves in the summer of 2010 and 2002. The growth of CO_2 at high (> 25÷28 °C) T is the result of both smoky haze, and the oppression of green trees. Smoky haze at extremely high T also affects SO_2 concentration.

Acknowledgments

The authors express their deep gratitude to their colleagues from the Meteorological Observatory of Lomonosov Moscow State University and Obukhov Institute of Atmospheric Physics for their help. This work was partially supported by the Russian Scientific Foundation (Project 21-17-00210).

References

[1] Brimblecombe P 1986 *Air Composition and Chemistry* (Cambridge, UK: Cambridge University Press) p 224

[2] Bezuglaya E Yu and Smirnova I V 2008 *Air in the Cities and Its Changes* (St. Petersburg, Russia: Asterion) p 254, in Russian

[3] Bezuglaya E Yu 1980 *Meteorological Potential and Climatic Features of Urban Air Pollution* (Leningrad, USSR: Gidrometeoizdat), p 184, in Russian

[4] Elansky N F, Mokhov I I, Belikov I B, Berezina E V, Elokhov A S, Ivanov V A, Pankratova N V, Postylyakov O V, Safronov A N, Skorokhod A I and Shumskii R A 2011 *Izvestiya, Atmospheric and Oceanic Physics* **47** 6 672–681

[5] Elansky N F, Lokoshchenko M A, Trifanova A V, Belikov I B, and Skorokhod A I 2015 *Izvestiya, Atmospheric and Oceanic Physics* On Contents of Trace Gases in the Atmospheric Surface Layer over Moscow **51** 1 30-41

[6] Lokoshchenko M A, Elansky N F, and Trifanova A V 2014 *Vestnik Ross. Akad. Estestv. Nauk* Influence of meteorological conditions on the air pollution in Moscow **14** 1 64–67, in Russian

[7] Slabkovich G I, Ed., 1985: Instructions for hydrometeorological stations and posts: **3** Part I (Leningrad, USSR: Gidrometeoizdat), p 300, in Russian

[8] Lokoshchenko M A, Elansky N F, Malyashova V P and Trifanova A V 2008 *Atmospheric and Oceanic Optics* Dynamics of sulfur dioxide surface concentration in Moscow **21** 5 384-391

[9] Lokoshchenko M A 2012 *Izvestiya, Atmospheric and Oceanic Physics* Catastrophic Heat of 2010 in Moscow from Data of Ground-Based Meteorological Measurements **48** 5 463-475

[10] Incropera F P 2016 *Climate Change: A Wicked Problem* (Cambridge, UK: Cambridge University Press) p 337

[11] Bardin M Yu et al. 2020 *A report on climate features on the Territory of the Russian Federation in 2019* Report of Russian Federal Service for Hydrometeorology and Environmental Monitoring (Moscow, Russia: Roshydromet), p 97, in Russian

[12] Teskey R, Wertin T, Bauweraerts I, Ameye M, McGuire M A, and Steppe K 2015 *Plant, Cell and Environment* Review Responses of tree species to heat waves and extreme heat events **38** 1699–1712

The parameters of internal gravity waves in the atmosphere from the amplitude fluctuations of radio occultation signals

V Kan[1], **M E Gorbunov**[1,2], **A V Shmakov**[1], **O V Fedorova**[1] **and V F Sofieva**[3]

[1] A M Obukhov Institute of Atmospheric Physics Russian Academy of Sciences, Moscow, Russia
[2] Hydrometeorological Research Center of Russia, Moscow, Russia
[3] Finnish Meteorological Institute, Helsinki, Finland

E-mail: valerykan@list.ru

Abstract We develop a method for the reconstruction of the statistical parameters of internal gravity wave (IGW) in the atmosphere from the amplitude fluctuations of signals acquired in radio occultation observations. We discuss the choice of the model of the IGW spatial spectrum, derive the relationships between the IGW spectra and the amplitude fluctuation spectra, develop the reconstruction algorithm and estimate its uncertainties. The IGW spectrum parameters to be retrieved are the outer (dominant) scale and the structure characteristic, the latter determining the power of saturated IGW. The method was tested on the COSMIC (Constellation Observing System for Meteorology, Ionosphere and Climate) observations acquired in the winter and summer seasons 2011. The IGW parameters were reconstructed in the height range from the upper boundary of the tropopause to 28 km. We show the altitude-latitudinal distributions of the IGW parameters and two integral characteristics: the variances of the temperature fluctuations and the IGW potential energy. We perform a detailed comparison of our results with those obtained from radiosonde and radio occultation observations. The developed method can be applied for the global monitoring of IGW parameters and activity in the middle atmosphere.

1. Introduction

Internal gravity waves (IGW) in the atmosphere are the main source of small-scale fluctuations of wind and temperature with characteristic vertical scales from several kilometers to a hundred meters and periods from five minutes to about ten hours. They play the key role in energy exchange and atmospheric global circulation. IGW breaking generates turbulence in the middle atmosphere [1]. In order to include IGW related processes into global atmospheric circulation models, their statistical parameterization based on the global distribution of IGW activity, is necessary. The importance of this problem is confirmed by a large number of publications in this area. The most accurate observations of wind and temperature are acquired by radiosondes [2–5]. However, it is only satellite constellations carrying corresponding instruments that can really provide the global monitoring of IGW. Widespread are studies based on radio occultation (RO) observations performed by low-Earth Orbiters (LEO) receiving the highly stable signals of global navigation satellite systems, such as GPS [6–10]. For example, the COSMIC constellation consisting of 6 LEOs, at its maximum capability provided about 3000 events per day, homogeneously distributed all over the globe. The GPS signals at wavelengths of λ_1=19.03 cm and

λ_2=24.42 cm are used for the measurements of the atmospheric excess phase, from which atmospheric temperature profiles including both regular and random components, are retrieved. A detailed analysis of the modern principles of RO sounding of the atmosphere can be found in the monograph [11].

The IGW parameters indicate a high spatiotemporal variability [1]. This fact, along with the importance of the role played by IGW in the atmospheric dynamics, motivates the development of new methods of their global monitoring. In our studies, we use the statistical description of the fluctuations in the amplitude of RO signals probing the atmosphere. In [12], we derived the main relationships connecting the measured spectra of the RO signal amplitude and phase fluctuations with the IGW spatial spectra. In [13], we developed the method for the reconstruction of the main statistical parameters of IGW from the amplitude fluctuation spectra; later the method was tested on COSMIC observations in 2011 [14].

This paper is a brief review of [12-14] complemented with a detailed comparison with other studies. The approximations adopted in our analysis impose weaker restrictions upon the spatial resolution, as compared to the standard geometric optical (GO) retrieval of temperature profiles [7, 8, 11]. From the COSMIC observations in January–February and June–July 2011, we reconstruct the altitude-latitudinal distributions of the main parameters of the IGW spectrum: the structure characteristic and the outer scale; we also obtain the variance of temperature fluctuations and the potential energy per unit mass in the height range from the tropopause to 28 km. We compare our results with other studies based on radiosonde, RO, and stellar occultation observations.

2. The reconstruction method

Fluctuations in the amplitude and phase of signals probing the atmosphere, are primarily caused by turbulence and IGW. In optical occultations, the contributions of these two sources of the light intensity fluctuations are nearly equal [15]. In RO observations, the main role is played by IGW, while the contribution of the Kolmogorov turbulence is 2–3 orders of magnitude smaller [12]. This is explained by the spectral slopes of IGW and turbulence and by a thousandfold difference in the main (Fresnel) scale for optical and radio waves [16]. Therefore, RO observations constitute a convenient tool for the study of IGW.

Our method is based on the statistical description of the random field of the atmospheric inhomogeneities and the RO signal fluctuations caused by them. In the limb observation geometry, a radio wave interacts with a representative ensemble of IGW. To take into account the effects of the atmospheric sphericity in RO observations, we assume that the random field of refractivity inhomogeneities is locally homogeneous in a spherical layer of the atmosphere.

The statistical description of wave propagation in the atmosphere is based on the 3-D spectrum of relative fluctuations of air refractivity, or temperature if humidity is negligible. We employ the 3-D power-law universal spectrum of the temperature fluctuations caused by IGW, developed by A.S. Gurvich [17].

$$\begin{aligned} \Phi_{\delta T/\overline{T}}(\kappa_x, \kappa_y, \kappa_z) &= C_W^2 \eta^2 (\kappa^2 + \mathrm{K}_W^2)^{-\frac{5}{2}}, \\ \kappa^2 &= \kappa_z^2 + \eta^2 \kappa_\perp^2, \\ \kappa_\perp^2 &= \kappa_x^2 + \kappa_y^2, \end{aligned} \tag{1}$$

where $\delta T/\overline{T}$ are the relative temperature fluctuations, $\overline{T}$ is the seasonal-regional mean temperature, C_W^2 is the structure characteristic determining the intensity of temperature fluctuations, η is the anisotropy coefficient characterizing the ratio of the characteristic horizontal to vertical scales (for IGW $\eta \gg 1$), κ_z is the vertical wavenumber, κ_x and κ_y are the horizontal wavenumbers, x axis coincides with the line of sight, $\mathrm{K}_W = 2\pi/L_W$, and L_W is the vertical outer (dominant) scale.

The single-sided 1-D vertical spectrum equals the integral of $\Phi_{\delta T/\overline{T}}$ over κ_x, κ_y:

$$V_{\delta T/\overline{T}}(\kappa_z) = \frac{4\pi}{3} C_W^2 (\kappa_z^2 + \mathrm{K}_W^2)^{-\frac{3}{2}}, \qquad \kappa_z \geq 0. \tag{2}$$

The main parameters of spectra (1) and (2), which have to be reconstructed, are the outer scale $L_W = 2\pi/\mathrm{K}_W$ and the structure characteristic C_W^2 [13, 14]. The outer scale L_W defines the transition from saturated to unsaturated waves at large scales $\kappa_z \leq \mathrm{K}_W$. The structure characteristic C_W^2 is the only parameter that defines the power of saturated waves with $\kappa_z > \mathrm{K}_W$. Spectrum Φ_W defined by equation (1) is the 3-D generalization of the known universal model of saturated IGW with the vertical spectrum [18, 19]:

$$V_{\delta T/\overline{T}} = \beta \frac{\omega_{B.V.}^4}{g^2} \kappa_z^{-3}, \tag{3}$$

where $\beta \approx 0.1$ is the IGW model coefficient that was first introduced theoretically and then confirmed experimentally, $\omega_{B.V.}$ is the Brunt–Väisälä frequency, and g is the gravity acceleration. The structure characteristic C_W^2 is related to the parameters of the universal spectrum (3) as follows:

$$C_W^2 = \frac{3\beta\omega_{B.V.}^4}{4\pi g^2}. \tag{4}$$

We adopt the approximations of phase screen and weak fluctuations, which are traditional for the statistical interpretation of occultation observations. Due to the averaging of the RO amplitude fluctuations over a large Fresnel zone, the condition of weak fluctuations is fulfilled for perigee heights down to several kilometers over the Earth's surface. These approximations greatly simplify the solution of the forward problem and result in a transparent analytical expression of the vertical spectrum of RO amplitude fluctuations though the vertical IGW spectrum in equation (2) [12, 13]:

$$V_{\delta A/\overline{A}}(\kappa_z) = \frac{k^2\overline{\psi}^2}{\sqrt{1+\kappa_z^2 H_0^2}} V_{\delta T/\overline{T}}(\kappa_z) \sin^2\frac{\kappa_z^2}{\kappa_F^2} = \frac{4\pi}{3}\frac{k^2\overline{\psi}^2}{\sqrt{1+\kappa_z^2 H_0^2}} C_W^2 (\kappa_z^2 + K_W^2)^{-\frac{3}{2}} \sin^2\frac{\kappa_z^2}{\kappa_F^2} \tag{5}$$

where $k = 2\pi/\lambda$, $\overline{\psi}$ is the regular component of the excess phase [17], $H_0 = 6-8$ км is the atmospheric vertical scale, $\kappa_F = 2\pi/\rho_F$, $\rho_F = \sqrt{\pi q \lambda D}$ is the vertical Fresnel scale, q is the refractive attenuation, $D = D_t D_r/(D_t + D_r)$ is the reduced observation distance ($D_r \approx$3,200 km and $D_t \approx$25,800 are the distances from LEO and GPS satellites to the ray perigee, respectively). The spectra in equation (5) are written down for wavenumbers $\kappa_z \geq 0$ in the phase screen plane. The sine factor describes the diffraction after the phase screen, and the square root term in the denominator takes into account the sphericity of the atmosphere. Due to the strong anisotropy of IGW, practically all occultation events can be considered as vertical ones [12]. The reconstructed parameters were used for the evaluation of the integral characteristics describing the IGW activity: the temperature fluctuation variance $\sigma_{\delta T}^2$ and the potential energy per unit mass E_p [1, 4]:

$$\sigma_{\delta T}^2 = \int_0^\infty V_{\delta T/\overline{T}}(\kappa_z) d\kappa_z = \frac{1}{3\pi}\overline{T}^2 C_W^2 L_W^2, \tag{6}$$

$$E_p = \frac{g^2}{2\overline{T}^2 \omega_{\mathrm{B.V.}}^2} \sigma_{\delta T}^2 \tag{7}$$

Our method fits the theoretical spectra of amplitude fluctuation in equation (5) to the experimental ones obtained from COSMIC observations, by varying the two parameters of the IGW model: the outer scale L_W and the structure characteristic C_W^2, and minimizing the residual. The residual is defined as the squared difference between the theoretical and experimental spectra integrated over the wavenumber.

The standard method of temperature profile retrieval from RO signals uses the approximations of the geometrical optics and the local spherical symmetry of the atmosphere [7, 8, 20]. The GO resolution is defined by the Fresnel zone scale ρ_F, which is about 1.5 km in the stratosphere [7, 20]. The approximation of the local spherical symmetry limits the horizontal resolution by wavelengths of about 250-300 km [21]. In our method, we take diffraction into account, and the vertical resolution, which is mainly limited by the receiver noise, is about 0.5 km [12]. Moreover, the condition of the spherical symmetry is imposed not upon the whole structure of the atmosphere, but upon the second moments of the random inhomogeneity field only [17]. This result in weaker restrictions upon the IGW anisotropy and, accordingly, upon the horizontal resolution of the method [12, 17]. A higher spatial resolution allows, in particular, separately reconstructing the two parameters, the outer scale and the structure characteristic. A detailed discussion of wave-optical methods of retrieving high-resolution temperature profiles from RO observations can be found in [11].

In order to estimate the uncertainties of the two reconstructed parameters, we used fragments of a large ensemble consisting of about 30,000 events from spring 2011, under the assumption of its statistical homogeneity. Our tests indicated that the reconstruction accuracy is 10–20% for the outer scale and 20–40% for the structure characteristic [13].

3. Altitude-latitudinal distribution of IGW parameters

For the validation of the method, we studied the altitude-latitudinal distribution of IGW parameters in latitude bands 0°–20°, 20°–40°, 40°–60°, 60°–90° in the Northern and Southern hemispheres for winter and summer COSMIC observations in 2011. For each season, we used 20,000 events. The IGW parameters were reconstructed in the height range from 4 km above the tropopause up to 28 km. Figure 1 presents the profiles of the outer scale for the local winter (panels A and C) and local summer (panels B and D) in both hemispheres. For the comparison, we present the multi-year measurements of the outer (dominant) scale obtained by other authors from radiosonde temperature observations [2–5, 22], stellar scintillations [23] and RO in the GO approximation [20, 21], the latter being justified by the fact that the outer scale exceeds the Fresnel zone size.

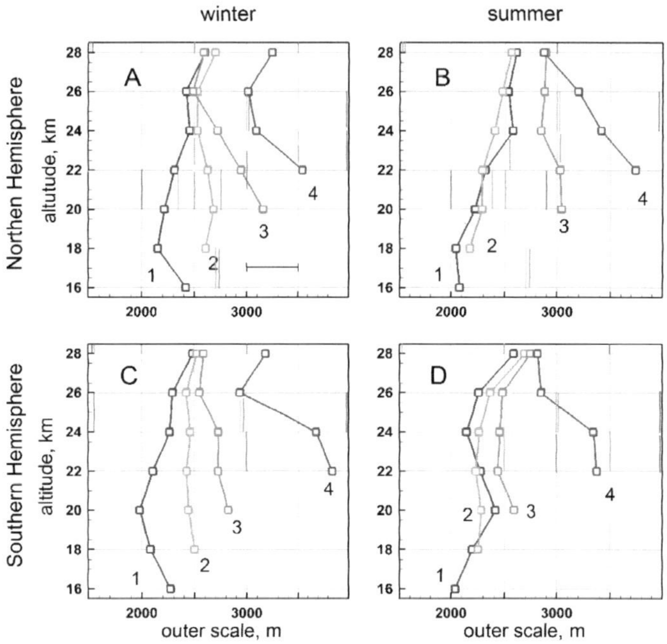

Figure 1. The outer scale profiles. Upper panels A and B: Northern hemisphere; lower panels C and D: Southern hemisphere. Left panels A and C: local winter; right panels B and D: local summer. The numeric labels of the curves correspond to the latitudinal bands: 1: 60°–90°; 2: 40°–60°; 3: 20°–40°; 4: 0°–20° in both hemispheres. The horizontal bar in panel A shows the uncertainty estimate. The vertical bars of the corresponding colors represent the data from other studies [2–5, 20–23].

The outer scale in our observations varies from 2 to 4 km. There is a noticeable increase of the outer scale from polar to tropical latitudes, especially in the lower latitude range. Our data are generally in close agreement with the other studies.

Figure 2 presents the profiles of the structure characteristic for the local winter (panels A and C) and local summer (panels B and D) in both hemispheres. Recall that the structure characteristic C_W^2 is the

IOP Conf. Series: Earth and Environmental Science 1040 (2022) 012008 doi:10.1088/1755-1315/1040/1/012008

only parameter of the saturated IGW spectrum. The structure characteristic varies from 10^{-11} m^{-2} to 10^{-10} m^{-2}. It indicates a noticeable increase from the poles to the tropics. For a comparison, we also show the results obtained from radiosonde measurements [2, 3, 24] and satellite stellar scintillation observations [23]. RO data, due to their insufficient vertical resolution, do not allow obtaining a reliable estimate of the structure characteristic. On the other hand, radiosonde observations lack the global coverage, which does not leave much data for the comparison. Optical and RO observations only overlap at the upper border of the altitude range used in this study [23].

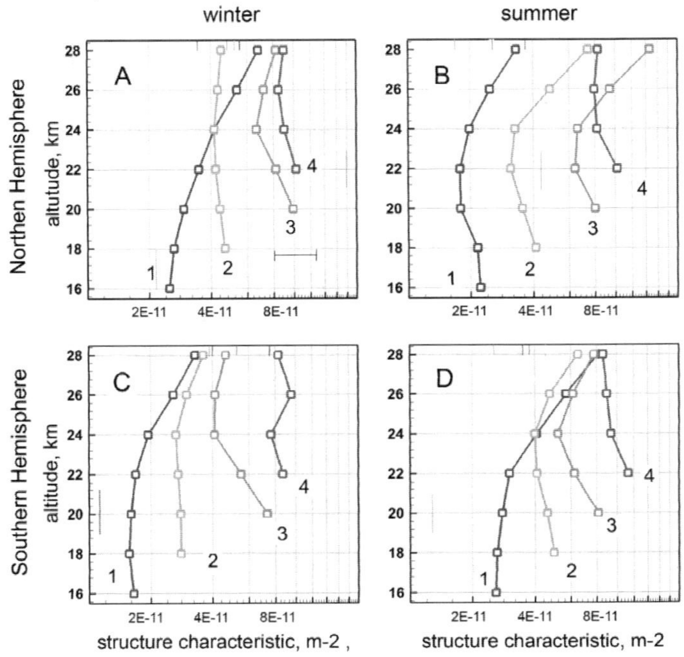

Figure 2. The structure characteristic profiles. The panel notations are the same as in Figure 1. The vertical bars represent the data from other studies [2, 3, 23, 24].

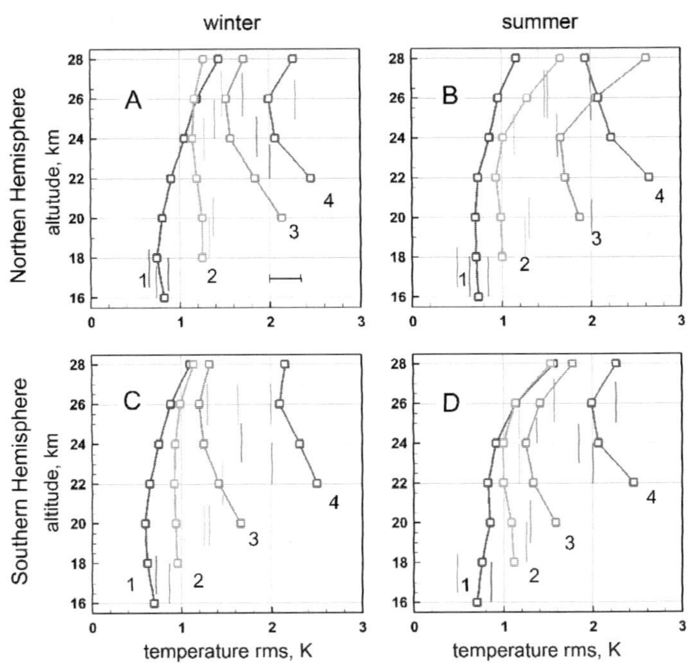

Figure 3. The temperature fluctuation profiles. The panel notations are the same as in Figure 1. The horizontal bar in panel A shows the variability estimate of the mean-square temperature deviations. The vertical bars represent the data from other studies [8, 21, 24–27].

The structure characteristic estimates indicate a high variability. Typical are the results obtained from the radiosonde measurements performed during 5 years at the same station located on a flat terrain [3]. The complete database was processed using the same algorithm. Still, the spectral amplitudes vary ten and more times for different measurements in close conditions.

Figure 3 presents the profiles of the mean-square temperature fluctuations for the local winter (panels A and C) and local summer (panels B and D) in both hemispheres. The vertical bars of the corresponding colors show the results obtained from radiosonde measurements [24–26] and RO retrieved temperature profiles [8, 21, 27]. The temperature fluctuation variance is determined by large-scale waves; therefore, it can be reliably estimated from both radiosonde data and RO observations processed in the GO approximation. The mean-square temperature deviations in our measurements are in the range from 1 to 3 K. As expected, the temperature fluctuations indicate a significant increase from the poles to the tropics. The mean-square temperature deviations obtained in our and other studies are in close agreement with each other.

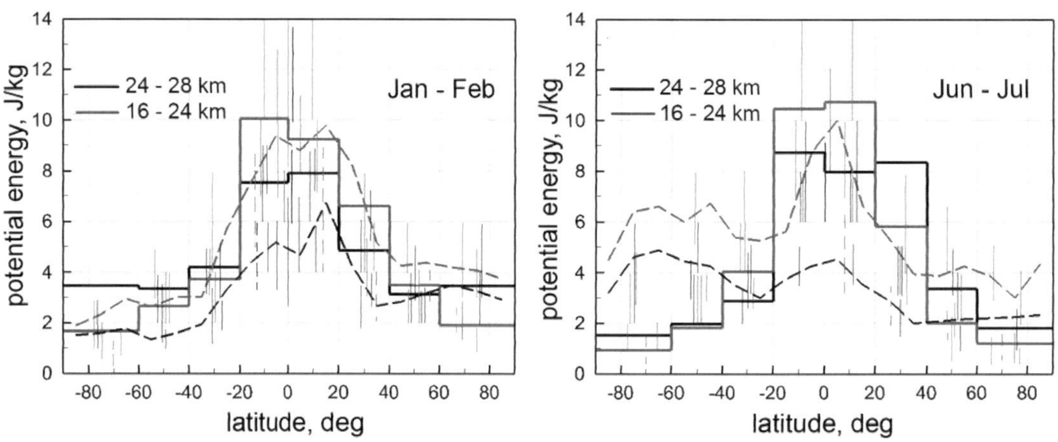

Figure 4. The stepwise solid lines show the latitudinal distribution of the integral potential energy of IGW for the altitude intervals of 16–24 km (red) и 24–28 km (black). The dashed lines show the measurement results [8] for 15–25 km (red) and 20–30 km (black). The solid vertical bars show the data of RO observations [9, 10, 21, 27–29, 31], stellar scintillation observations [30], dashed bars show the radiosonde measurement data [4, 24, 29]. The height of the bars is the estimate of the measurement variability, including the uncertainties.

A very important characteristic of the IGW activity is their energy per unit mass. The radiosonde observations allow the determination of both energy components: the kinetic one from the pulsation power of the horizontal wind component and the potential one from the temperature fluctuation intensity. In RO monitoring, the IGW activity indicator is the potential energy per unit mass [1, 4, 21], because the kinetic energy is linked to the potential one by the polarization relationships: in the linear theory their ratio equals the modulus of the exponent of the temporal fluctuation spectrum of temperature [1, 2]. Figure 4 shows the latitudinal distributions of the integral potential energy defined by equation (7), averaged over the two altitude intervals: 16–24 km and 24–28 km, for January–February (left panel) and June–July (right panel). Noticeable are the high values of E_p in the equatorial zone, reaching 10–11 J/kg in the altitude range 16–24 km and 8–9 J/kg in the altitude range 24–28 km. This indicates the leading role of the convective processes for IGW generation in the tropics [1, 8, 20]. An enhanced IGW activity in the equatorial zone may be partially explained by the Kelvin wave contribution into large-scale temperature fluctuations [8, 9, 27]. The IGW activity indicates a fast decay with increasing latitudes, reaching values E_p of 3–4 J/kg in the middle latitudes and 1–2 J/kg in the high latitudes. In these zones, the potential energy E_p for altitudes 24–28 km is slightly higher and has a flatter distribution as compared to the lower altitude interval. The global potential energy distribution is approximately symmetric for the two hemispheres, however, in Northern latitudes over 60° the winter values of E_p are approximately twice as high as the summer ones.

As noticed above, the temperature fluctuation intensity and IGW potential energy defined by equations (6) and (7), respectively, are determined by large-scale waves close to dominant ones. This justifies the estimates based on RO temperature profiles obtained in the GO approximation, which are intensively used during the two last decades. Below we compare our results with pioneering study [8], which analyzed the global distribution of the IGW potential energy inferred from the GPS/MET RO data 1995–1997. As follows from Figure 4, our results mostly agree with [8], although there are some differences. In [8], an enhanced IGW activity at high latitudes in local winter is noticed; in the Northern hemisphere (left panel) this effect is modest, while in the Southern hemisphere (left panel) for 50°S–80°S latitude band it is noticeable. Besides, in the altitude interval 24–28 km our estimates of E_p are higher. Figure 4 also shows the vertical bars of the corresponding colors representing the results of other studies, obtained both from RO, stellar scintillation observations, and radiosonde measurements. We conclude that our results agree with those from other studies, within the high variability of IGW potential energy.

4. Conclusions

Here we summarize the main statements of our study.

We developed a method of the reconstruction of the main statistical parameters of IGW in the atmosphere: the outer scale and the structure characteristic, from the measurements of amplitude fluctuations in RO experiments.

The approximations of the phase screen and weak fluctuations allowed the derivation of simple relations between the vertical spectra of IGW and RO signal amplitude fluctuations. A higher spatial resolution, as compared to the traditional GO approximation used in temperature profile retrieval from RO observations, allows the separate reconstruction of the IGW parameters.

Our validation indicated that the method can be successfully applied for the global monitoring of the statistical parameters and activity of IGW in the middle atmosphere.

Acknowledgments

The work of V. Kan, M.E. Gorbunov, A.V. Shmakov and O.V. Fedorova was supported by the Russian Foundation for Basic Research, project no. 20-05-00189 A. The work of V.F. Sofieva was supported by the Academy of Finland (Centre of Excellence of Inverse Modelling and Imaging and the TT-AVA project).

References

[1] Fritts D C and Alexander M J 2003 *Rev. Geophys.* **41** 3–1–64
[2] Tsuda T, Van Zandt T E, Mizumoto M, Kato S and Fukao S 1991 *J. Geophys. Res.* (96) 17265–78.
[3] Nastrom G D, Van Zandt T E and Warnock J M 1997 *J. Geophys. Res.* **102** 6685–701
[4] Wang L, Geller M A, Alexander M J 2005 *J. Atmos. Sci.* **62** 125–42
[5] Murphy D J, Alexander S P, Klekociuk A R, Love P T and Vincent R A 2014 *J. Geophys. Res.* **119** 11,973–96
[6] Ware R et al. 1996 *Bull. Amer. Soc.* **77** 19–40
[7] Kursinski E R, Hajj G A, Schofield J T, Linfield R P and Hardy K R 1997 *J. Geophys. Res.* **102** 23429–65
[8] Tsuda T, Nishida M, Rocken C and Ware R H 2000 *J. Geophys. Res.* **105** 7257–73
[9] Schmidt T, Alexander P and de la Torre A 2016 *J. Geophys. Res. Atmos.* **121** 4443–67
[10] Rapp M, Dörnbrackl A and Kaifler B 2018 *Atmos. Meas. Tech.* **11** 1031–48

IOP Conf. Series: Earth and Environmental Science 1040 (2022) 012008 doi:10.1088/1755-1315/1040/1/012008

[11] Gorbunov M E 2019 *Physical and Mathematical Principles of Satellite Radio Occultation Sensing of the Earth's Atmosphere* (GEOS, Moscow, 2019) p 288 [in Russian]
[12] Kan V, Gorbunov M E and Sofieva V F 2018 *Atmos. Meas. Tech.* **11** 663–80
[13] Kan V, Gorbunov M E, Shmakov A V and Sofieva V F 2020 *Izvestiya, Atmospheric and Oceanic Physics* **56** 435–47
[14] Kan V, Gorbunov M E, Fedorova O V and Sofieva V F 2020 *Izvestiya, Atmospheric and Oceanic Physics* **56** 564–75
[15] Sofieva V F, Gurvich A S, Dalaudier F and Kan V *J. Geophys. Res.* 2007 **112** D12113
[16] Tatarskii V I 1971 *The Effects of the Turbulent Atmosphere on Wave Propagation* (US Dept. Commerce: Springfield) p 417
[17] Gurvich A S and Brekhovskikh V L 2001 *Waves in Random Media* **11** 163-81
[18] Smith S A, Fritts D C and Van Zandt T E 1987 *J. Atmos. Sci.* **44** 1404–10
[19] Dewan E M and Good R F 1986 *J. Geophys. Res.* **91** 2742–8
[20] Steiner A K and Kirchengast G 2000 *J. Atmos Ocean. Technol.* **17** 495–503
[21] Tsuda T and Hocke K 2002 *J. Meteor. Soc. Japan.* **80** 925–938
[22] Shimizu 1997 *J. Geophys. Res.* **102** 26159–71
[23] Sofieva V F, Gurvich A S and Dalaudier F 2009 *Geophys. Res. Lett.* **36** L05811
[24] Yoshiki M and Sato K 2000 *J. Geophys. Res.* **105** 17995–8011
[25] Sofieva V F, Dalaudier F, Kivi R and Kyro E 2008 *Geophys. Res. Lett.* **35** L23808
[26] Alexander S P and Tsuda T 2008 *J. Geophys. Res. Atmos.* **113** D04109
[27] Schmidt T, de la Torre A and Wickert J 2008 *Geophys. Res. Lett.* **35** L16807
[28] Hei H, Tsuda T and Hirooka T 2008 *J. Geophys. Res.* **113** D04107
[29] Tsuda T, Ratnam M V, May P T, Alexander M J, Vincent R A and MacKinnon A 2004 *J. Geophys. Res.* **109** D20S03
[30] Sofieva V F, Dalaudier F, Hauchecorne A and Kan V 2019 *Atmos. Meas. Tech.* **12** 585–98
[31] Ratnam M V, Tetzlaff G and Jacobi C 2004 *J. Atmos. Sci.* **61** 1610–20

IOP Conf. Series: Earth and Environmental Science 1040 (2022) 012009 doi:10.1088/1755-1315/1040/1/012009

Methodological approach to assessing environmental risk (health risk) from air pollution in the Baikal region in a changing climate

A A Makosko[1], **A V Matesheva**[1,2]

[1] Obukhov Institute of Atmospheric Physics of the RAS, Moscow, Russia
[2] Russian University of Transport, Moscow, Russia

E-mail: aamacosco@mail.ru, matesheva@ ifaran.ru

Abstract. The methodological approach to assessing the environmental risk (health risk) from atmospheric pollution in the Baikal region taking into account the changing climate was formulated. It is based on the US EPA's health risk assessment methodology and impurity concentration estimates by solving the adjoint equation for impurities transport and diffusion. The dynamics of health risk for the population of Irkutsk from atmospheric pollution PM10, PM2.5 due to emissions from possible sources located both in the Baikal region and beyond including sources of transboundary pollution in 1980-2050 under various scenarios of climate change was researched. A tendency for a moderate decrease in the considered risk over the past several decades was shown. In the forecast period until 2050, under both climate scenarios, significant changes in the risk were not noted. The spatio-temporal dynamics of the areas of PM10, PM2.5 sources location that create the increased health risk in relation to Irkutsk was studied. The results indicate the moderate narrowing of the increased risk zone for Irkutsk in the first half of the 21st century. In the period until 2050, the main impact on the population health in the Irkutsk area from emission sources located in the southwest, south and southeast is expected. In case of transboundary pollution, the main danger from sources in Kazakhstan, Mongolia, and China is expected. The results are important to develop proposals for ensuring safety for public health, developing health care and planning environmental protection measures in the Baikal and neighboring regions.

1. Introduction

The Baikal region is a strategically important natural area. At the same time, air pollution here is one of the acute environmental problems [1, 2] and presents a potential hazard to the health of the population [3] and natural ecosystems.

Observed and predicted climate change can change the level of danger of atmospheric pollution. The change in the general circulation of the atmosphere in a changing climate can lead to a significant redistribution of the incoming pollutants and areas that affect the Baikal region in the event of long-range pollution [4]. In this regard, studying environmental risks associated with air pollution is of great interest to the region today. It is especially important to study health risks as one of the most important criteria for assessing the environmental situation.

The goal of this work is to develop a methodological approach to assessing environmentally related health risks (trends in these risks) taking into account climate change. The area of application is studying of the dynamics of risk to public health due to atmospheric pollution in regions, in particular, in the Baikal region, in the 21st century. The methodological approach at solving the following tasks is aimed [5]:

1. Assessment of the pollution potential hazard for the region from all possible emission sources located both in the considered area and beyond, including sources of transboundary pollution, under various scenarios of climate change;
2. Determination of sources or areas of sources location that present the greatest potential health hazard to the population of the region under various scenarios of climate change.

2. Materials and research methods

The methodological approach on the methodology for assessing the health risk from environmental pollution by the US Environmental Protection Agency is based [6] and involves an early assessment of carcinogenic and non-carcinogenic risks.

According to [6], the assessment of carcinogenic risk from air pollution is based on using the unit risk factor [6, 7, 8] that characterizes the probability of cancer throughout a person's life due to exposure to a chemical substance with the annual average concentration in the air equal to 1 $(mg/m^3)^{-1}$.

Assessment of non-carcinogenic risk is determined by dividing the annual average concentration in the air by the appropriate reference exposure level (reference concentration, mg/m^3). A reference exposure level is used as an indicator of potential noncancer health impacts and is defined as the concentration at which no adverse noncancer health effects are anticipated [6, 8, 9].

Thus, the indicators of the sensitivity of the human body to the pollutants concentration in the atmospheric air are known a priori from the results of biomedical research. The main difficulty in early health risk assessment from air pollution in the Baikal region is the impurity concentration assessment in its ecologically significant territories, which has a number of features.

The first feature is a significant number of potential sources, including transboundary ones, and the presence of various mechanisms of atmospheric pollution in ecologically significant areas of Lake Baikal.

The second feature is uncertainty about the future impact on the atmosphere: changes in the number, location and power of emission sources.

The assessment of the concentration can be obtained in at least two ways. The first is the multiple solution of the impurity transport equation with sources of different intensity and different locations. Another method is more science-intensive, but it requires only a single solution of the adjoint problem [10, 11], by means of which it is possible to assess the degree of potential danger of atmospheric pollution in a given zone from all sources located in the problem solution area under given scenarios of the meteorological regime of the atmosphere. At the same time, no limitations are imposed on the location of the sources and their intensity. This method is used below.

The evolution of impurity q in the atmosphere over the Northern Hemisphere Ω within $0 \leq z \leq H$ (H – tropopause height) with a velocity characterizing the average transport in the troposphere is considered. For definiteness, we will assume that this is the air velocity on the surface of 500 hPa, which characterizes, first of all, long-range transport. Changes in impurity concentration due to transport in the near region are considered to be due to sedimentation. The generalization to the three-dimensional case is realized trivial.

With sufficient accuracy, long-range transport in the middle troposphere by a two-dimensional equation for impurities transport and diffusion is described [10]. At the same time, background pollution of the atmosphere is neglected as not interesting for solving this problem.

On basis of the Lagrange identity, the foundational problem is corresponded with the adjoint problem [10]. Its solution is the adjoint function q^*, which is a weight function that determines the contribution of each pollution source I to the amount of air pollution in an ecologically significant zone (given region) G.

The effect of atmospheric pollution from sources in the area Ω integrated over G region over time T will characterize the functional [10]

$$Q = \int_0^T \mathrm{d}t \int_\Omega I q^* \mathrm{d}\Omega \quad (1)$$

Thus, we obtain an expression for solving the first task indicated above in the Introduction.

When solving the second task to determine the areas of sources location that present the greatest hazard to the population in the G region the impurity concentration from a source located in a separate zone k belonging to the area Ω will characterize the functional

$$Q_k = \int_0^T \mathrm{d}t \int_{\Omega_k} I_k q_k^* \mathrm{d}\Omega_k . \quad (2)$$

On basis of the pollutant concentrations averaged over a certain period (for a year) the environmental risk research is executed. Therefore, functionals (3), (4) take the form

$$Q = \frac{1}{T} \int_0^T \mathrm{d}t \int_\Omega I q^* \mathrm{d}\Omega , \quad (3)$$

$$Q_k = \frac{1}{T} \int_0^T \mathrm{d}t \int_{\Omega_k} I_k q_k^* \mathrm{d}\Omega_k . \quad (4)$$

Taking into account [6, 7, 9], let us set down expressions for calculating the health risk caused by carcinogenic (i) and non-carcinogenic (j) impurities:

- for solving first task

$$r_i = \frac{R_i}{T} \int_0^T \mathrm{d}t \int_\Omega I_i q_i^* \mathrm{d}\Omega \quad \text{(carcinogenic risk),} \quad (5)$$

$$h_j = \frac{1}{H_j T} \int_0^T \mathrm{d}t \int_\Omega I_j q_j^* \mathrm{d}\Omega \quad \text{(non-carcinogenic risk),} \quad (6)$$

- for solving second task

$$r_{ik} = \frac{R_i}{T} \int_0^T \mathrm{d}t \int_{\Omega_k} I_{ik} q_{ik}^* \mathrm{d}\Omega_k \quad \text{(carcinogenic risk),} \quad (7)$$

$$h_{jk} = \frac{1}{H_j T} \int_0^T \mathrm{d}t \int_{\Omega_k} I_{jk} q_{jk}^* \mathrm{d}\Omega_k \quad \text{(non-carcinogenic risk).} \quad (8)$$

Here R_i is the unit risk factor for the substance i, H_i is reference concentration of the substance j, I is the source intensity.

Let us also write in a general form of expressions for calculating environmental risks from impurities emitted by sources of the area Ω:

$$r_i = \sum_{i=1}^{N} R_i Q_i \quad \text{(carcinogenic risk),} \quad (9)$$

$$h_j = \sum_{j=1}^{M} \frac{Q_i}{H_j} \quad \text{(non-carcinogenic risk).} \quad (10)$$

Accordingly, the expressions for calculating risks from the source of the zone k of the area Ω:

$$r_{ik} = \sum_{i=1}^{N} R_i Q_{ik} \quad \text{(carcinogenic risk),} \quad (11)$$

$$h_{jk} = \sum_{j=1}^{M} \frac{Q_{ik}}{H_j} \quad \text{(non-carcinogenic risk).} \quad (12)$$

In long-term estimates, information on future emissions and sources may not be known. In this case, the methodological approach allows considering a source with a unit emission. We put I = const (for convenience, below we will assume $I = 1$). In this case, expressions (5) - (12) will characterize the unit normalized risks from atmospheric pollution only due to the influence of weather and climatic processes. If the power of the source becomes known (for example, from the region development plan or it will be consider the impact of a specific source on the region), it will be enough (in the linear approximation of the source power influence, which is permissible for long-term planning) multiply the corresponding integrand terms in (3) - (8) or the corresponding summands in (9) - (12) by the actual value of emissions.

To solve the adjoint equation of impurity transport and diffusion and to obtain the fields of adjoint functions, a numerical model is used, which was previously used in other works of the authors including the solution of the adjoint problem in a three-dimensional formulation [5, 12].Therefore, it will be noted only some features of its construction below.

The region of the solution of the problem is the Northern Hemisphere. The grid area of the numerical model is 90×360 knots. The grid spacing of the model is 1°. For the numerical solution of the adjoint turbulent diffusion equation, the splitting method is used [13]. The photochemical processes, coagulation, absorption by droplets of fog and precipitation, radioactive decay include implicitly. Fields of wind and other meteorological quantities are taken from either reanalysis or from the results of scenario calculations of climate change. This provides an opportunity to assess trends of environmental risks in a changing climate.

3. Results

On basis of the developed methodological approach, the dynamics of the environmentally related health risk in Irkutsk city due to atmospheric pollution with PM10, PM2.5 particulate matter in 1980-2050 taking into account climate change has been investigated.

The research of the risk to public health from air pollution by microparticles is associated with the increased potential hazard of these substances. They are a mixture with different physical and chemical characteristics depending on their origin and the possibility of their transfer over thousands of kilometers from the source from urbanized and industrial zones, areas of natural fires, oil and gas production, neighboring countries. The health effects of microparticles are varied. In addition to oncological pathology (for example, lung cancer), fine impurities promote to the manifestation of non-cancer diseases of the respiratory and cardiovascular systems, for example, exacerbation of asthma and respiratory symptoms, and an increase in the number of hospitalizations [14].

Within the framework of the study, two tasks were solved.

Task 1. Assessment of the dynamics of health risks for the population of Irkutsk from air pollution PM10, PM2.5 due to emissions from possible sources located both in the Baikal region and beyond, including sources of transboundary pollution.

For this, the annual average values of unit normalized non-carcinogenic risks (since information on future sources and emissions is unknown) were calculated in the form of hazard indices for non-carcinogenic diseases for two substances (PM10, PM2.5) from 1980 to 2050 with a step of 5 years. In this case, two series of calculations were performed taking into account two characteristic scenarios of climate change in the 21st century adopted by the Intergovernmental Panel on Climate Change (IPCC) – "moderate" (RCP4.5) and "hard" (RCP8.5) [15]. These scenarios are based on different pathways for greenhouse gas emissions and concentrations in the atmosphere, pollutant emissions and land use in the 21st century that affect the climate system. Thus, the "moderate" scenario assumes a peak in greenhouse gas emissions around 2040, while the "hard" scenario shows a gradual increase in emissions in the period under review.

The following data as meteorological fields were used:

- 1980-1995 – NCEP Climate Forecast System Reanalysis (CFSR) [16];
- 2000-2020 – NCEP FNL Operational Model Global Tropospheric Analyses [17];
- 2025-2050 – data of calculations on the climatic model of the Marchuk Institute of Numerical Mathematics RAS (scenarios RCP4.5 and RCP8.5 [18, 19]).

The figure 1 shows graphs of the dynamics of unit normalized non-carcinogenic risks in 1980-2050 with an interval of 5 years, taking into account various climatic scenarios.

Calculations for Irkutsk a tendency for a noticeable, but quite moderate decrease in the considered risk (about 15%) over the past several decades have shown. In the forecast period until 2050, under both climate scenarios, significant changes in the hazard to humans from air pollution with particulate matter in the area of Irkutsk were not noted (the trend is several percent).

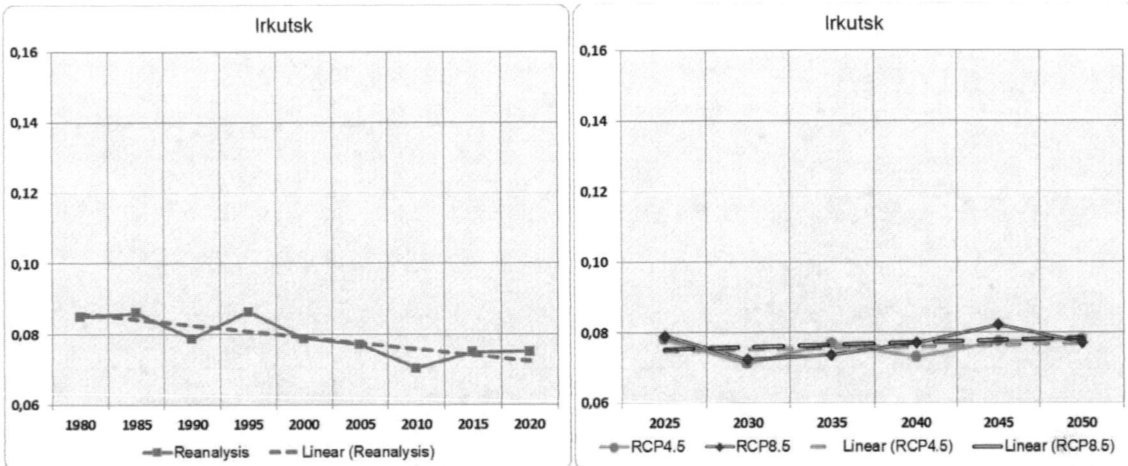

Figure 1. Dynamics of unit normalized non-carcinogenic risk (hazard index of non- carcinogenic diseases) in Irkutsk due to atmospheric pollution with PM10, PM2.5 particles from potential nearby and remote sources in 1980-2020 (left) and in 2025-2050 (on the right).

Task 2. Study of the spatio-temporal dynamics of the areas of sources location of atmospheric pollution that create the increased health risk in relation to Irkutsk under various scenarios of climate change.

For this purpose, the annual average values of unit normalized non-carcinogenic risks for Irkutsk from potential sources of PM10, PM2.5 located in different areas relative to Irkutsk (in the *k*-zones of the *Ω* region) from 1980 to 2050 with a step of 5 years under various climatic scenarios were calculated. Maps of the dynamics of these risks have been built. However, the presentation in the article of all maps for the period under consideration requires a significant reduction in scale and thus leads to extremely low information content. Therefore, below, as an example, two maps are shown that demonstrate one of the largest and smallest areas of the sources location of increased risk for the Irkutsk city for 1995 and 2010 respectively (figure 2). Areas of different colors show the level of risk created by the source in Irkutsk, when the source is located in this area (area of this color).

According to calculations, taking into account both climatic scenarios, in the first half of the 21st century (after 2005), there is some narrowing of the zone of the location of sources that create increased levels of risk in the Irkutsk region in the first half of the 21st century, in certain directions. In fact, the area of increased risk (for example, the area of the 0.0001 level in figure 2) changes in some directions relative to Irkutsk by tens and even hundreds of kilometers (taking into account the grid spacing on the map of 5 degrees). Thus, the results of solving second task indicate a decrease in the hazard to public health in the Irkutsk area from some remote sources of microparticles in the first half of the 21st century due to the influence of weather and climatic processes.

As in the 1980s and 1990s, in the first half of the 21st century, the main impact on the health of the population of the Baikal region (in the Irkutsk area) from emission sources located in the southwestern, southern and southeastern directions relative to Irkutsk is expected. In case of transboundary pollution, the main danger from sources in Kazakhstan, Mongolia, and China is expected.

It is important to emphasize that the resulting risk assessments are much lower than the acceptable level (unity). This is due to the fact that the unit normalized risk was assessed - the risk due to emissions of 1 g / s intensity produced by only one source, which, however, can be located at any point (zone) of the territory under consideration. The actual value of the risk can be obtained by multiplying the unit normalized risk by the actual value of the source intensity.

Given the uncertainties inherent in long-term assessments, the presented results should be considered as trends in environmental risks for the Baikal region.

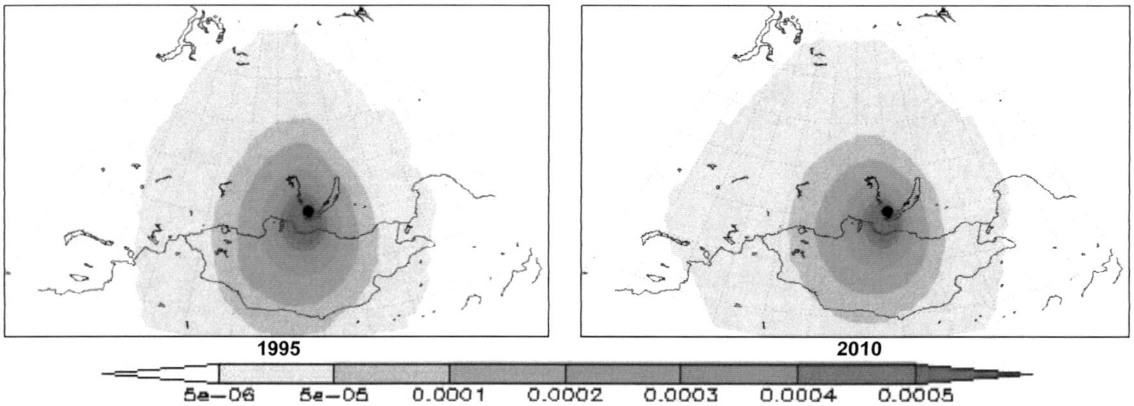

Figure 2. An example of dynamics of unit normalized non-carcinogenic risk for Irkutsk from potential sources of PM10, PM2.5 located in different areas relative to Irkutsk in 1995 and 2010 (Irkutsk is shown with a black dot).

4. Conclusion

The methodological approach to assessing the environmental risk (health risk) from atmospheric pollution in the Baikal region, taking into account the changing climate, was formulated.

The dynamics of health risk for the population of Irkutsk from atmospheric pollution PM10, PM2.5 due to emissions from possible sources located both in the Baikal region and beyond including sources of transboundary pollution under various scenarios of climate change was researched. Calculations for Irkutsk a tendency for a noticeable, but quite moderate decrease in the considered risk over the past several decades have shown. In the forecast period until 2050, under both climate scenarios, significant changes in the hazard to humans from air pollution with particulate matter in the area of Irkutsk were not noted.

The spatio-temporal dynamics of the areas of sources location of air pollution that create the increased health risk in relation to Irkutsk under various scenarios of climate change was studied. The results indicate the presence of a certain spatial-temporal dynamics: the moderate narrowing of the zone of increased risk for Irkutsk in the first half of the 21st century due to the influence of weather and climatic processes. An inter-scenario variability of the zone of increased risk until 2050 is insignificant.

In the period until 2050, the main impact on the population health in the Irkutsk area from emission sources located in the southwest, south and southeast is expected. In case of transboundary pollution, the main danger from sources in Kazakhstan, Mongolia, and China is expected.

The results are important to develop proposals for ensuring safety for public health, developing health care and planning environmental protection measures in the Baikal and neighboring regions.

Acknowledgments
The authors are grateful to the referee for carefully reading the manuscript and valuable comments. The work was supported partly within the state task as well as with partial funding by RFBR, project number 17-29-05102.

IOP Conf. Series: Earth and Environmental Science 1040 (2022) 012009 doi:10.1088/1755-1315/1040/1/012009

References

[1] *On the state and environmental protection of the Russian Federation in 2018* 2019 (Moscow: Ministry of nature of Russia, NPP Kadastr), p 844
[2] *On the state of the Lake Baikal and measures for its protection in 2017* 2018 (Irkutsk: ANO KC Expert), p 340
[3] *On the state of sanitary and epidemiological well-being of the population in the Russian Federation in 2018* 2019 (Moscow: Federal Service for Supervision of Consumer Rights Protection and Human Welfare), p 254
[4] Makosko A A and Matesheva A V 2020 *IOP Conf. Ser.: Earth Environ. Sci.* **606** 012034
[5] Makosko A A and Matesheva A V 2020 *Air pollution and the quality of population life in the XXI century: threats and prospects* (Moscow: Russian Academy of Sciences) p 257
[6] Air Toxics Hot Spots Program, 2015: Risk Assessment Guidelines. The Air Toxics Hot Spots Program Guidance Manual for Preparation of Health Risk Assessments (California EPA) p 231
[7] Air Toxics Hot Spots Program, 2009: Air Toxics Hot Spots Risk Assessment Guidelines Part II: Technical Support Document for Cancer Potency Factors (California EPA) p 89
[8] Integrated Risk Information System (IRIS) Database (U.S. EPA) Available at: http://www.epa.gov/IRIS/
[9] Air Toxics Hot Spots Program, 2008: Air Toxics Hot Spots Risk Assessment Guidelines: Technical Support Document for the Derivation of Noncancer Reference Exposure Levels (California EPA) p 131
[10] Marchuk G I 1982 *Mathematical modeling in the environmental problem* (Moscow: Nauka) p 320
[11] Aloyan A E 2002 *Dynamics and kinetics of gas impurities and aerosols in the atmosphere* (Moscow: IVM RAN) p 201
[12] Makosko A A and Matesheva A V 2016 An experience of the identification of air pollution sources in the Moscow region *Russian Chem. J.* LX **3** pp 113-120
[13] Marchuk G I et al 1987 *Mathematical models in geophysical hydrodynamics and numerical methods for their implementation* (Moscow: Gidrometeoizdat) p 296
[14] Health effects of particulate matter: *Policy implications for countries in eastern Europe, Caucasus and central Asia* 2013 (WHO: Regional Office for Europe), p 15
[15] IPCC, 2014: Climate Change 2014: Synthesis Report. Contribution of Working Groups I, II and III to the Fifth Assessment Report of the Intergovernmental Panel on Climate Change (IPCC, Geneva, Switzerland) p 151
[16] Saha S et al 2010 *NCEP Climate Forecast System Reanalysis (CFSR) Selected Hourly Time-Series Products. January 1979 to December 2010* (Research Data Archive at the National Center for Atmospheric Research: Computational and Information Systems Lab.) Available at: https://doi.org/10.5065/ D6513W89
[17] *NCEP FNL Operational Model Global Tropospheric Analyses continuing from July 1999* (Research Data Archive at the National Center for Atmospheric Research: Computational and Information Systems Lab.) Available at: https://doi.org/10.5065/D6M043C6
[18] Volodin E and Diansky N 2013 *INMCM4 model output prepared for CMIP5 RCP4.5. served by ESGF* (World Data Center for Climate (WDCC) at DKRZ) Available at: https://doi.org/10.1594/WDCC/CMIP5.INC4r4
[19] Volodin E and Diansky N 2013 *INMCM4 model output prepared for CMIP5 RCP8.5. served by ESGF* (World Data Center for Climate (WDCC) at DKRZ) Available at: https://doi.org/10.1594/WDCC/CMIP5.INC4r8

IOP Conf. Series: Earth and Environmental Science 1040 (2022) 012010 doi:10.1088/1755-1315/1040/1/012010

Research of synoptic processes in the south-east of the Russian plain during different climatic periods

S V Morozova[1], **K E Denisov**[2], **E A Polyanskaya**[1], **S I Pryakhina**[1], **E I Ormeli**[1], **N K Kononova**[3, 4]

[1]Saratov State University, Saratov, Russia
[2]Saratov State Agrarian University named after N.I. Vavilov, Saratov, Russia
[3]Institute of Geography, Russian Academy of Sciences, Moscow, Russia
[4] A.O. Kovalevsky Institute of Biology of the Southern Seas of RAS, Sevastopol, Russia

E-mail: swetwl@yandex.ru

Abstract. This article examines the annual and seasonal (winter and summer) changes in the frequency of occurrence of the main types of synoptic processes in the Lower Volga region in two natural climatic periods of the state of the earth climatic system - the stabilization period and the second wave of global warming. The frequency of occurrence of some types of synoptic processes revealed both climatic and seasonal differences. In the second wave of global warming, in comparison with the stabilization period, the frequency of cyclone impact on the Lower Volga region developed on the Arctic front and the number of Arctic anticyclones increase, and the frequency of polar-front cyclones and cases of the impact of the western periphery of the winter Asian anticyclone decreases. Seasonal differences in the frequency of occurrence were revealed in the Arctic anticyclones. During the second wave of global warming, these anticyclones penetrate into the Lower Volga region in summer more often than in winter. It is concluded that the contribution of the transformation factor to the observed warming increases and the role of summer seasons in the observed climatic changes increases, which is not typical for processes in the earth climatic system. It is indicated that obtained result is of great climatological significance and will serve the further development of the physical theory of climate.

1. Introduction

The main trend in the global climate change is a steady progressive increase in the surface air temperature with some fluctuations [2, 14]. It is known that in the Northern Hemisphere over the past 100–120 years, during the mean surface air temperature, climatic periods such as the first wave of global warming, the period of stabilization, and the second wave of global warming are distinguished [2, 11, 13, 14]. However, at the regional level, climatic changes may have their own characteristics and do not always coincide with global trends. For example, in the Lower Volga region, the first wave of global warming did not manifest itself at all, and the development of the second wave of global warming began in the mid-1980s, and not from the mid-1970s, as in the hemisphere as a whole [4, 5, 7, 9].

Undoubtedly, regional climatic variability is formed under the influence of the general circulation of the atmosphere, namely, synoptic processes developing over the region. The recurrence of certain synoptic processes forms the regional climatic background.

The purpose of this work was to objectively identify the boundaries of climatic periods and identify the prevailing types of synoptic processes in different climatic intervals. The study area was the southeast of the European part of Russia, namely the Middle and Lower Volga regions.

2. Materials and methods

Since the most striking indicator of modern climatic changes is air temperature, the initial material of the study was the data on the anomalies of the near-surface air temperature in the Northern Hemisphere, taken from the website (http://www.cru.uea.ac.uk/cru/data/temperature/#datdow).

For an objective identification of the boundaries of climatic periods, the stepwise trend method was applied to the available time series using the Kolmogorov statistics [3, 6, 8]. To further substantiate the statistical significance of changes in the near-surface air temperature to the selected climatic intervals, the method of confidence intervals was used using Student's statistics (95% significance level). The calculations were carried out according to the following formula (1)

$$\bar{x} \pm t\gamma \frac{s}{\sqrt{n}}\ , \qquad (1)$$

where $\bar{x}$ is an average value; $t\gamma$ - Student's statistics, γ - Significance level indicator (95 %); s - mean square; n - Number of members of the series.

The regional typification of synoptic processes developed by professor V.L. Arkhangelsky [1], and refined by E.A. Polyanskaya [12] was chosen as a characteristic of the circulation conditions. According to the research of V.L. Arkhangelsky and E.A. Polyanskaya, all the diversity of synoptic processes in the Lower Volga region is divided into seven main types.

I. Cyclonic activity in the Arctic front.
II. The impact of the Arctic anticyclone.
III. Impact of the Asian winter anticyclone.
IV. Impact of the subtropical anticyclone.
V. Low-gradient fields of high and low pressure.
VI. Cyclonic activity in the polar front.
VII. Deformation field.

The type of synoptic process is determined according to the methodological recommendations of V.L. Arkhangelsky [1] based on a set of synoptic material, including daily weather maps, baric topography maps and thermobaric maps. For each type of process, V.L. Arkhangelsky and E.A. Polyanskaya give examples of weather maps, the distribution of the geopotential values characteristic of this type, the location of altitudinal zones, and the thermobaric field of the middle troposphere [1, 12]. The type of process is determined separately for each selected item, taking into account the history of the development of the process and its transformation over time. It is a visual analysis maps.

To analyze synoptic processes during the stabilization period, the authors used an archive of types of synoptic processes of V.L. Arkhangelsky based on synoptic material received by the forecast center in Saratov via facsimile channels. Facsimile weather maps, baric topography maps and other maps were compiled at the Hydrometeorological Center (Moscow) and broadcast by fax. The archive of types of processes based on the available synoptic material covers the period from 1949 to 1969. This time interval falls on the stabilization period. The authors of this study believe that this working sample is quite representative for the characterization of processes in this climatic interval.

The sample of synoptic material characterizing atmospheric processes during the second wave of global warming was formed by daily downloading of synoptic maps, which are regularly posted on the *ftp*-server of the Hydrometeorological Center of Russia in the GRIB code. Since the definition of the type of synoptic process is very difficult, the Calendar of synoptic processes was compiled by E.A. Polyanskaya only for 2011. Therefore, to analyze synoptic processes during the second wave of global warming, we used a sample containing information on the circulation conditions from 1998 to 2013, which the authors consider to be representative for characterizing the circulation conditions during the second wave of global warming.

Let us point out that in Tables below, the stabilization period is designated as SP, the second wave of global warming as SW. For each period, the frequency of synoptic processes was calculated, expressed in the number of days with a process of this type, i.e. absolute frequency. Samara, Saratov, Volgograd and Astrakhan were taken as reference points in which the type of synoptic process was determined. The type of synoptic process was determined by the surface weather map (or 00 hours GMT).

3. Results

The study of the time series of surface air temperature anomalies in the Northern Hemisphere for uniformity revealed years of disturbance in uniformity. These years were 1908, 1944, 1975. The significance level was assumed to be 95%. The procedure made it possible to objectively determine the boundaries of climatic intervals. Table 1 shows the boundaries of climatic intervals and an assessment of the statistical significance of temperature changes.

Table 1. Assessment of the statistical significance of changes in the average annual temperature anomalies in the Northern Hemisphere

Period, year	Statistical characteristics of changes significance			Confidence interval
	α	$\bar{x}$	σ	
1) Little Ice Age (1850-1907)	-0.0014	-0.283	0.147	[-0.508; -0. 315]
2) first wave (1908 – 1943)	0.0163	-0.185	0.194	[-0.240; -0.130]
3) stabilization (1944 – 1974)	-0.0058	-0.020	0.124	[-0.078; 0.038]
4) second wave (1975 – 2016)	0.0240	0.356	0.320	[0.273; 0.439]

The alternation of positive and negative values of the coefficients of linear trends, as well as the absence of overlapping of the boundaries of the confidence intervals, allows the changes to be considered statistically significant. S.V. Morozova [10, 15, 16] proposed to name these time intervals the natural climatic periods of the state of the Earth climatic system.

Figure 1 shows a graph of the change in the surface air temperature of the Northern Hemisphere with the drawn trend lines. This graph clearly shows the natural climatic periods of the state of the Earth climatic system.

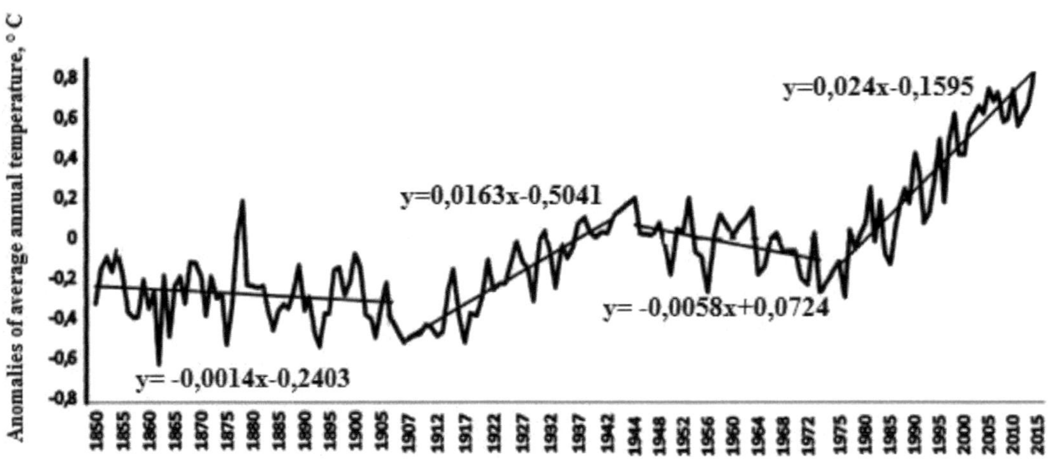

Figure 1. Variability of the average hemispheric air temperature

The characteristics of the circulation conditions in two natural climatic periods are given in Table 2. This Table shows the average annual number of days with processes of each type at four points of the Lower Volga region - Samara, Saratov, Volgograd, and Astrakhan during the stabilization period and during the second wave of global warming.

According to Table 2, three types of processes are decisive in the formation of weather and climatic instability at all reference points of the Lower Volga region. These are cyclonic activity in the Arctic and polar fronts (types I and VI) and the impact of Arctic anticyclones (type II).

For the entire studied period, including two natural climatic periods, in Samara and Saratov, cyclonic activity on the Arctic front and the impact of Arctic anticyclones prevail of the above three types of synoptic processes. In Volgograd and Astrakhan, located to the south, the weather is most often determined by Arctic invasions and polar front cyclones. It can be explained by the fact that polar front depressions are pumping fresh portions of cold Arctic air into their rear to the extreme southeast of the European Russia, while a new cyclonic series has already formed in the Arctic front to the north, in the Middle Volga region. The arctic air is stationary over the territories for a long time according to the type of anticyclonic circulation. A detailed description of such phenomena with an analysis of daily synoptic situations can be found in [9].

Table 2. Frequency of synoptic processes (year)

Process type	Period	Point			
		Samara	Saratov	Volgograd	Astrakhan
I	SP	84.2	74.1	56.5	42.1
	SW	95.4	98.6	63.5	43.6
II	SP	73.2	69.9	61.9	58.8
	SW	105.6	99.5	103.0	100.0
III	SP	40.9	38.3	33.8	35.6
	SW	36.2	27.5	30.5	32.1
IV	SP	38.0	45.6	56.5	64.2
	SW	21.0	26.4	32.5	35.9
V	SP	23.7	28.5	25.9	30.6
	SW	51.4	53.7	64.4	87.3
VI	SP	84.5	90.8	111.4	95.4
	SW	34.3	44.5	51.5	42.5
VII	SP	7.8	17.7	19.2	8.8
	SW	21.0	18.5	20.2	23.2

Comparison of the frequency of synoptic processes during the stabilization period and the second wave of global warming revealed a change in the frequency of synoptic processes in these two climatic intervals.

In all points of the Lower Volga region, with the exception of Astrakhan, during the second wave of global warming, compared with the stabilization period, the frequency of cyclones developed in the Arctic front significantly increased. The absolute frequency of this process has increased by an average of 1.5 times. In Astrakhan, no change in the frequency of this type of process (I) from one natural climatic period to another was observed.

During the second wave of global warming, the influence of the Arctic anticyclones on the region almost doubled in comparison with the stabilization period. The difference in the frequency of these invasions between two natural climatic periods increases from north to south (from Samara to Astrakhan). Another feature associated with this type of process is that during the stabilization period, the frequency of Arctic invasions decreased from north to south (from Samara to Astrakhan), and during the second wave of global warming, the number of Arctic invasions in all points turned out to be approximately the same.

It should be noted that while the influence of the Arctic anticyclones on the region increased during the second wave of global warming, other anticyclones determined the weather in the Lower Volga region much less often. In the second studied climatic interval (the second wave of global warming), the territory was less often influenced by the western periphery of the winter Asian anticyclone and the eastern periphery of the Azores pressure maximum.

During the second wave of global warming, in comparison with the previous period (stabilization), the frequency of cyclones developed in the polar front decreased by almost two times. Moreover, this tendency is typical for all points. Besides, deformation (type VII process) and low-gradient (type V process) fields over the Lower Volga region were much more often than during the stabilization period.

According to Table 1, an increase in the number of impacts on the region of the Arctic anticyclones was found out. Undoubtedly, the question arises about the apparent discrepancy between the observed warming and the increase in the number of Arctic invasions. To resolve this paradox, let us consider the recurrence of processes in winter and summer (Table 3).

Table 3. Frequency of various types of synoptic processes in winter and summer.

Process type	Period	Reference point							
		Samara		Saratov		Volgograd		Astrakhan	
		winter	summer	winter	summer	winter	summer	winter	summer
I	SP	25.1	14.4	24.5	11.4	19.3	7.6	15.5	5.0
	SW	30.5	18.6	29.6	22.0	19.9	14.2	13.8	9.2
II	SP	15.9	15.3	15.3	13.5	14.4	11.2	13.5	10.3
	SW	12.3	30.0	10.2	26.8	12.9	27.3	12.6	26.2
III	SP	22.3	-	20.6	-	17.5	-	18.8	-
	SW	22.2	-	16.6	-	18.7	-	19.7	-
IV	SP	2.7	16.9	3.1	20.0	5.0	26.5	5.9	26.2
	SW	0.6	10.7	2.3	10.5	3.1	14.1	2.8	17.8
V	SP	4.4	11.4	5.6	11.9	4.8	11.4	5.0	13.1
	SW	10.6	20.6	11.9	18.6	13.9	22.9	21.0	28.7
VI	SP	14.8	29.5	16.8	30.8	24.8	34.2	21.8	30.0
	SW	8.6	9.0	13.1	9.5	12.4	11.5	11.5	7.5
VII	SP	1.8	1.2	4.5	4.4	4.5	3.9	3.5	1.5
	SW	5.3	4.8	4.4	3.3	4.7	3.9	4.9	4.9

According to Table 3, both in winter and summer, from the stabilization period to the second wave of global warming, at three points in the Lower Volga region (with the exception of Astrakhan), the number of cyclones developed in the Arctic front increases and the number of polar-front cyclones decreases. In Astrakhan, seasonal differences in the frequency of occurrence of this type of process are noticeable. In winter, from the first climatic period to the second one, the frequency of occurrence of the Arctic-front cyclones increases, in summer it decreases.

Strong seasonal differences were found out in the frequency of incursions into the region by arctic anticyclones. In winter, during the second wave of global warming, the frequency of penetration of the Arctic cores into the region is less than during the stabilization period. The increased temperature background of the winter seasons is quite explainable by the intensification of cyclonicity and a decrease in the number of intrusions of cold air masses from the Arctic.

In summer, the frequency of the invasion of Arctic anticyclones into the region during the second wave of global warming is more than twice as high as during the stabilization period. Let us recall that sharp and intense cold waves in the continental regions in summer are manifested by short-term cold snaps, followed by long periods of very hot weather. According to the research of V.L. Arkhangelsky and E.A. Polyanskaya [1, 12], in summer under the conditions of the Lower Volga region, fresh Arctic air is transformed into local tropical air within one or two days.

Circulation near the earth's surface cannot but manifest itself in the middle troposphere. Based on the NCEP / NCAR reanalysis data on the geopotential values at the AT-500 GPa level, the fields of geopotential anomalies were constructed in two climatic periods (stabilization and the second wave of global warming) for the central months of winter (Figure 2) and summer (Figure 3).

According Figures 2 and 3, climatically stable areas of negative and positive geopotential anomalies are distinguished on all maps, which indicates the predominance of certain types of processes. In January (Figure 2 a) during the stabilization period, two closed centers of negative geopotential anomalies indicate active cyclonic activity over eastern and southern Europe in the Arctic and polar fronts. During the second wave of global warming (Figure 2b), the disappearance of these foci and the formation of a single band of negative geopotential anomalies is an indicator of a change in the frequency of circulation forms, which was mentioned above.

IOP Conf. Series: Earth and Environmental Science 1040 (2022) 012010 doi:10.1088/1755-1315/1040/1/012010

In July (Figure 3a), during the stabilization period, the area of negative geopotential anomalies over southern Europe indicates the predominance of cyclonicity in this region. In the next climatic period (Figure 3 b), almost all of Europe is occupied by a center of positive geopotential deviations, which indicates the predominance of anticyclones in this region. An independent focus with the highest values of positive geopotential deviations is located in the south of ER. Such a geopotential field indicates an increase in the intensity and frequency of arid events in the south of ER.

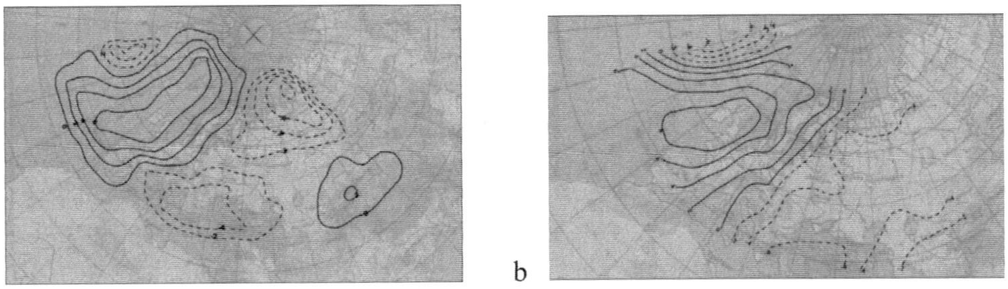
a b

Figure 2. Average long-term field of isanomalous of the surface geopotential of 500 GPa in January: a – stabilization period; b - second wave of global warming.

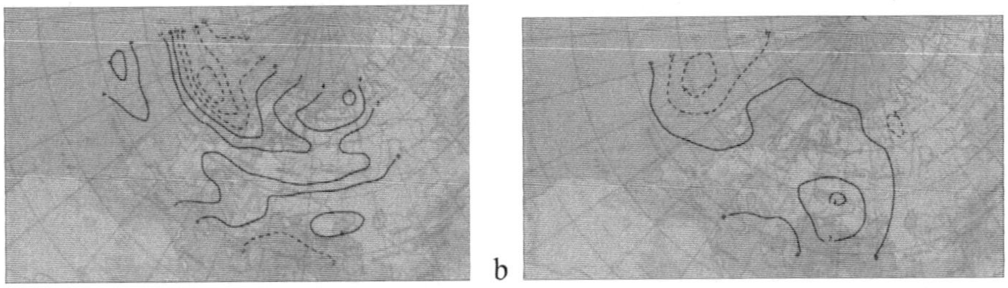
a b

Figure 3. Average long-term field of isanomalous of the surface geopotential of 500 GPa in July: a – stabilization period; b - second wave of global warming.

Taking into account the above, we can formulate a following conclusion: in the future, with a possible increase in the average global temperature, the contribution of summer seasons to an increase in the general temperature background may become more significant as a result of transformation processes - heating of dry cold air masses under anticyclone. If in the second wave of global warming, the increase in temperatures was mainly determined by milder winter seasons (advective factor), then in the future, possibly, summer seasons will influence the development of warming due to an increase in the meridionality of processes and an increase in the role of the transformation factor.

An increase in average annual temperatures due to an increase in temperatures in summer seasons is not typical for processes in the earth climatic system, at least in the era of regular meteorological observations. We believe that this provision can be of great importance for the development of the physical climate theory.

It also should be noted (Table 3) that during the period of stabilization both in winter and in summer, the number of Arctic invasions decreased from north to south (from Samara to Astrakhan). During the second wave of global warming in winter, Arctic anticyclones do not extend as far southward than in summer. The number of summer incursions of Arctic cores in Saratov, Volgograd and Astrakhan is the same, that is, summer sharp cold waves propagate further south than winter ones. This phenomenon is also unusual for the development of synoptic processes according to "climatic" rules.

Let us note that the preliminary climatic generalization of synoptic material by the types of synoptic processes (2014 - 2019) showed the preservation of the above trends.

4. Conclusions

In the course of the study, the following conclusions were obtained.

1. The number of cases of impact on the region of some types of synoptic processes differ significantly during the period of stabilization and the second wave of global warming. During the second wave of global warming, in comparison with the stabilization period, cyclones developed on the Arctic front and Arctic anticyclones began to be observed more often in the Lower Volga region. The repeatability of low-gradient and deformation fields has also increased.

During the second climatic period, the western periphery of the winter Asian anticyclone less often settled in the region, subtropical ridges spread less often, and there were fewer polar front cyclones.

2. There are seasonal differences in the frequency of occurrence of some types of synoptic processes in the Lower Volga region. In the second wave of global warming, compared with the stabilization period in winter, the number of Arctic invasions into the region increases, and in summer decreases. This feature allows one to assume an increase in the impact of the thermal features of summer seasons due to an increase in the role of the transformation factor, which is uncharacteristic for the development of processes in the earth climatic system. We believe that obtained result can be of independent importance for the development of the physical climate theory.

3. The features of the distribution of the Arctic cores in the Lower Volga region in two natural climatic periods have been revealed. During the stabilization period, the number of incidents of Arctic invasions into the Lower Volga region decreased from north to south. During the second wave of global warming, the Arctic cores spread throughout the entire territory.

5. Acknowledgements

The work is performed within the Program Fundamental Research Institute Geography RAS, project No. 0148-2019-0009.

The study was carried out as a part of the IBSS research (registration No.: AAAA-A19-119061-190081-9).

6. References

[1] Arkhangelsky V L 1968 Regional forecaster of the Lower Volga region (Saratov)
[2] Second Assessment Report of Roshydromet on Climate Changes and Their Consequences on the Territory of the Russian Federation 2014 (Moscow)
[3] Gmurman V E 2003 Theory of Probability and Mathematical Statistics (Moscow)
[4] Ivanova G F , Levitskaya N G and Shatalova O V 2008 Climate change and anomalous winters in the Saratov region in the late XX - early XXI century (Saratov)
[5] Ivanova G F , Sklyarov Yu A and Levitskaya N G 2006 *Izvestiya of the Saratov University* 10-15
[6] Isaev A A 1988 Statistics in meteorology and climatology (Moscow)
[7] Levitskaya N G Morozova S V and 2015 *Izvestiya of the Saratov University* **1** 21-24
[8] Malinin V N 2007 Statistical methods for the analysis of hydrometeorological information (Saint-Petersburg)
[9] Morozova S V 2013 Study of synoptic processes by the method of standards (Saratov)
[10] Morozova S V 2015 The role planetary circulation objects in global climate processes (Saratov)
[11] Perevedentsev Yu P 2009 Climate theory (Kazan)
[12] Polyanskaya E A 1986 Synoptic processes and weather phenomena in the Lower Volga region (Saratov)
[13] Sherstyukov B G 2008 Regional and seasonal patterns of modern climate change (Obninsk)
[14] Intergovernmental Panel on Climate Change, Climate change 2013 The physical science basis. (Cambridge)
[15] Morozova S V , Polyanskaya E A , Ivanova G F , Levitskaya N G , Denisov K E and Molchanova N P 2018 *IOP Conference Series: Earth and Environmental Science* **107**
[16] Morozova S V, Polyanskaya E A, Kononova N K, Molchanova N P and Solodovnikov A P 2019 *IOP Conference Series: Earth and Environmental Science* **381**

IOP Conf. Series: Earth and Environmental Science 1040 (2022) 012011 doi:10.1088/1755-1315/1040/1/012011

Application of the numerical model TSUNM3 to study the urban heat island and the intensity of precipitation over the Siberian city of Tomsk

AV Starchenko[1,2], L I Kizhner[1], S L Odintsov[2,1], E A Danilkin[1,2] and A A Bart[1]

[1]National Research Tomsk State University, Tomsk, 634050, Russia
[2]V.E. Zuev Institute of Atmospheric Optics, Tomsk, 634055, Russia

E-mail: starch@math.tsu.ru

Abstract. The relevance of the study is associated with the development and validation of the TSUNM3 model for numerical local weather prediction. The results of its application over a limited territory of the Siberian region are presented. A study of the characteristics of an urban "heat island" and the analysis of the reliability of forecasting precipitation have been carried out. The calculation results are compared with observations of the atmospheric boundary layer parameters obtained with meteorological instruments of the Joint Use Center "Atmosphere" at the Institute of Atmospheric Optics of the Siberian Branch of the Russian Academy of Sciences.

1. Introduction

Nowadays, numerical models of high spatial resolution (with a horizontal resolution from several hundred meters to several kilometers) for short- and ultra-short-term weather forecasting and research, are being developed and improved, which make it possible to predict local mesoscale atmospheric phenomena for different regions of the globe [1].

In many countries, the most widespread model is the Weather Research and Forecasting (WRF) mesoscale model [2], which is widely used to solve many practical problems in the numerical study of weather and air quality. For the territory of the Russian Federation the COSMO-Ru model [3], created by international cooperation between the countries of Europe and Russia, is one of the main regional models currently used at the Hydrometeorological Center of Russia for forecasting weather elements. For the European territory of the Russian Federation, versions of the COSMO model with a grid step from 1.1 to 13.2 km have been implemented, and for the Siberian region, the COSMO-Ru14-Sib technology is used, where numerical forecasts are carried out along with forecasts using the ICON model in the German Meteorological Service. The recent COSMO model version reasonably reproduces “urban heat island” effect in high-resolution simulations for selected European cities, namely Turin, Naples and Moscow [4].

The aim of this work is to simulate numerically the atmospheric boundary layer parameters over a particular territory of the Siberian region as a result of the application of the mesoscale meteorological model TSUNM3 (Tomsk State University Nonhydrostatic Mesoscale Meteorology Model) [5], and to make a comparative analysis of the calculation results with observational data reached using the meteorological instruments of "Atmosphere".

2. TSUNM3 numerical model

The basic equations of the TSUNM3 atmospheric boundary layer model are the Reynolds-averaged Navier–Stokes equations [6] in the coordinate system associated with the Earth's surface, using the following assumptions:

1. Mesoscale density variations are quasistationary. The Boussinesq approximation is used to represent the buoyancy force in the vertical velocity equation.
2. Molecular diffusion is assumed to be negligible in relation to turbulent exchange.
3. Phase transitions of moisture in the atmospheric boundary layer, short- and long-wave radiation heat transfer with an explicit representation in the atmosphere are taken into account.

The mathematical model of TSUNM3 includes the following equations:

$$\frac{\partial(\rho u)}{\partial x}+\frac{\partial(\rho v)}{\partial y}+\frac{\partial(\rho w)}{\partial z}=0 \quad (1)$$

$$\rho\left(\frac{\partial u}{\partial t}+u\frac{\partial u}{\partial x}+v\frac{\partial u}{\partial y}+w\frac{\partial u}{\partial z}\right)=-\frac{\partial p}{\partial x}+\rho f v+\frac{\partial}{\partial x}\left(K_H\frac{\partial u}{\partial x}\right)+\frac{\partial}{\partial y}\left(K_H\frac{\partial u}{\partial y}\right)+\frac{\partial}{\partial z}\left(K_Z^m\frac{\partial u}{\partial z}\right) \quad (2)$$

$$\rho\left(\frac{\partial v}{\partial t}+u\frac{\partial v}{\partial x}+v\frac{\partial v}{\partial y}+w\frac{\partial v}{\partial z}\right)=-\frac{\partial p}{\partial y}-\rho f u+\frac{\partial}{\partial x}\left(K_H\frac{\partial v}{\partial x}\right)+\frac{\partial}{\partial y}\left(K_H\frac{\partial v}{\partial y}\right)+\frac{\partial}{\partial z}\left(K_Z^m\frac{\partial v}{\partial z}\right) \quad (3)$$

$$\rho\left(\frac{\partial w}{\partial t}+u\frac{\partial w}{\partial x}+v\frac{\partial w}{\partial y}+w\frac{\partial w}{\partial z}\right)=-\frac{\partial p}{\partial z}-\rho g+\frac{\partial}{\partial x}\left(K_H\frac{\partial w}{\partial x}\right)+\frac{\partial}{\partial y}\left(K_H\frac{\partial w}{\partial y}\right)+\frac{\partial}{\partial z}\left(K_Z^m\frac{\partial w}{\partial z}\right) \quad (4)$$

$$\rho\left(\frac{\partial \theta}{\partial t}+u\frac{\partial \theta}{\partial x}+v\frac{\partial \theta}{\partial y}+w\frac{\partial \theta}{\partial z}\right)=\frac{\partial}{\partial x}\left(K_H\frac{\partial \theta}{\partial x}\right)+\frac{\partial}{\partial y}\left(K_H\frac{\partial \theta}{\partial y}\right)+\frac{\partial}{\partial z}\left(K_Z^h\frac{\partial \theta}{\partial z}\right)+\frac{\theta}{c_p T}\left(Q_{rad}-\rho L_w\Phi\right) \quad (5)$$

$$p=\rho RT,\quad R=R_0\left[\frac{1-q_V}{M_{air}}+\frac{q_V}{M_{H_2O}}\right] \quad (6)$$

Here t is time, u, v, w are the longitudinal, transverse and vertical components of the averaged wind velocity vector in the direction of the Cartesian coordinates x, y, z, respectively, ρ is the density, f is the Coriolis parameter, K_H is the horizontal diffusion coefficient, K_z^m is the momentum vertical diffusion coefficient, g is the gravity acceleration, p is the pressure, T is the absolute temperature, θ is the potential temperature, $\theta=T(p_0/p)^{R_0/c_p}$, c_p is the heat capacity of air at constant pressure, p_0=101300 N/m^2, R_0 is the universal gas constant, Q_{rad} is the heating (cooling) of the atmosphere due to long- and short-wave radiation heat fluxes propagating in a humid atmosphere, $\rho L_w\Phi$ is the temperature change due to moisture phase transitions in the atmosphere, K_z^h is the vertical diffusion coefficient of heat and moisture, L_w is the heat of vaporization, M_{air} is the molar air mass, M_{H2O} is the molar mass of water vapor, and q_v is the specific density of water vapor in the atmosphere.

To simulate the processes of phase transformations of water moisture in the atmosphere, we use the 6-class scheme of microphysics of moisture WSM6 [7], developed by Korean scientists Hong and Lim for the well-known mesoscale meteorological model WRF [2]. It considers six states of atmospheric moisture (water vapor, cloud water, rain water, ice particles, snow, graupel (hail)). The transport equation is used for each of the moisture state parameters in the atmosphere. For each transport equation, along with advective transport, various parameterizations of physical processes are included that lead to a change in the phase state of the considered forms of moisture state.

The basic equations of the WSM6 scheme are as follows:

$$\rho\left(\frac{\partial q_j}{\partial t}+u\frac{\partial q_j}{\partial x}+v\frac{\partial q_j}{\partial y}+w\frac{\partial q_j}{\partial z}\right)+\frac{\partial\left(\rho V_j q_j\right)}{\partial z}=\frac{\partial}{\partial x}\left(K_H\frac{\partial q_j}{\partial x}\right)+\frac{\partial}{\partial y}\left(K_H\frac{\partial q_j}{\partial y}\right)+\frac{\partial}{\partial z}\left(K_Z^h\frac{\partial q_j}{\partial z}\right)+\rho\Phi_j \quad (7)$$

$$j=V,C,R,S,I,G$$

Here q_V, q_C, q_R, q_S, q_I, q_G are mass concentrations of water vapor (Vapor), cloud moisture (Cloud), rainfall moisture (Rain), snow (Snow), ice crystals (Ice) and graupel (Graupel) in the atmosphere. V_j is the deposition rate of component j ($V_V=V_C=0$) [7]. For non-gaseous components, diffusion processes are not taken into account.

The source terms of the equations Φ_j for j class are a mathematical notation for the parametrization of the atmospheric moisture transitions from one state to another. Some of the transitions are carried out at a positive air temperature, and some at a negative air temperature. The WSM6 scheme [7] considers such processes as capture of some components by others (accretion), melting of ice crystals, snow, graupel, evaporation/condensation of rain drops or cloud moisture, deposition/sublimation of graupel or snow, cloud-to-rain (ice crystals-to-snow or snow-to-graupel) autoconversion, evaporation/melting of snow, rain drop freezing with the formation of graupel, etc.

To complete the system of equations (1)-(7), a turbulence model, which consists of an equation for kinetic energy [8], and algebraic relations for determining the coefficients of turbulent diffusion are used:

$$\begin{aligned}&\rho\frac{\partial k}{\partial t}+\rho u\frac{\partial k}{\partial x}+\rho v\frac{\partial k}{\partial y}+\rho w\frac{\partial k}{\partial z}=\\&=\frac{\partial}{\partial x}\left(\rho\sigma_k\sqrt{kl}\frac{\partial k}{\partial x}\right)+\frac{\partial}{\partial y}\left(\rho\sigma_k\sqrt{kl}\frac{\partial k}{\partial y}\right)+\frac{\partial}{\partial z}\left(\rho\sigma_k\sqrt{kl}\frac{\partial k}{\partial z}\right)+P+B-\frac{\rho C_D k^{3/2}}{l}\end{aligned} \quad (8)$$

Here k is the kinetic energy of turbulent pulsations; l is the integral scale of turbulence; $\sigma_k=1.2$ is the numerical coefficient; the term $C_D k^{3/2}/l$ is responsible for the viscous dissipation of turbulence energy ($C_D=0.189$), and the P and B terms express the generation of turbulence due to shear stresses and the action of the buoyancy force:

$$P=K_z^m\left[2\left(\left(\frac{\partial u}{\partial x}\right)^2+\left(\frac{\partial v}{\partial y}\right)^2+\left(\frac{\partial w}{\partial z}\right)^2\right)+\left(\frac{\partial u}{\partial y}+\frac{\partial v}{\partial x}\right)^2+\left(\frac{\partial u}{\partial z}+\frac{\partial w}{\partial x}\right)^2+\left(\frac{\partial w}{\partial y}+\frac{\partial v}{\partial z}\right)^2\right]$$

$$B=-K_z^h\frac{g}{\theta}\left(\frac{\partial\theta}{\partial z}-\gamma_\theta\right)$$

where $\gamma_\theta = 0.00065$ K/m.

The coefficients of vertical diffusion of the momentum and heat are calculated as [9]:

$$K_z^m=\rho c^m\sqrt{kl},\ K_z^h=\rho c^h\sqrt{kl}$$

The horizontal diffusion coefficient is estimated using the Smagorinsky formula [10]:

$$K_H=\alpha_{Sm}\left(\Delta x\Delta y\right)\left(\frac{1}{2}\left(\frac{\partial u}{\partial y}+\frac{\partial v}{\partial x}\right)^2+\left(\frac{\partial u}{\partial x}\right)^2+\left(\frac{\partial v}{\partial y}\right)^2\right)^{1/2}$$

where α_{Sm} is the coefficient depending on the choice of horizontal grid steps Δx and Δy.

The TSUNM3 mesoscale model predicts the components of wind velocity, temperature and humidity in the boundary layer of the atmosphere at 50 vertical levels (up to H=10000 m) for an area of 200×200 km and enclosed in it 50x50 km (grid step 1 km with the center in the city of Tomsk) for 24 hours. The initialization of the model is carried out on the basis of the results of a numerical forecast using the operational SL-AV global model [11] of the Hydrometeorological Center of the Russian Federation.

The boundary conditions for equations (1)-(7) have the following form:

$$z=H:\ \frac{\partial u}{\partial z}=\frac{\partial v}{\partial z}=\frac{\partial q_j}{\partial z}=\frac{\partial k}{\partial z}=0,\ w=0,\ \frac{\partial \theta}{\partial z}=\gamma_\theta \tag{9}$$

$$x=0,\ x=L_x:\ \begin{cases}\dfrac{\partial \phi}{\partial t}+c_\phi\dfrac{\partial \phi}{\partial x}=\dfrac{\partial \phi_0}{\partial t}+c_\phi\dfrac{\partial \phi_0}{\partial x}\\ \dfrac{\partial w}{\partial x}=0,\dfrac{\partial q_j}{\partial x}=0,\dfrac{\partial k}{\partial x}=0\end{cases}\quad y=0,\ y=L_y:\ \begin{cases}\dfrac{\partial \phi}{\partial t}+c_\phi\dfrac{\partial \phi}{\partial y}=\dfrac{\partial \phi_0}{\partial t}+c_\phi\dfrac{\partial \phi_0}{\partial y},\ \phi=u,v,\theta,q_V\\ \dfrac{\partial w}{\partial y}=0,\dfrac{\partial k}{\partial y}=0,\dfrac{\partial q_j}{\partial y}=0\end{cases} \tag{10}$$

The index $(\ )_0$ corresponds to the dynamic and thermodynamic parameters of the synoptic scale, c_φ is the phase velocity. Boundary conditions of the form (10) are often called "radiation" type conditions [12]. The phase velocity c_φ is calculated numerically from the spatial and temporal trends within the grid region near its boundaries.

Near the underlying surface, the conditions are set to correspond to the basic relations of the Monin-Obukhov similarity theory [13, 14].

When determining the temperature of the Earth's surface θ_g , the heat flux through the surface is used, due to radiation heating (cooling) and turbulent fluxes of heat and moisture on the surface; the change in temperature and moisture content of the lower soil layers is also taken into account (depth $\approx$2m, the simulation period is several days).

The ISBA (Interaction Soil Biosphere Atmosphere) parameterization scheme developed by Noilhan and Planton [15] is used to simulate the processes of heat and moisture exchange between the lower level of the atmosphere, vegetation, and soil. This scheme takes into account changes in the thermal regime of the soil, moisture content in the soil, precipitation accumulating on vegetation, and aerodynamic transport processes in the surface layer of the atmosphere. The scheme uses a soil heat and moisture recovery model and an evaporation model.

When the effect of cloudiness is taken into account, Stephens' assumption is used [16], according to which it is assumed that clear and cloudy areas of the sky can separately contribute to the total radiation balance. The short- and long-wave components of radiation are modified by cloudy areas according to empirical relationships that determine the degree of transmission and scattering of radiation depending on the content of condensed moisture in the atmosphere [17, 18].

When modeling the magnitude of anthropogenic heat flux from the surface of an urban area (land use category "urban development"), information from [19] was used. In [19], the most universal profile of changes in the normalized value of anthropogenic heat flow during the day is recommended, which is also used in the WRF modeling system [2]. As the maximum values of anthropogenic heat flow for the city of Tomsk (a city with a population of 0.5 million inhabitants from a region with a sharply continental climate), according to the recommendation [19], 25 W/m^2 was chosen for the warm period of the year (from May 1 to October 1) and 90 W/m^2 - for cold.

3. Numerical method

Before solving the system of equations by the method of grids, the coordinate transformation for this system was used:

$$\begin{aligned}\xi&=x,\\ \eta&=y,\\ \sigma&=H\frac{z-h(x,y)}{H-h(x,y)}\end{aligned} \tag{11}$$

The approximation of the above-constructed differential problem with the performed transformation (11) is carried out on the basis of the finite volume method. For discretization, a grid that is uniform along the horizontal directions is used with a refinement of the grid planes when approaching the Earth's surface. The values of the velocity components are determined at the edges of the finite volumes, and the scalar characteristics are determined at their centers. The approximation of the convective terms of the transport equations is performed using the monotonized linear upstream MLU Van Leer scheme [20]. For discretization in time, the explicit Adams-Bashforth method and the implicit Crank-Nicolson method of the second order of approximation are used:

$$\Phi_h^{n+1} = \Phi_h^n + \Delta t_n \left(\frac{3}{2} L_h\left(\Phi_h^n\right) - \frac{1}{2} L_h\left(\Phi_h^{n-1}\right) \right) + \\ + \Delta t_n \left(\frac{1}{2} \Lambda_h\left(\Phi_h^{n+1}\right) + \frac{1}{2} \Lambda_h\left(\Phi_h^n\right) \right) + \Delta t_n \left(\frac{3}{2} S_h\left(\Phi_h^n\right) - \frac{1}{2} S_h\left(\Phi_h^{n-1}\right) \right) \quad (12)$$

Here L_h is the finite-difference analogue of the convective-diffusion operator in equations (2)-(5), (7), (8) with the exception of vertical diffusion along the Oz axis, Λ_h is the difference analogue of the differential operator of vertical diffusion, $S(\Phi)$ are the source terms of equations (2)-(5), (7), (8). Note that the chosen approximation method for convective-diffusion equations makes it possible to use a tridiagonal matrix algorithm along vertical grid lines in their numerical solution.

In the hydrodynamic part of the model, a predictor-corrector method was used to match the velocity and pressure fields. According to this method the explicit-implicit scheme (12) for the equations of motion (2)-(4) was performed the function of a predictor for the velocity components. The correction to the velocity field for the purpose of satisfying the grid analogue of the continuity equation (1) was carried out on the basis of the iterative solution of the discrete Poisson's equation for the correction to the non-hydrostatic part of the pressure $p_h' = p_h^{n+1} - p_h^n$. In this paper, to solve the difference equation for the pressure correction p_h', the linear Seidel method with red-black ordering of the computational grid nodes [21] is used. The parallel implementation of such a method shows the independence of the convergence rate of the iterative process on the number of used processor elements during calculations. It is important that such an implementation of the algorithm on a multiprocessor computer system wholly preserves the property of a sequential algorithm and scales well to any reasonable number of computational nodes.

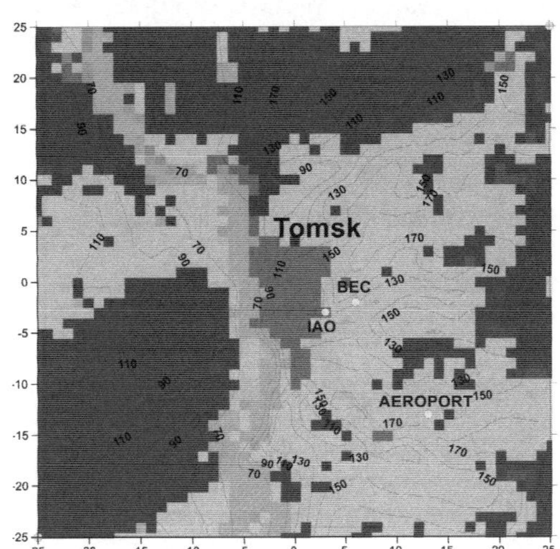

Figure 1. Distribution of land-use categories (red for urban, green for forest, gray for low vegetation, blue for water) and surface elevations above sea level for the study area. The yellow circles mark the positions of observation points IAO, BEC and Aeroport.

4. Results and discussion

The developed mesoscale model was applied to the study of meteorological conditions over the city of Tomsk (85.0°E 56.5°N, center of the city) and Bogashevo airport (Aeroport, 85.21°E, 56.38°N) (Figure 1).

The results of calculations using the TSUNM3 model [5] were compared with measurements obtained using meteorological instruments (ultrasonic meteorological stations «Meteo-2», temperature profiler MTP-5, sodar «Volna-4M») of the Joint Use Center "Atmosphere" IAO SB RAS for various seasonal conditions [22]. Ultrasonic meteorological stations "Meteo-2" are installed at points IAO and BEC, and the temperature profiler MTP-5 at point IAO. Point IAO is located on the eastern border of the city, and point BEC is located to the east, in a suburban area surrounded by forest plantations. The airport is located 12 km southeast of the city (Figure 1).

The ability of the TSUNM3 model to predict precipitation in the Siberian region was estimated for several selected case studies in the period of 2016–2018 (11.07.2016, 24.11.2016, 27.06.2016, 5.08.2017, 6.08.2017, 25.01.2017, 29.12.2017, 2.05.2018, 8.05.2018, 21.05.2018), when intense precipitation was observed on the territory of Bogashevo airport and which was recorded by standard meteorological observations [23].

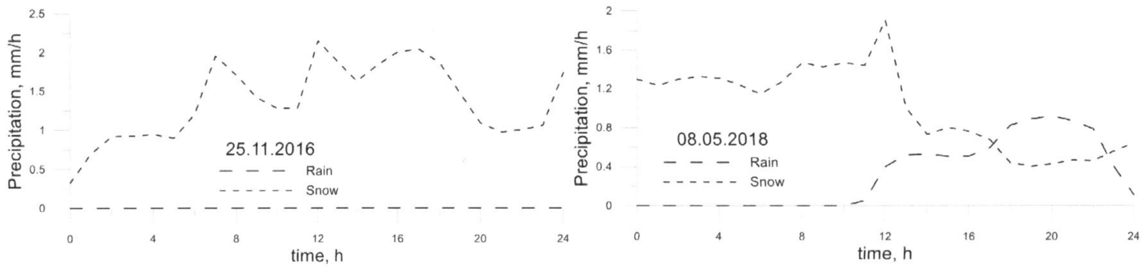

Figure 2. The predicted values of precipitation accumulated over 1 hour on November 25, 2016 and May 8, 2018 in the vicinity of the airport.

Figure 2 shows the precipitation calculated using the TSUNM3 model in the area of Bogashevo airport from November 25, 2016 and May 8, 2018. The figure shows that the first date (25.11.2016) is characterized by snowfall. Observations at the airport [23] indicate continuous snowfall during the day with an increase during the time intervals: 7:00-9:30, 17:00-18:30 (local time). In the second date (8.05.2018) until 10:30 meteorologists recorded snowfall [23], then until 17:00 only liquid precipitation was observed. After 17:00, light snow and rain were recorded. In general, the TSUNM3 model satisfactorily predicted the observed pattern of precipitation changes during the day for each selected period.

Table 1. Characteristics of the quality of precipitation forecast based on precipitation.

№	Characteristics	Rain forecast	Snow forecast	Overall precipitation forecast
1	General accuracy U	77	68	74
2	Accuracy of the forecast of precipitation events U_+	34	68	57
3	Forewarning of precipitation events P_+	59	98	87
4	Accuracy of the forecast of events without precipitation U_-	92	–	91
5	Forewarning of events without precipitation P_-	80	–	67
6	The Pearcy-Obukhov criterion T	39	–	54

The results of assessing the quality of the numerical forecast of precipitation based on precipitation for the selected case studies in the period 2016–2018, are given in Table 1. The last column represents

IOP Conf. Series: Earth and Environmental Science 1040 (2022) 012011 doi:10.1088/1755-1315/1040/1/012011

the characteristics for all precipitation events (rain, snow). The quality of numerical precipitation forecasts is established by their presence or absence and the amount of precipitation by comparing the calculated data on precipitation with actual observations. Assessing the precipitation forecast in fact, the following characteristics are calculated: the general accuracy U, the accuracy of the forecast of precipitation events U_+ and without precipitation events U_- forewarning of precipitation events P_+ and without precipitation events P_- (all values are calculated as %) and also the Pearcy-Obukhov quality criterion T. To obtain these characteristics, a contingency table is constructed showing the relationship between the number of forecast cases and the fact of precipitation. Assessing the quality of the precipitation forecast based on the fact of precipitation, hourly forecasts and actual precipitation data were used for the dates considered above.

Table 1 concludes the following:

- general accuracy for all types of precipitation according to the TSUNM3 model is 74%;
- for all precipitation, characteristics 1, 2, 4 and 6 correspond to the quality of modern mesoscale models;
- the forewarning of precipitation is higher, and the absence of precipitation is slightly lower than for the known models;
- general accuracy for rain is slightly higher than accuracy for snow;
- the TSUNM3 model better predicts the absence of rain, and its presence is better predicted from snow, which is possibly related to the peculiarities of the samples;
- the model accurately predicts the phase state of precipitation: there was rain given as snow only in one case in the forecast;
- out of 6 cases of hail, the model did not predict it.

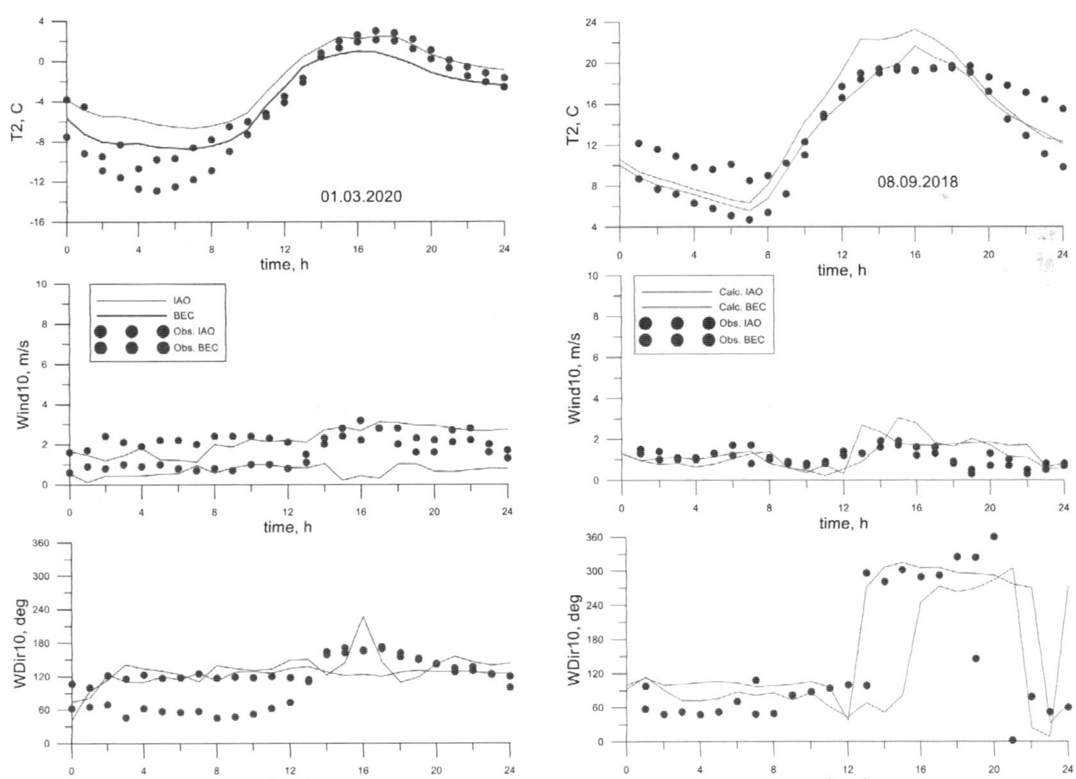

Figure 3. Numerical and experimental values of surface temperature, wind velocity and direction for March 1, 2020 (left) and September 8, 2018 (right).

As for the estimation of the model for forecasting the amount of precipitation, due to the lack of actual hourly data on its amount, it is impossible to obtain numerical characteristics. However, it could be preliminarily stated that the model predicts "rain" and "heavy rain" for the corresponding hour in the amount of precipitation from 0.2 to 1.0 mm. In the case of snowfall based on "snow", the calculations by the TSUNM3 model give the amount of precipitation up to 1 mm/h, in the case of "heavy snow" precipitation, the model forecasts from 1 to 5 mm/h.

Also, the TSUNM3 numerical weather forecast model considered in this work was applied to study the "urban heat island" phenomenon in the city of Tomsk. Figure 3 below shows the results of calculations and observations for two selected dates: March 1, 2020 and September 8, 2018. The results of the numerical calculations are compared with the experimental data.

Figure 3 shows the hourly experimental and numerical results of calculations of the surface wind velocity, and wind direction and temperature for the selected day. The first date refers to the cold season, when the earth's surface is completely covered with a uniform layer of dry crumbly snow and the river is under a meter layer of ice. The meteorological situation in the area under consideration on March 1, 2020 was characterized by [23] as weak wind and cloudy weather conditions. The temperature during the night hours dropped to -10÷-12°C. During the daytime, positive values of air temperature were observed. Figure 3 shows that due to the prevailing quiet, cloudless dry weather, both calculations and observations (during most of the day) show a weak southeast wind, which changes slightly in velocity and direction during the day. Note that the wind velocity and direction also change insignificantly for the IAO and BEC points located in the city and outside the city.

As follows from the results of calculations and measurements, the lowest air temperature values near the surface on March 1, 2020 are observed at night. This fact occurred due to the cooling of the underlying snow surface in calm, clear weather. The increase in air temperature near the earth's surface during the day is associated with solar radiation. In this case, the "urban heat island" effect is observed - the values of the surface air temperature are lower outside the city than in the city. The temperature difference reaches, as follows from calculations and observations, several degrees. Due to the prevailing eastern transport, albeit weak, the area of higher urban temperatures shifts slightly to the west.

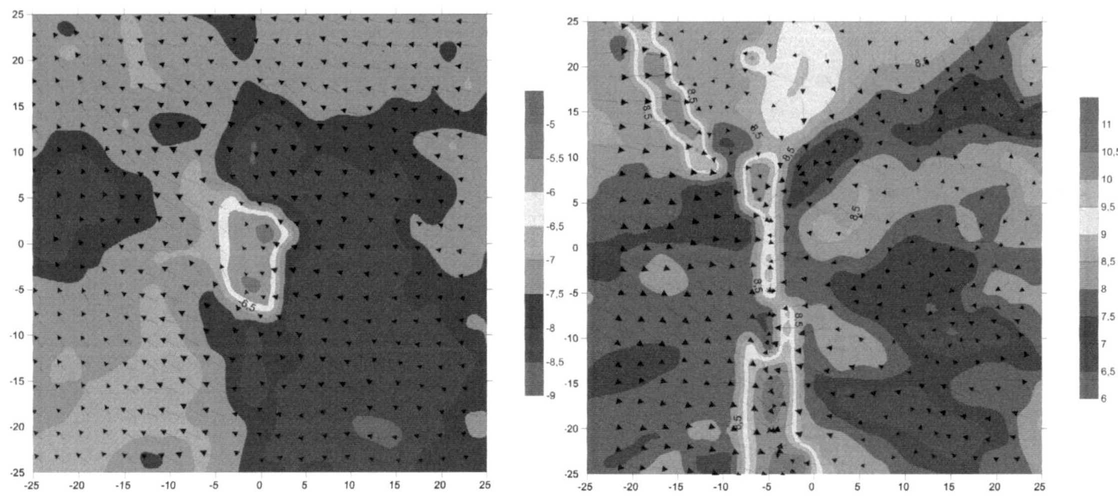

Figure 4. Colored surface temperature maps (scale to the right in °C) and surface wind vector field calculated using the TSUNM3 model at 04:00 on March 1, 2020 and 04:00 on September 8, 2018, in the center of Tomsk. Area size in km.

The second date, September 8, 2018, is typical for the warm season. During this period, the thermal regime of the surface air was warmed significantly by the nearby Tom River, about 1 km wide, in the city area. Therefore, the effect of a small anthropogenic heat flux is insignificant. In this case, the air temperature at the BEC point was predicted accurately, but in the IAO it turned out to be underestimated by about 2 °; the "urban heat island" did not appear in the calculations. In a warmer area along the river bed, according to calculations, convergence of air flows is observed.

Figure 4, to confirm the findings mentioned above, shows the isolines of the surface air temperature and the wind field calculated by the TSUNM3 model for 4 hours on March 1, 2020 and September 8, 2018. The figure shows that at night during the cold period of the year the "urban heat island" meteorological phenomenon manifests itself much more strongly, although a significant increase in temperature does not have a significant effect on the horizontal velocity of the surface wind.

5. Conclusion

The paper considers the mathematical formulation and numerical weather prediction in the Siberian region using the developed high-resolution mesoscale meteorological model TSUNM3. This model, as the "leading" global weather forecast model, provides calculations to the operational weather forecast model SL-AV of the Hydrometeorological Center.

The results of numerical forecasting of the meteorological parameters: temperature, wind velocity and direction, precipitation in different seasons are presented. The model was applied for pre-selected case studies characterized by intense precipitation in the form of rain, snow or graupel in the Tomsk region, as well as the experimentally established effect of an urban heat island.

The results were as follows. In general, the improved TSUNM3 model adequately simulates the time of precipitation and the intensity of precipitation, however, in some cases, the start and end times do not always coincide with observations, and the difference between predictions and observations can reach several hours. The phase state of precipitation is reliably displayed. More than 70% of precipitation cases are confirmed by numerical calculations. The model predicts the temperature and humidity characteristics satisfactorily. The quality of the precipitation forecast model is comparable to modern mesoscale models [3].

The capabilities of the model are shown to predict the state of the atmosphere over the city, including the "heat island" and its characteristics (intensity and horizontal dimensions, lifetime, air circulation).

Acknowledgments

The work was supported by the Russian Science Foundation (project no.19-71-20042).

References

[1] Sokhi R, Baklanov A, Schlünzen H. et al. 2018 *Mesoscale Modelling for Meteorological and Air Pollution Applications* (New York: Anthem Press)
[2] Skamarock W C, Klemp J B, Dudhia J, Gill D O, Liu Z, Berner J, Huang X 2021 *NCAR/TN-556+STR* 165 (Boulder, Colorado)
[3] Vil'fand R M, Rivin G S and Rozinkina I A 2010 *Russ. Meteorol. Hydrol.* **35** 503
[4] Garbero V, Milelli M, Bucchignani E, Mercogliano P, Varentsov M, Rozinkina I, Rivin G, Blinov D, Wouters H, Schulz J-P, Schättler U, Bassani F, Demuzere M, Repola F 2021 *Atmosphere* **12(2)** 237
[5] Starchenko A V, Bart A A, Kizhner L I and Danilkin E A 2020 *Vestnik Tomskogo gosudarstvennogo universiteta. Matematika i mekhanika* **66** 35
[6] Pielke R 2002 *Mesoscale meteorological modeling* (Academic Press, San Diego, California)
[7] Hong S-Y and Lim J-O 2006 *J. of Korean meteorological society* **42** 129
[8] Andren A 1990 *J. Appl. Meteorol.* **29** 224
[9] Yamada T 1983 *J. of Atmospheric Sciences* **40(1)** 91
[10] Smagorinsky J 1963 *Monthly Weather Review* **91** 99
[11] Tolstykh M A, Fadeev R Yu, Shashkin V V, Goyman G S, Zaripov R B, Kiktev D B,

Makhnorylova S V, Mizyak V G and Rogutov V S 2018 *Russ. Meteorol.Hydrol.* **43** 773
[12] Carpenter K 1982 *J. of Royal Meteorology Society* **108** 717
[13] Monin A S and Obukhov A M 1954 *Trudy Akademii Nauk SSSR Geofizika* **24** 163
[14] Dyer A J and Hicks B B 1970 *Quart. J. Roy. Meteorol. Soc.* **96** 715
[15] Noilhan J and Mahfouf J-F 1996 *Global and Planetary Change* **13** 145
[16] Stephens G 1978 *J. of the Atm. Sci.* **35** 2123
[17] Mahrer Y and Pielke R A 1977 *Beitr. Phys. Atmosph.* **50** 98
[18] Hurley P 2002 *CSIRO Atmospheric Research* **55** 49
[19] Saylor D J, Georgescu M, Milne J M and Hart M A 2015 *Atmospheric Environment* **118** 7
[20] Van Leer B 1974 *J. of Computational Physics* **14** 361
[21] Ortega J 1988 *Introduction to Parallel and Vector Solution of Linear Systems* (Springer)
[22] Odintsov S L 2020 *Atmospheric and Oceanic Optics* **33** 104
[23] http://rp5.ru

IOP Conf. Series: Earth and Environmental Science 1040 (2022) 012012 doi:10.1088/1755-1315/1040/1/012012

Estimating the parameters of wind turbulence from spectra of radial velocity measured by a pulsed Doppler lidar

V A Banakh, I N Smalikho

Institute of Atmospheric Optics SB RAS, Tomsk, 634055, Russia

E-mail: banakh@iao.ru, smalikho@iao.ru

Abstract. The paper proposes the method for estimating the turbulence energy dissipation rate and the variance of the vertical component of the wind velocity vector from the spectral density of radial velocity measured by a pulsed coherent Doppler lidar. The method takes into account the averaging of the radial velocity over the probing volume. It is shown that neglecting the spatial averaging leads to the understating the dissipation rate.

1. Introduction

Turbulence in the stably stratified atmosphere is a subject of study for many decades, but still remains poorly understood despite numerous publications. At the stable stratification in the atmospheric boundary layer (ABL), low level jet streams (LJS) and internal gravity waves (IGW) are formed, and turbulence is characterized by intermittence and not always obeys the Kolmogorov-Obukhov law.

The use of pulsed coherent Doppler lidars (PCDL) allows obtaining the information about wind, turbulence, and IGW in the whole ABL. Different PCDL measurement strategies (geometries) and methods for estimation of wind and turbulence parameters from raw lidar data have been developed [1-7]. In particular, the method of azimuthal structure function (ASF) for estimating the turbulence energy dissipation rate ε and the variance $\sigma_r^2 = <V_r^2> - <V_r>$ of radial velocity V_r from PCDL data obtained with conical scanning by a lidar probing beam around the vertical axis was proposed [6]. However, during IGW this method appears to provide highly overstated values of ε and σ_r^2. The modified ASF method was proposed in [8]. For the first time, it allowed obtaining unbiased estimates of the dissipation rate ε at heights of jet streams during IGW propagation. However, the following experiments have shown that this modified ASF method not always gives correct results.

Other scenarios of lidar measurements are also used to study wind turbulence. For example, in [4] it was proposed to determine the turbulence energy dissipation rate from temporal spectra of the vertical component of wind velocity vector measured by PCDL at the probing vertically upward. This method was applied to examine variations of the dissipation rate in the atmospheric boundary layer [7]. However, when obtaining estimates of ε, the authors of [4,7] ignored the spatial averaging of the radial velocity along the probing beam axis, which can lead to significant underestimation of the dissipation rate.

This paper proposes a strategy of lidar measurements, which allows us to determine both the wind velocity vector and the temporal and spatial (along the mean wind direction) spectra of turbulent fluctuations of the vertical component of wind velocity vector from data of a single PCDL. The theoretical relations presented allow the averaging over the probing volume to be taken into account when estimating the spectra of fluctuations of the vertical velocity measured by the lidar. These relations form the basis for the new method for estimation of the turbulence energy dissipation rate ε and the variance of vertical component of wind velocity $\sigma_w^2 = \sigma_r^2$ from the spectra of the vertical velocity measured by the lidar.

2. Measurement strategy and methodology of estimating the wind velocity vector and spectra of vertical velocity

For the data of a single lidar could be used for retrieving the wind velocity vector and calculating the temporal and spatial spectra of turbulence fluctuations of the vertical velocity, the following measurement strategy is applied. First, one conical scan by the probing beam at an elevation angle φ = 60° for T_{scan} = 1 min is performed. Then the probing beam is directed vertically upward (elevation angle φ = 90°), and the vertical probing is conducted for T_{vert}. Then the elevation angle is changed from 90° to 60° and the procedure is repeated. The duration of one measurement cycle is $T_c = T_{\text{scan}} + \delta t + T_{\text{vert}} + \delta t$, where $\delta t \sim 10$ s is the time needed to change the angle φ. In the experiment, the time of vertical probing was taken T_{vert} = 500 s. The time for measurement of the radial velocity Δt is determined by the number of accumulated echo signals N_a, which is set equal to 7500. At the pulse repetition frequency of StreamLine PCDL (Halo Photonics, Brockamin, Worcester, United Kingdom) f_p = 15 kHz, this corresponds to $\Delta t = N_a / f_p = 0.5$ s. Correspondingly, the number of measurements of the radial velocity was $M_1 = T_{\text{vert}} / \Delta t = 1000$ for the time T_{vert} = 500 s, and $M_s = T_{\text{scan}} / \Delta t = 120$ per one scam.

As a result, we obtained two arrays of data for estimating the signal-to-noise ratio $\text{SNR}(m'\Delta t; R_k, \varphi_{\text{scan}}, n)$ and the radial velocity $V_L(m'\Delta t; R_k, \varphi_{\text{scan}}, n)$ from measurements at the conical scanning and estimates of $\text{SNR}(m\Delta t; R_k, \varphi_{\text{vert}}, n)$ and $V_L(m\Delta t; R_k, \varphi_{\text{vert}}, n)$ from measurements in the vertical direction, where $m' = 0, 1, 2, ..., M_s - 1$; $m = 0, 1, 2, ..., M_1 - 1$ $R_k = R_0 + k\Delta R$ is the distance from the lidar; $k = 0, 1, 2, ..., K - 1$; ΔR is the range step; φ_{scan} = 60° and φ_{vert} = 90° are the elevation angles of the conical and vertical scanning, respectively, and $n = 1, 2, 3, ...$ is the number of cycle with duration T_c.

The sine-wave fitting method [9] was applied to find the horizontal wind velocity $U_1(h_k, n)$, wind direction angle $\theta_V(h_k, n)$, and the velocity of vertical wind component $V_z(h_k, n)$ averaged over the circle of the scanning cone at the height $h_k = R_k \sin\varphi_{\text{scan}}$ for the time T_{scan} from the $V_L(m'\Delta t; R_k, \varphi_{\text{scan}}, n)$ arrays obtained at the conical scanning. The $V_L(m\Delta t; R_k, \varphi_{\text{vert}}, n)$ arrays obtained at the vertical scanning were used to determine the vertical component of the wind velocity vector w and to estimate the temporal $\hat{S}_L(f_l)$ and spatial $\tilde{S}_L(\kappa_l)$ spectra of turbulent fluctuations of the vertical velocity, where f_l and $\kappa_l = f_l / U$ are the temporal and spatial frequencies, respectively, and U is the average horizontal velocity at the height of spectral measurements.

To obtain estimates of wind turbulence parameters, data measured for half an hour are usually used. At T_{vert} = 500 s, the cycle duration is $T_c \approx 10$ min. Therefore, to estimate the spectral density of the vertical velocity measured by the lidar $\hat{S}_L(f_l)$, the data of three consecutive measurement cycles were used. Correspondingly, the average velocity was estimated from four values of U_1 as $U(h_k, n) = (1/4)\sum_{i=0}^{3} U_1(h_k, n-1+i)$. The spectra $\hat{S}_L(f_l)$ are estimated from the arrays of the radial velocity $V_L(m\Delta t; R_k, \varphi_{\text{vert}}, n)$ at heights $h_k = R_k \sin\varphi_{\text{vert}} = R_k$, which differ from the heights of measurement of the average velocity $h_k = R_k \sin\varphi_{\text{scan}}$. Therefore, to find the average velocity at the height of spectral measurements, the interpolation between the measured values of U was applied. Experimental temporal spectra of the radial (vertical) velocity $\hat{S}_L(f_l)$ were calculated as

$$\hat{S}_L(f_l) = \frac{1}{3}\sum_{i=0}^{2}\frac{\Delta t}{M_1}\left|\sum_{m=0}^{M_1-1} V_L(m\Delta t; R_k, \varphi_{\text{vert}}, n-1+i)\exp\left(-2\pi j\frac{lm}{M_1}\right)\right|^2 . \qquad (1)$$

In Eqs. (1), $f_l = l\Delta f$; $l = 1,2,3,...,M_1/2$ is the number of the spectral channel; M_1 = 1000; $\Delta f = (M_1\Delta t)^{-1}$ = 0.002 Hz is the width of the spectral channel; 0.002 Hz $\le f_l \le$ 1 Hz, and $j = \sqrt{-1}$ is the imaginary unit. The summation of spectral components over all frequencies gives the estimate of the variance of lidar estimate of the radial velocity

$$\hat{\sigma}_L^2 = 2\Delta f \sum_{l=1}^{M_1/2} \hat{S}_L(f_l). \qquad (2)$$

The Taylor hypothesis of frozen turbulence allows us to pass from the temporal $\hat{S}_L(f_l)$ to spatial $\tilde{S}_L(\kappa_l) = U\hat{S}_L(U\kappa_l)$ spectra, where $\kappa_l = l\Delta\kappa$, $\Delta\kappa = \Delta f / U$. Whence we can find $\hat{S}_L(f_l) = U^{-1}\tilde{S}_L(f_l/U)$ and $\hat{\sigma}_L^2 = 2\Delta\kappa\sum_{l=1}^{M_1/2}\tilde{S}_L(\kappa_l)$.

3. Consideration of spatial averaging of the radial velocity

To take into account the averaging of radial velocity over the probing volume when estimating the dissipation rate ε from the spectrum $\hat{S}_L(f_l)$ (or $\tilde{S}_L(\kappa_l)$), it is necessary to know the low pass filter transfer function $H(\kappa_l)$ [5]. The equation for $H(\kappa_l)$ was derived in [5] for the case of estimating the radial velocity from the position of centroid of the Doppler spectrum. In the StreamLine PCDL with the low pulse energy, the signal-to-noise ratio SNR in the 50-MHz frequency bandwidth of the receiver usually does not exceed -10 dB. That is why the radial velocity is estimated from the position of maximum of the Doppler spectrum.

To make sure that the known equation for the transfer function $H(\kappa_l)$ [5] can be used to take into account the spatial averaging when estimating the radial velocity from the maximum of the Doppler spectrum, we have used numerical simulation to compare the transverse fluctuation spectra of the radial velocity estimated from Doppler spectra of the StreamLine lidar and the transverse spectra of the radial velocity averaged over the probing volume with the same averaging function as in [5]. The simulation was performed in the following way.

Assume that the longitudinal dimension of the probing volume Δz exceeds its transverse dimension Δy ($\Delta y = U\Delta t$ at the vertical probing). Then, according to [5], the equation for the normalized correlation function of the useful component of the complex lidar signal $C_S(l'T_s, z, y)$ can be written as

$$C_S(l'T_s; z, y) = \int_{-\infty}^{+\infty} dz' A(l', z')(-1)^{l'} \exp[2\pi j l' B_V^{-1} V_r(z+z', y)] , \qquad (3)$$

where $l' = 0,1,2,...,6$; $T_s = B^{-1}$; B = 50 MHz is the receiver frequency bandwidth; z is the distance from the lidar to the center of the probing volume; $y = Ut$ is the transverse coordinate; t is time; $A(l',z') = \frac{1}{7-l'}\sum_{m'=0}^{6-l'} F(z'-(m'-3)\delta R)F(z'-(m'+l'-3)\delta R)$; $\delta R = cT_s/2 = 3$ m; c is the speed of light; $F(z') = (\sqrt{\pi}\Delta p)^{-1/2}\exp[-(z'/\Delta p)^2/2]$; $\Delta p = c\sigma_P/2 = 15.3$ m; $B_V = \lambda/(2T_s) = 37.5$ m/s; $\lambda = 1.5$ µm is probing beam wavelength, and $V_r(z,y)$ is the 2D random distribution of the radial velocity over the plane $\{z,y\}$. Random realizations of $V_r(z,y)$ at the nodes of the 1024×1024 computational grid with 3-m cells were simulated with the use of the von Karman model for the two-dimensional spectrum of wind velocity $S_r(\kappa_z,\kappa_y)$ based on the algorithm described in [9].

The Doppler spectrum is calculated from $C_S(l'T_s;z,y)$ with the Fast Fourier Transform (FFT) as

$$S_D(f_{k'};z,y) = \mathrm{Re}\left\{\sum_{l'=0}^{6}(2-\delta_{l'})(1-l'/7)C_s(l'T_s;z,y)\exp\left(-2\pi j\frac{l'k'}{N}\right)\right\}, \quad (4)$$

where $f_{k'} = k'\delta f$ is the frequency; $k' = 0,1,2,...,N$ is the spectral channel number; $\delta f = B/N$ is the spectral channel width, $N = 1024$ is the number of spectral channels, and $\delta_{l'}$ is the Kronecker delta ($\delta_0 = 1$ and $\delta_{l'\neq 0} = 0$). In Eq. (4), we can pass from the frequency distribution to the velocity distribution: $V_{k'} = k'\delta V - B_V/2$; $\delta V = B_V/N = 0.03662$ m/s. The position of maximum in the Doppler spectrum is used to estimate the radial velocity $V_{\max}(z,y)$ and to calculate the transverse spectrum of fluctuations of the radial velocity as

$$S_{\max}(\kappa) = \int_{-\infty}^{+\infty} dr B_{\max}(r)\exp(-2\pi j\kappa r), \quad (5)$$

where $B_{\max}(r) = \langle V'_{\max}(z,y+r)V'_{\max}(z,y)\rangle$ is the transverse correlation function of velocity fluctuations $V'_{\max} = V_{\max} - \langle V_{\max}\rangle$, and angular brackets denote the averaging over an ensemble of realizations.

The radial velocity $V_a(z,y)$ averaged over the lidar probing volume is given by the integral [5]

$$V_a(z,y) = \int_{-\infty}^{+\infty} dz' Q_s(z')V_r(z+z',y), \quad (6)$$

where the filtering function (or averaging function) is described as

$$Q_s(z') = \frac{1}{2\Delta R}\left[\mathrm{erf}\left(\frac{z'+\Delta R/2}{\Delta p}\right) - \mathrm{erf}\left(\frac{z'-\Delta R/2}{\Delta p}\right)\right]. \quad (7)$$

In Eq. (7), $\mathrm{erf}(x) = (2/\sqrt{\pi})\int_0^x d\xi\exp(-\xi^2)$ is the probability integral. In our case, the range step is $\Delta R = 6\delta R = 18$ m, and the longitudinal dimension of the probing volume [5] is $\Delta z = Q_s^{-1}(0) = \Delta R/\mathrm{erf}(\Delta R/(2\Delta p)) \approx 30$ m. The transverse spectra $S_a(\kappa)$ are calculated from the transverse correlation functions $B_a(r) = \langle V'_a(z,y+r)V'_a(z,y)\rangle$, where $V'_a = V_a - \langle V_a\rangle$, by the equation similar to Eq. (5).

Equations (3), (4), (6), and (7) are applied to calculate $V_{\max}(z, y)$ and $V_a(z, y)$ as functions of the transverse coordinate y at fixed z from the simulated distributions $V_r(z, y)$ and then to determine the transverse spectra $S_{\max}(\kappa)$ and $S_a(\kappa)$. The averaging over an ensemble was performed with the use of no less than 1000 independent random realizations of $V_r(z, y)$. For comparison, we also calculated the transverse spectra of the radial velocity without spatial and temporal averaging $S_r(\kappa) = \int_{-\infty}^{+\infty} dr B_r(r) \exp(-2\pi j \kappa r)$, where $B_r(r) = \langle V_r'(z, y+r) V_r'(z, y) \rangle$, $V_r' = V_r - \langle V_r \rangle$. The simulation of V_r was carried out with the use of the von Karman model (see equation (17) in [9]).

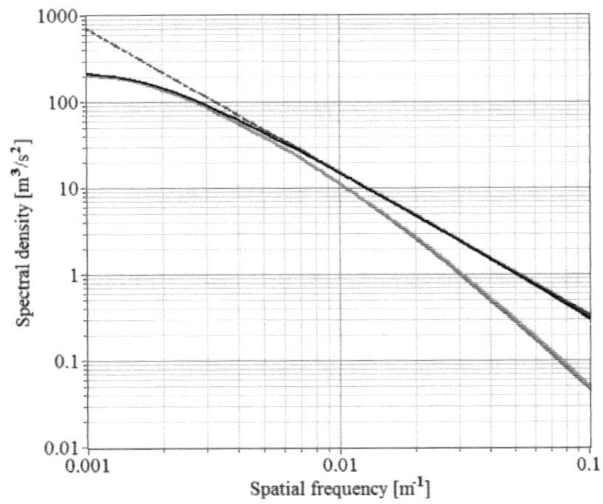

Figure 1. Transverse spectra of the radial velocity $S_r(\kappa)$ (black curve), $S_{\max}(\kappa)$ (green curve), and $S_a(\kappa)$ (red curve). The blue dashed curve is for the Kolmogorov-Obukhov "-5/3" spectrum.

Figure 1 shows an example of the spectra $S_r(\kappa)$, $S_{\max}(\kappa)$, and $S_a(\kappa)$ obtained in the numerical experiment with the following wind turbulence parameters given: variance of the radial velocity $\sigma_r^2 = 1$ (m/s)^2 and the integral scale of turbulence $L_V = 100$ m. In addition, the transverse Kolmogorov-Obukhov "-5/3" spectrum [10]

$$S_r(\kappa) = \frac{4}{3} 0.073 C_K \varepsilon^{2/3} \kappa^{-5/3}, \quad (8)$$

where $C_K = 2$ is the Kolmogorov constant, is shown by the dashed curve. Whence for the von Karman model, we have [9]:

$$L_V = 0.6973 \sigma_r^3 / \varepsilon, \quad (9)$$

where σ_r^2 is the variance of radial velocity. It can be seen from Fig. 1 that the spectra $S_{\max}(\kappa)$ and $S_a(\kappa)$ completely coincide. This means that Eqs. (6) and (7) can be used to take into account the averaging of the radial velocity over the probing volume when estimating wind turbulence parameters from measurements by the StreamLine lidar. With the use of Eq. (7), the spatial low pass filter transfer function $H_{\parallel}(\kappa_z)$ in the longitudinal direction has the form [5]

$$H_{\parallel}(\kappa_z) = \left| \int_{-\infty}^{+\infty} dz' Q_s(z') \exp(-2\pi j \kappa_z z') \right|^2 = \left[\exp\{-(\pi \Delta p \kappa_z)^2\} \mathrm{sinc}(\pi \Delta R \kappa_z) \right]^2, \quad (10)$$

where $\mathrm{sinc}(x) = \sin(x)/x$.

One can also see from Fig.1 that at the frequencies $\kappa \geq 1/(2L_V) = 0.005$ m^{-1} corresponding to the inertial range of turbulence, where the blue dashed curve and the black solid curve coincide, the averaging of the radial velocity within the probing volume along the axis z significantly overstates the spectral density. Thus, at the frequency $\kappa = 0.1$ m^{-1} the amplitude of the spectrum $S_a(\kappa)$ is sevenfold smaller than the amplitude of $S_r(\kappa)$. This should necessarily be taken into account when estimating the dissipation rate ε from turbulent fluctuations of the velocity measured by the lidar.

We have conducted similar numerical experiments at different values of σ_r^2 (from 0.01 to 4 m^2/s^2) and L_V (from 30 to 300 m). In all the cases, very close results were obtained for $S_{\max}(\kappa)$ and $S_a(\kappa)$, which confirms that Eqs. (6), (7), and (10) can be used for a good reason to take into account the averaging of the radial velocity measured by the StreamLine lidar over the probing volume.

4. Estimation of wind turbulence parameters

To estimate parameters of wind turbulence from the spectrum of radial velocity measured by the lidar (1), it is necessary to have the theoretically calculated estimate of the spectrum averaged over the ensemble of realizations, $S_L(f_l) = <\hat{S}_L(f_l)>$, with allowance made for the averaging over the probing volume and the instrumental error of measurement of the radial velocity. According to [3, 5], the unbiased lidar estimate of the radial velocity $V_L(m\Delta t)$ can be represented as

$$V_L(m\Delta t) = V_a(m\Delta t) + V_e(m\Delta t), \tag{11}$$

where

$$V_a(m\Delta t) = \frac{1}{\Delta t}\int_{-\Delta t/2}^{\Delta t/2} dt' \int_{-\infty}^{+\infty} dz' Q_s(z') V_r(R_k + z', (m\Delta t + t')U) \tag{12}$$

is the radial velocity averaged over the probing volume with the longitudinal dimension Δz and the transverse dimension $\Delta y = U\Delta t$; $V_r(z, y)$ is the radial velocity at the point with coordinates $\{z, y\}$; $V_e(m\Delta t)$ is the random instrumental error (caused mostly by the noise component of the Doppler spectrum), which is characterized by the following statistical properties: $<V_e> = <V_a V_e> = 0$ and $<V_e(m'\Delta t)V_e(m''\Delta t)> = \sigma_e^2 \delta_{m'-m''}$ $\sigma_e^2 = <V_e^2>$.

Upon the substitution of Eq. (11) into Eq. (1) and averaging over the ensemble of realizations, with allowance made for the random instrumental error we obtain the equation

$$S_L(f_l) = S_a(f_l) + S_N, \tag{13}$$

where

$$S_a(f_l) = \frac{\Delta t}{M_1} < \left| \sum_{m=0}^{M_1 - 1} V_a(m\Delta t) \exp\left(-2\pi j \frac{lm}{M_1}\right) \right|^2 > \tag{14}$$

is the spectral component carrying the information about wind turbulence, and

$$S_N = \Delta t \sigma_e^2 \tag{15}$$

is the noise component (white noise) caused by the instrumental error of estimation of the radial velocity ($\sigma_e \neq 0$).

On the assumption of isotropic turbulence for $S_a(f_l)$ at frequencies $f_l \geq U/(2L_V)$ corresponding to the inertial range, we can obtain from Eqs. (12) and (14) that

$$S_a(f_l) = \int_{-\infty}^{+\infty} d\kappa_z \int_{-\infty}^{+\infty} d\kappa_y S_r(\kappa_z, \kappa_y) H_{\parallel}(\kappa_z)[\mathrm{sinc}(\pi U \Delta t \kappa_y)]^2 P(f_l - U\kappa_y), \quad (16)$$

where

$$S_r(\kappa_z, \kappa_y) = 0.0163\varepsilon^{2/3}(\kappa_z^2 + \kappa_y^2)^{-4/3}\left[1 + \frac{8}{3}\frac{\kappa_y^2}{\kappa_z^2 + \kappa_y^2}\right] \quad (17)$$

is the two-dimensional spatial spectrum of the radial velocity in the inertial range of turbulence (Kolmogorov-Obukhov spectrum) [9] and

$$P(f_l - U\kappa_y) = \frac{\Delta t}{M_1} \frac{1 - \cos[2\pi\Delta t(f_l - U\kappa_y)M_1]}{1 - \cos[2\pi\Delta t(f_l - U\kappa_y)]} . \quad (18)$$

In addition to low-pass filtering, Eq. (16) takes into account the phenomenon of frequency substitution, which is observed in experiments when the spectrum is nonzero at frequencies higher than the Nyquist frequency $f_{\mathrm{N}} = (2\Delta t)^{-1}$ [11]. If the frequency substitution is neglected, then the function $P(f_l - U\kappa_y)$ can be represented as

$$P(f_l - U\kappa_y) = U^{-1}\delta(\kappa_y - f_l / U), \quad (19)$$

where $\delta(x)$ is the delta function. Upon the substitution of Eq. (19) into Eq. (16), we obtain the approximate equation for the spectrum with the neglected frequency substitution

$$S_a^{(1)}(f_l) = 0.0326(\varepsilon U)^{2/3} f_l^{-5/3} \int_0^{\infty} d\xi (1+\xi^2)^{-4/3}[1 + (8/3)/(1+\xi^2)] H_{\parallel}(f_l\xi / U)[\mathrm{sinc}(\pi \Delta t f_l)]^2 . \quad (20)$$

The frequency substitution can be taken into account, if we use the equation [11]

$$S_a(f_l) = S_a^{(1)}(f_l) + \sum_{n'=1}^{N'} \left[S_a^{(1)}(2n'f_{\mathrm{N}} - f_l) + S_a^{(1)}(2n'f_{\mathrm{N}} + f_l) \right]. \quad (21)$$

The calculations of $S_a(f_l)$ by Eqs. (16)-(18) and by Eqs. (20)-(21) demonstrate that already at $N' = 1$ the results coincide almost completely.

The spectrum $S_a(f_l)$ is proportional to $\varepsilon^{2/3}$, where ε is the sought parameter. Therefore, it is convenient to introduce the fitting function of the frequency f_l and the average wind velocity U as

$$G(f_l) = S_a(f_l) / \varepsilon^{2/3} . \quad (22)$$

Equations (16) - (22) are valid for frequencies $f_l \geq U/(2L_V)$ (inertial range of turbulence). Let U = 10 m/s and L_V = 100 m. Then $f_l \geq 0.05$ Hz. Consider the functions $G(f_l)$, $G_1(f_l) = S_a^{(1)}(f_l) / \varepsilon^{2/3}$ (fitting function neglecting the frequency substitution), and

$$G_{\mathrm{K}}(f_l) = 0.0974 U^{2/3} f_l^{-5/3} \quad (23)$$

(fitting function neglecting both the averaging of the radial velocity over the probing volume and the frequency substitution) in the frequency range 0.05 Hz $\le f_l \le$ 1 Hz.

Figure 2 exemplifies the fitting functions $G_K(f_l)$, $G_1(f_l)$, and $G(f_l)$ calculated by Eqs. (20)-(23) at different values of the average wind velocity U. One can see that the effect of frequency substitution manifests itself mostly in the range 0.8 - 1 Hz (compare the red and green curves), where the decisive contribution to the spectrum $S_L(f_l)$ measured by the lidar is often due to the noise component of the spectrum S_N. Nevertheless, at the strong wind (and turbulence) and low noise ($S_N < S_a(f_l)$) the consideration of the frequency substitution can be important in the range from 0.8 to 1 Hz for correct determination of the instrumental error of estimation of the radial velocity σ_e. It also follows from Figure 2 that if the dissipation rate ε is determined from the spectrum $S_L(f_l)$ measured by the lidar within the frequency range 0.05 Hz $\le f_l \le$ 0.2 Hz at wind velocities U higher than 10 m/s, then the use of $G_K(f_l)$, as in [4], can provide quite an acceptable result, since in this frequency range the difference between $G_K(f_l)$ and $G(f_l)$ is small (see Fig. 2(c,d)).

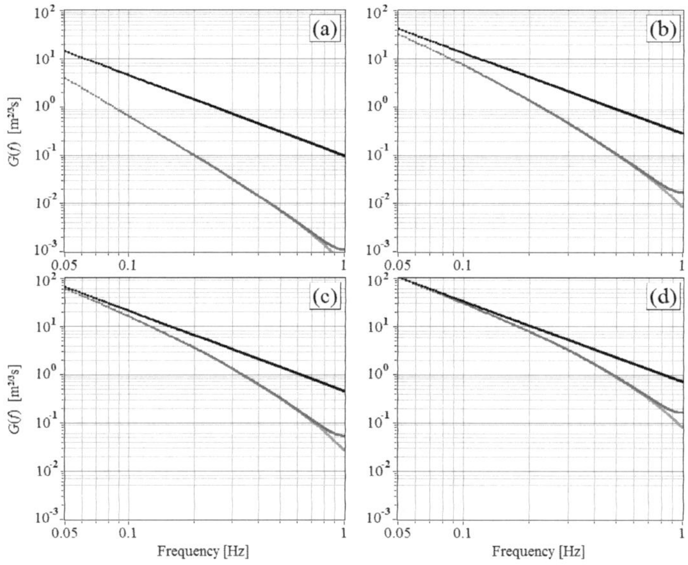

Figure 2. Fitting functions $G_K(f_l)$ (black curves), $G_1(f_l)$ (green curves), and $G(f_l)$ (red curves) at the average wind velocity U = 1 m/s (a), 5 m/s (b), 10 m/s (c), and 20 m/s (d).

Using equation $\eta = \left(\sum_l G_K(f_l) / \sum_l G(f_l)\right)^{3/2}$, where the summation is performed over frequencies from 0.05 to 0.2 Hz, we have calculated how many times the dissipation rate is underestimated when the effect of averaging the radial velocity over the probing volume is neglected. The results calculated are given in Table 1.

Table 1. Ratio of the averaged estimates of the turbulence energy dissipation rate obtained with and without regard for the averaging of the radial velocity over the probing volume.

U [m/s]	1	5	10	20
η	14.1	2.2	1.45	1.16

As follows from Table 1, the estimates of the dissipation rate with the neglected averaging of the radial velocity over the probing volume at weak wind (U = 1 m/s) turn out to be underestimated 14 times. With an increase of the average wind velocity, the bias of the estimate of dissipation rate decreases. However, even at U = 10 m/s, it remain significant.

According to [12], the dissipation rate ε can be found from the lidar estimate of the spectrum of radial velocity $\hat{S}_L(f_l)$ by the maximal likelihood method from the maximum of the function

$$\Phi(\varepsilon, S_N) = -\sum_{l=l_3}^{M_1/2}\left[\ln\left(\varepsilon^{2/3}G(f_l) + S_N\right) + \frac{\hat{S}_L(f_l)}{\varepsilon^{2/3}G(f_l) + S_N}\right], \tag{24}$$

where $\Delta f l_3 = 0.05$ Hz and $l_3 = 25$. To find the dissipation rate ε and the instrumental error $\sigma_e = \sqrt{S_N / \Delta t}$ based on Eq. (24), one can use the following iterative procedure. Assuming that in the frequency range from $\Delta f l_1 = 0.8$ Hz (l_1 = 400) to $\Delta f M_1/2 = 1$ Hz ($M_1/2$ = 500) the noise component is a major contributor to the spectrum $\hat{S}_L(f_l)$ measured by the lidar, we can find the estimate S_N as

$$\hat{S}_N^{(1)} = \frac{1}{M_1/2 - l_1 + 1}\sum_{l=l_1}^{M_1/2}\hat{S}_L(f_l). \tag{25}$$

Then we obtain the following estimate of the dissipation rate from the difference $\hat{S}_L(f_l) - \hat{S}_N^{(1)}$ in the frequency range from $\Delta f l_3 = 0.05$ Hz to $\Delta f l_2 = 0.2$ Hz (l_2 = 100):

$$\hat{\varepsilon}^{(1)} = \left[\frac{1}{l_2 - l_3 + 1}\sum_{l=l_3}^{l_2}\frac{\hat{S}_L(f_l) - \hat{S}_N^{(1)}}{G(f_l)}\right]^{3/2}. \tag{26}$$

At the second step of the iterative procedure, S_N is estimated by the equation

$$\hat{S}_N^{(2)} = \frac{1}{M_1/2 - l_1 + 1}\sum_{l=l_1}^{M_1/2}[\hat{S}_L(f_l) - \mu G(f_l)], \tag{27}$$

where $\mu = \left(\hat{\varepsilon}^{(1)}\right)^{2/3}$. Having substituted $\hat{S}_N^{(2)}$ into Eq. (26) in place of $\hat{S}_N^{(1)}$, we find the estimate of the dissipation rate at the second step of the iterative process as

$$\hat{\varepsilon}^{(2)} = \left[\frac{1}{l_2 - l_3 + 1}\sum_{l=l_3}^{l_2}\frac{\hat{S}_L(f_l) - \hat{S}_N^{(2)}}{G(f_l)}\right]^{3/2}. \tag{28}$$

The experiments have shown that one to two iterations are quite sufficient.

With allowance for Eqs. (11) - (23), the obtained estimates of the variance $\hat{\sigma}_L^2$ given by Eq. (2), dissipation rate $\hat{\varepsilon}$, and the noise component of the spectrum $\hat{S}_N$ allow us to estimate the variance of radial velocity $\hat{\sigma}_r^2$ as

$$\hat{\sigma}_r^2 = \hat{\sigma}_L^2 - \hat{S}_N / \Delta t + \hat{\varepsilon}^{2/3}2\Delta f\sum_{l=1}^{M_1/2}[G_K(f_l) - G_1(f_l)]. \tag{29}$$

At vertical probing, the estimate of variance of radial velocity is the estimate of variance of the vertical component of the wind velocity vector $\sigma_w^2 = <w^2> - <w>^2$. In contrast to the results obtained by other authors, Eq. (29) takes into account not only the instrumental error (second term from the right), but also the effect of averaging of the radial velocity over the probing volume on the estimate σ_w^2 (third term). With the obtained estimates $\hat{\varepsilon}$ and $\hat{\sigma}_r^2$, it is possible to calculate the integral scale of turbulence L_V (L_w at vertical probing) by Eq. (9).

5. Conclusions

The paper proposes the strategy of measurements by PCDL, which allows us to obtain spatial (along the mean wind direction) spectra of turbulent fluctuations of the vertical component of the wind velocity vector. The proposed measurement strategy forms the basis of the method for estimating the turbulence energy dissipation rate and the variance of the vertical component of the wind velocity vector from the spectral density of radial velocity measured by PCDL. For the first time, the method takes into account the averaging of the radial velocity over the probing volume. It is shown that if the spatial averaging of the radial velocity is neglected, the lidar estimate of the dissipation rate is understated.

6. References

[1] Eberhard W L, Cupp R E and Healy K R 1989 Doppler lidar measurement of profiles of turbulence and momentum flux *J. Atmos. Ocean. Technol.* **6** 809-819

[2] Frehlich R G, Hannon S M and Henderson S W 1998 Coherent Doppler lidar measurements of wind field statistics *Bound.-Layer Meteorol.* **86** 223-256

[3] Frehlich R G, Meillier Y, Jensen M L, Balsley B and Sharman R 2006 Measurements of boundary layer profiles in urban environment *J. Appl. Meteorol. Climatol.* **45** 821-837

[4] O'Connor E J, Illingworth A J, Brooks I M, Westbrook C D, Hogan R J, Davies F, Brooks B J 2010 A method for estimating the kinetic energy dissipation rate from a vertically pointing Doppler lidar, and independent evaluation from balloon-borne in situ measurements *J. Atmos. Ocean. Technol.* **27** 1652-1664

[5] Banakh V A and Smalikho I N 2013 *Coherent Doppler Wind Lidars in a Turbulent Atmosphere* (Artech House: Boston MA USA, London UK) p 248

[6] Smalikho I N and Banakh V A 2017 Measurements of wind turbulence parameters by a conically scanning coherent Doppler lidar in the atmospheric boundary layer *Atmos. Meas. Tech.* **10** 4191-4208

[7] Bodini N, Lundquist J K and Newsom R K 2018 Estimation of turbulence dissipation rate and its variability from sonic anemometer and wind Doppler lidar during the XPIA field campaign *Atmos. Meas. Tech.* **11** 4291-4308

[8] Banakh V A and Smalikho I N 2018 Lidar Studies of wind turbulence in the stable atmospheric boundary layer *Remote Sens.* **10** 1219

[9] Smalikho I N and Banakh V A 2013 Accuracy of estimation of the turbulent energy dissipation rate from wind measurements with a conically scanning pulsed coherent Doppler lidar. Part I. Algorithm of data processing *Atmos. Ocean. Optics* **26** 404-410

[10] Monin A S and Yaglom A M 1971 *Statistical Fluid Mechanics, Volume II: Mechanics of Turbulence* (M I T Press: Cambridge MA USA)

[11] Vinnichenko N K, Pinus N Z, Shmeter R M and Shur G N 1976 *Turbulence in the Free Atmosphere* (Leningrad: Gidrometeoizdat)

[12] Smalikho I N 1997 Accuracy of the turbulent energy dissipation rate estimation from the temporal spectrum of wind velocity fluctuations *Atmos. Ocean. Optics*. **10** 559-563

IOP Conf. Series: Earth and Environmental Science 1040 (2022) 012013 doi:10.1088/1755-1315/1040/1/012013

Investigation of wind shear structure and turbulence characteristics in a warm front cloud system using a research aircraft

V V Volkov[1], A M Strunin[1], D V Kirin[1], G E Kolokutin[1] and M A Strunin[2]

[1]Central Aerological Observatory, 3, Pervomayskaya Str, Dolgoprudny, Moscow Region, 141701, Russia

[2] Moscow Institute of Physics and Technology, 9, Institutskiy per., Dolgoprudny, Moscow Region, 141701, Russia

Abstract. The manuscript presents the results of study vertical and horizontal wind shears, the spectral characteristics of turbulence in the warm front cloud system with using the Yak-42D aircraft-laboratory. Within the As-Cs clouds at altitudes of 3 - 8 km, vertical wind shear increased sharply to 1.5 ms^{-1} per 100 m. At altitudes of 8 - 9 km, in the jet stream zone, vertical wind shear increased up to significant values of 3.0 ms^{-1} per 100 m. The maximum horizontal wind shear of 0.3 ms^{-1} per km was reached in As-Ns clouds at altitudes of 2 - 4 km. Anisotropy of turbulence with a predominance of horizontal wind fluctuations was observed in the stable layer at the altitude of 6000 m. The normalized spectra of the wind speed fluctuations in the lower part of the cloud system (at an altitude of about 3 km) corresponded to the stable layers of the atmosphere. The maximum value of 0.8 at the coherence spectra between the temperature and the vertical wind speed fluctuations was observed at the scales of 1200 m. It is noted that a solid Cs-As-Ns cloud system could be divided into sub-layers, depending on structure of wind shear and turbulence.

1. Introduction

Wind shear, as well as turbulence, accompanies many meteorological processes: development of a planetary boundary layer, atmospheric jet streams, convective motions, breeze circulation, etc. Wind shear and turbulence affect the piloting of aircraft and can cause serious flight accidents. According to ICAO law regulations [7], vertical wind shears are considered to be weak, if they are 0 - 2 ms^{-1}, moderate, if they are within 2 - 4 ms^{-1}, strong, if they are 4 - 6 ms^{-1}, and very strong, if they are over 6 ms^{-1} (wind shear is calculated with a 30-m height interval). Changes in the longitudinal wind speed component are considered to be safe for aircraft flights, if they are over 5 ms^{-1} per 600 m of a horizontal flight path.

The currently employed research aircraft Yak-42D "Roshydromet" is equipped with instruments that allow defining the wind vector with a high accuracy and calculating wind shears and turbulent wind speed fluctuations in the atmosphere [1]. The principle of defining the wind speed vector aboard the research aircraft is based on calculating the difference between two vectors, those of the airspeed vector and the aircraft ground speed one. Absolute values of these variables are usually one order larger than the measured difference between them, and therefore, their accurate measurements are especially important. The following instruments are used to define wind speed and direction onboard the Yak-42D research aircraft:

- Inertial navigation system LaserRef-VI which is used to measure heading, roll and pitch angles, and three components of the aircraft ground speed;
- Satellite global navigation system BPSN-2 which is used to determine the geographic coordinates, the track angle and ground speed of the aircraft;
- Hemispherical air pressure probe with a system of pressure transducers, which is used to obtain the value of the airspeed vector, i.e. an absolute magnitude of the airspeed as well as the angles of attack and slidslip.

The aircraft measuring system provides wind speed measurements with an error of not larger than 0.3 ms^{-1} at horizontal flight conditions and not larger than 1 ms^{-1} during the aircraft maneuvers [9]. This accuracy was verified during numerous field flight and numerical experiments [4], [3]. Random and systematic temperature-wind measurement errors were determined, taking into account various disturbing factors (aerodynamic disturbances, aircraft roll and pitch maneuvers, the impact of cloud particles on the sensors).

The air temperature was measured using a high-frequency temperature sensor (HFTS) which is a hot-wire platinum sensitive element wound on a cruciform frame with a protective Teflon cone. Calculations of true air temperature were made, taking into account the dynamic heating of the air flow during its deceleration on the sensor [1]. The error of true air temperature determination was no more than 0.3 °C during flight segments in clear air or in purely crystalline clouds, and no more than 0.5 °C in clouds with a liquid-droplet fraction. The temperature sensor, as the most inertial one in the system, determined the frequency response of the aircraft measuring system, which was thus about 10 Hz. For a typical speed of the research aircraft of 100 - 150 ms^{-1}, this corresponds to a spatial scale of wind speed and air temperature fluctuations of 10 - 15 m. Measurements of wind speed and air were supplemented with data from the Russian BPSN-2 satellite navigation system, which made it possible to relate all the obtained data to the geographic coordinates with an accuracy of 0.0003 ° and to the geometric height with an error of not more than 10 m.

All the aircraft measuring systems received signals from a unified time server. The clocks of the on-board computers were synchronized according to these signals with an accuracy of 0.001 s, and a corresponding time was assigned to all measured parameters during the registration [3]. The well-known methods of calculating wind speed components and air temperature, which are described in detail in the textbook [9], were used for the analysis in this paper.

High accuracy aircraft sensors and designed accurate methods for calculating air temperature and wind speed components gave possibility to study thin structure wind speed and air temperature fields at different heights in atmosphere and investigate characteristics of wind shear and atmospheric turbulence. Clouds of different forms (types) severely distort the structure of dynamic structure of the atmosphere (see, for example, [8]), which is reflected in the occurrence of wind shears and turbulence. Thus, the main goal of the present manuscript was revealing of some features of wind speed shear and turbulence structure in cloudy atmosphere.

2. Calculation of wind shear and energy of turbulent fluctuations

The horizontal wind shear along the selected direction is determined through the expression:

$$\frac{\partial \vec{V}}{\partial L} = \vec{e}_x\left(\frac{\partial V_x}{\partial L}\right) + \vec{e}_y\left(\frac{\partial V_y}{\partial L}\right) \tag{1}$$

The vertical wind shear is given by:

$$\frac{\partial \vec{V}}{\partial Z} = \vec{e}_x\left(\frac{\partial V_x}{\partial Z}\right) + \vec{e}_y\left(\frac{\partial V_y}{\partial Z}\right) \tag{2}$$

where $\vec{e}_x$, $\vec{e}_y$ are the horizontal unit vectors of the coordinate system for representing the wind components, L is the distance along the selected horizontal direction, Z is the vertical distance (height), V_x is the latitudinal component of the wind speed, V_y is the longitudinal component of the wind speed.

The modulus of the horizontal wind shear $|S|$ is determined based on the wind shear components, using the formula:

$$|S|=\left(\left(\frac{dV_x}{dL}\right)^2+\left(\frac{dV_y}{dL}\right)^2\right)^{\frac{1}{2}}, \tag{3}$$

and the modulus of the vertical wind shear $|W|$ is determined though the expression:

$$|W|=\left(\left(\frac{dV_x}{dZ}\right)^2+\left(\frac{dV_y}{dZ}\right)^2\right)^{\frac{1}{2}} \tag{4}$$

The energy of turbulent wind speed fluctuations is calculated through the relations:

$$E_h=\frac{\sigma_u^2+\sigma_v^2}{2}, \qquad E_w=\frac{\sigma_w^2}{2} \tag{5}$$

where the variances of the horizontal u' and v' and vertical w' wind speed fluctuations are:

$$\sigma_u^2=\frac{1}{N}\sum_{j=1}^{N}(V_{x,j}-\tilde{V}_{x,j})^2, \quad \sigma_v^2=\frac{1}{N}\sum_{j=1}^{N}(V_{y,j}-\tilde{V}_{y,j})^2, \quad \sigma_w^2=\frac{1}{N}\sum_{j=1}^{N}(V_{z,j}-\tilde{V}_{z,j})^2 \tag{6}$$

In formulas (5) and (6), $j=1,2,\ldots N$ is the number of the values in sequences of wind components for horizontal measuring flight legs, σ_u^2, σ_v^2, σ_w^2 are the dispersions of turbulent fluctuations of the wind speed components, $\tilde{V}_{x,j}$, $\tilde{V}_{y,j}$, $\tilde{V}_{z,j}$ are the calculated regression curves for correspondent wind speed components, E_h and E_w are the energies of the horizontal and vertical of wind speed components fluctuations, respectively.

To calculate the wind shear, parameters of regression lines for wind speed components along the flight path were calculated based on the least squares method. The accuracy of this method was 0.007 ms^{-1} per km for horizontal wind shears and 0.07 ms^{-1} per 100 m for vertical ones. The results of the errors estimating will be published soon.

3. Synoptic situation and flight tracks during investigations

During the period of flight investigations (December 17, 2019), Moscow region was under the influence of two vast pressure systems, a multicenter cyclone over the northern part of Europe and the Scandinavian Peninsula and an anticyclone centered over the Caucasus and Asia Minor. The location of the synoptic vortexes contributed to the appearance of large horizontal pressure gradients and a powerful westerly air transport in the boundary layer and in the middle troposphere. The research aircraft Yak-42D took off from Ramenskoye airport on December 17, 2019 at 12:15 Moscow local time (MLT). The surface air temperature during the taking off was +1.6 °C, and the zero isotherm at the airport area was located at an altitude of 550 m. During the aircraft climb over the area of Voskresensk city and Yegoryevsk city, a continuous As-Cs layer was observed within an altitude range from 3000 to 9000 m (see figure 1). From 12:45 to 13:18 MLT, the aircraft was carrying out observations along the flight lines between Yegoryevsk and Voskresensk at an altitude of 9200 m, above the upper boundary of cirrostratus clouds (Cs). At this time, an onboard measuring system

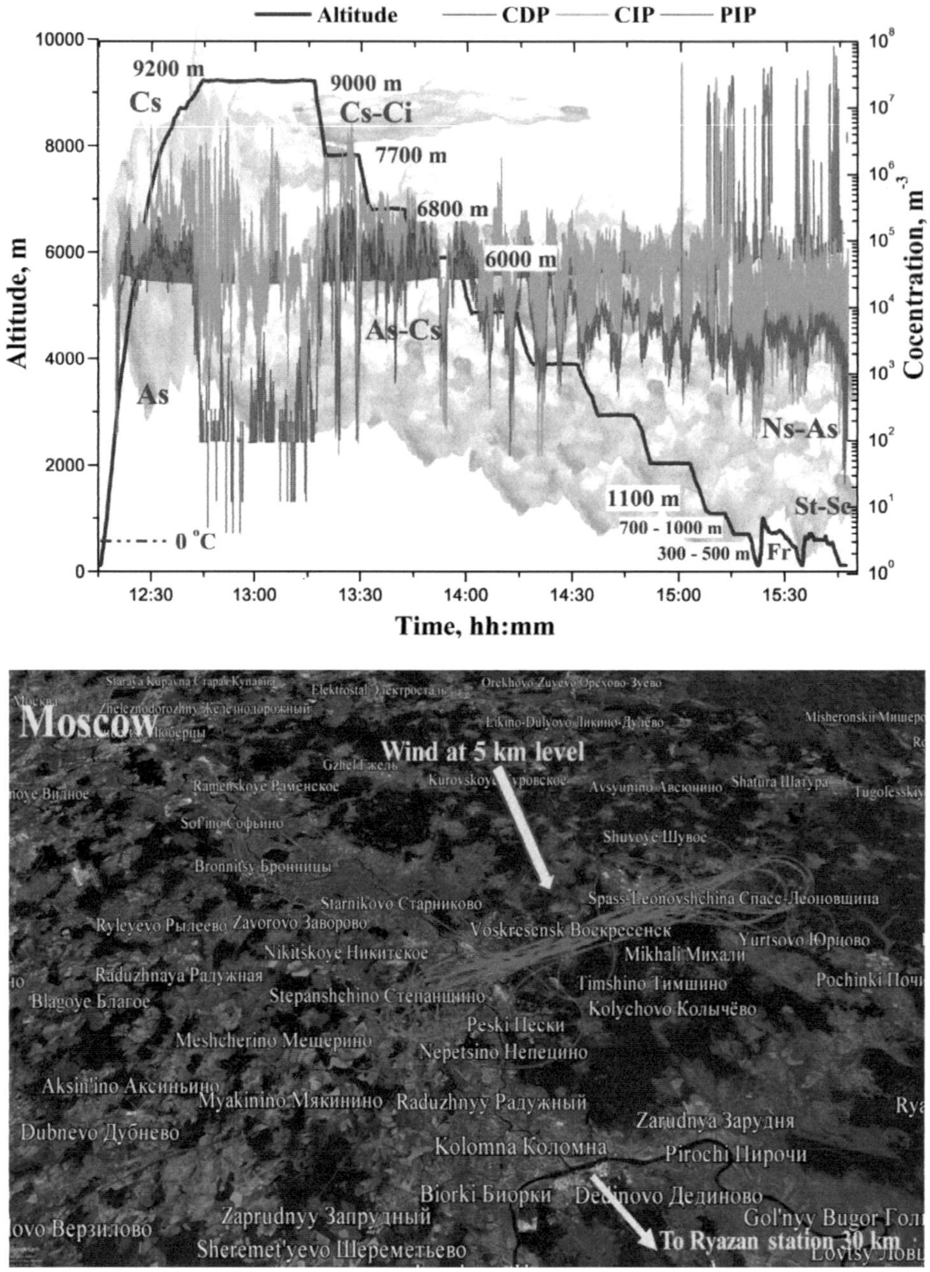

a)

b)

Figure 1. **a**) Location of cloud layers, vertical profile of flight and microphysical data, obtained during investigations. CDP is the concentrations of cloud droplets in the range from 2 to 50 mkm; CIP is the concentration of cloud particles in the range from 25 up to 1550 mkm; PIP is the concentration of cloud and precipitation particles in the range of 100 up to 6200 mkm. **b**)Trajectory of Yak-42D research aircraft flight at the map of locality, on 2019, December, 17. Average wind direction at the level of 5000 m and direction to Ryazan measuring site are also shown.

recorded a horizontal wind speed of up to 250 kmh^{-1}, related to an atmospheric jet stream axis directed from northwest to southeast. The jet was also found in radio-sounding data obtained at Ryazan radiosounding station (direction and distance to the station are shown at the map in figure 1), which was the closest to the aircraft measuring site. According to the radio-sounding data, there was mainly a stable stratification up to an altitude of 7500 m, and the axis of a jet stream with a 80 ms^{-1} wind speed was located at an altitude of 9000 - 10000 m. Between 13:20 and 15:10 MLT, the aircraft was descending from a 9000-m to a 600-m levels, carrying out observations over horizontal sampling flight legs oriented along the line between Voskresensk and Yegoryevsk. The main observations were carried out along horizontal sampling legs located approximately 1000 m apart, with flight duration along each of them being 8 - 10 minutes (see figure 1). Overall, measurements were fulfilled along 10 horizontal sampling legs.

During the aircraft descent after 13:30 of MLT, the cloud layering was visually observed below the 8000 m. At the same time, at heights of 8 - 9 km, a thin layer of cirrus and cirrostratus clouds (Ci-Cs) was clearly seen, and below the As cloud layer was located, but at 14:00 MLT upper boundary of the cloudy system has dropped up to level of 6 km. After 14:00 MLT, according to the data of ground-based meteorological radar, precipitation with an intensity of up to 1 mm per hour was recorded, the intensity tending to increase and reaching 3 - 5 mm per hour by 15:00 MLT. This indicated a change in the structure of cloudiness and the appearance of stratus clouds (Ns). During the aircraft landing in the area of Ramenskoye airport, Ns clouds were visually observed, with the lower boundary of 1100 m. Below, at an altitude of 700 - 1000 m, there was observed an inhomogeneous layer of stratocumulus clouds (Sc) and, at an altitude of 300 - 500 m, a layer of ragged rain clouds (Frnb).

Thus, during the period of observations, a dynamic process of cloudiness and weather conditions changing was observed. It was characterized by a passage of a classical warm atmospheric front.

4. Wind shears at different heights in the cloudy zone of the warm front

Calculations of the vertical shear of horizontal wind were made by the regression method, based on the wind speed data obtained during the aircraft climb to an altitude of 9000 m. The value of wind

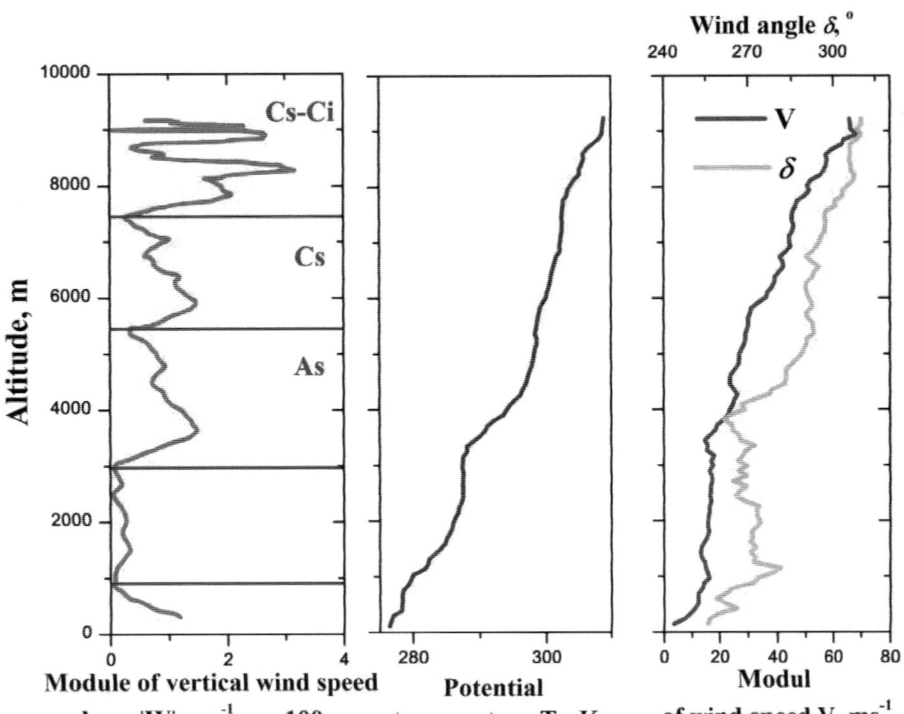

Figure 2. Vertical profiles of vertical shear of the horizontal wind speed module $|W|$, potential temperature T_p, wind speed V and wind angle δ measured during continuous aircraft climb.

shear for two horizontal wind speed components and module of wind speed $|W|$ was calculated for each level up to that height with a spatial interval of 500 m.

Figure 2 demonstrates variability of vertical shear of horizontal wind in cloudy system of warm front. The figure also shows vertical distribution of potential temperature T_p, wind speed V and wind angle δ, obtained during climb of aircraft. As it clear seen from the figure 1 the aircraft ascended through the As-Cs cloud system. The results of measurements showed that in the sub-cloud layer (practically without any clouds) up to a height of 3 km, the vertical shears of the horizontal wind were rather small, being about 0.10 - 0.30 ms^{-1} per 100 m. In the cloud layer at altitudes of 3 – 5.5 km (in the As clouds), the wind shear values sharply increased to significant values of 1.5 ms^{-1} per 100 m, becoming negligible at the lower boundary of Cs. In the Cs cloud zone wind shear become significant again, up to values of 1.5 ms^{-1} per 100 m. At altitudes of 8 - 9 km, in the jet stream zone, vertical wind shears reached significant values of 3.0 ms^{-1} per 100 m. Thus, vertical distribution of horizontal wind shear clearly correspondent to cloudy layers, which visually identified as cloud of different types. As also mentioned in literature (see textbook [8] for example) clouds have the influence on the atmosphere around them, which is reflected in the structure of the turbulence. The present studies have shown that cloud systems have also significant effect on the structure of wind shears, which is important both for solving the problems of cloud physics and for ensuring flight safety.

5. Turbulence intensity and horizontal wind shears in warm front cloud system

The descent of the research aircraft was fulfilled inside a slightly different air mass (a Cs-As-Ns cloud system, see figure 1). Measurements at horizontal sampling legs allowed calculating the horizontal

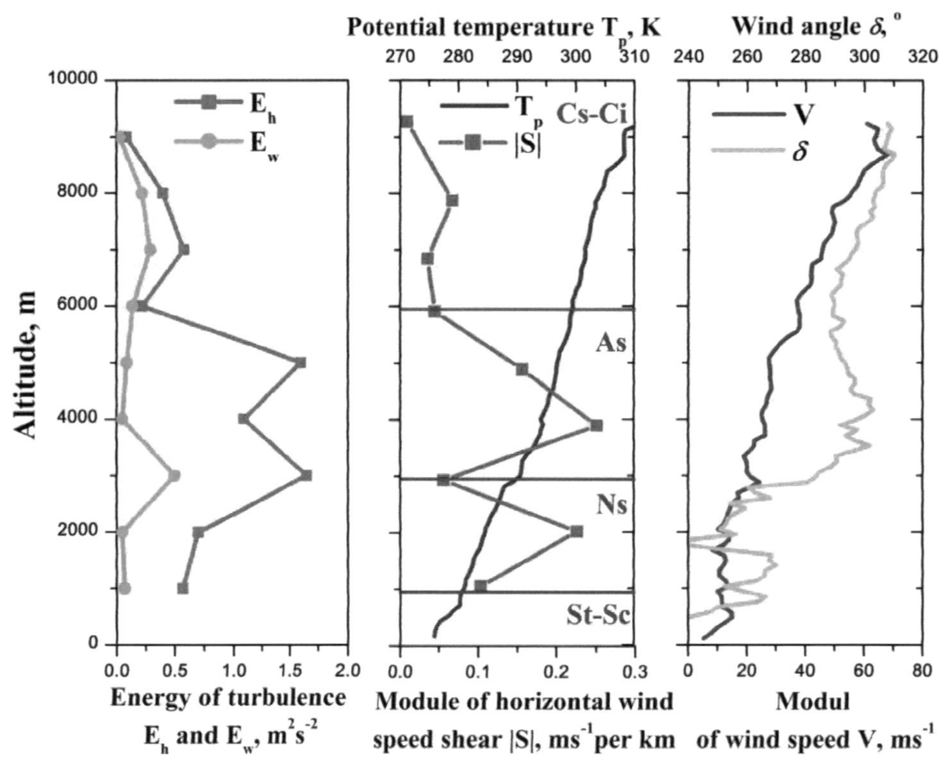

Figure 3. Energy of turbulent fluctuations of horizontal E_h and vertical wind components E_w, potential temperature T_p, modulus of horizontal wind shear $|S|$ determined from measurements at horizontal sampling legs, as well as wind speed V and direction δ obtained during the continuous aircraft descent.

shear of horizontal wind speed components and absolute wind shear values as well as estimating the intensity of turbulent fluctuations of wind speed components.

The results shown in figure 2 and 3 give reason to suggest a possibility of detecting some cloud layers inside an almost continuous Cs-As-Ns cloud system. Importantly, this assumption is based on the peculiarities of the wind shear distributions. The horizontal wind shears were the most significant in the Ns clouds (at heights from 1 km to 4 km), where wind shear values were about 0.25 ms^{-1} per km. The minimal horizontal wind shear was observed at an altitude of 3000 m in the field of temperature inversion. At this height there was the boundary of a sharp turn of wind direction, the wind angle changing from 240° to 300°. All this has led us to assume that the upper boundary of the layer was at an altitude of 3000 m.

In the overlying layers, inside the As cloud system, the horizontal wind shears increased again up to 0.27 ms^{-1} per km. Above 6000 m, in the Cs-Ci cloud layer, the horizontal wind shears diminished. At the same altitude, a sharp decrease of the intensity of both horizontal and vertical wind speed fluctuations was noted. Moreover, this made it possible to assume that dynamically (in terms of the structure of wind and fields of turbulence) the clouds of the upper layer located in the zone of the jet stream, significantly differed from the underlying As clouds. It is known [8] that clouds of various forms (types) differ significantly in their dynamics and, especially, turbulent structure. Therefore, presumably, differences in the structure of horizontal wind shears and turbulence in clouds make it possible to reveal individual cloud layers inside the general Cs-As-Ns cloud system.

Variances of wind speed fluctuations, determined for the horizontal sampling legs at different altitudes, were used as characteristics of turbulence intensity. For the Ns cloud layer, which was evidently located at levels between 1 and to 3 km, a maximum of the turbulent energy of the vertical and horizontal wind speed components was observed at an altitude of 3 km. At this level there was an inversion layer, and the corresponding energy maxima for horizontal and vertical components were 1.8 m^2s^{-2} and 0.5 m^2s^{-2}, respectively. Therefore, the turbulent energy maxima were located near the presumed upper boundary of the Ns layer.

It is well-known [6] that fluctuations of the longitudinal wind speed component receive energy directly from the main flow, but their lateral components get energy as a result of the redistribution one the longitudinal component. For a stable air-mass layer extending up to an altitude of 6000 m, the intensity of vertical wind speed fluctuations is observed to decrease, somewhat increasing in clouds of the upper tier in a jet stream zone. It is evident that under neutral stable conditions, or even stable stratification, the development of convection is suppressed by buoyancy forces, thus the main contribution to turbulence is made by fluctuations of the horizontal wind speed component, which remaining high nearly throughout the entire cloud thickness. A clearly pronounced anisotropy of turbulent fluctuations can also be noted. Usually, turbulence anisotropy is characterized by the parameter $\frac{2E_w}{E_h}$ [2], [8]. In our experiments, this parameter varied from 0.11 (at an altitude of about 5000 m) up to 0.6 (at an altitude of 3000 m). In an air layer with neutral temperature stratification above 6000 m, this ratio was close to 1 and varied from 1.27 at an altitude of 6000 m to 0.7 at an altitude of 9000 m, while at an altitude of 7000 m it was approximately equal to 1.

6. Spectral characteristics of turbulence in clouds of a warm atmospheric front

Calculations of longitudinal u', lateral v' and vertical w' wind speed components fluctuations and air temperature fluctuations T' were carried out using formulas (5) based on the data obtained at the sections of the horizontal steady flight (sampling legs) without significant changes in speed and altitude and with minor changes in the roll and pitch angles (changes in the airspeed of the aircraft did not exceed 20 km per hour, flight altitude - 10 m, and roll and pitch angles - 5°). The typical lengths of the sampling legs were 8 - 10 km, which were enough to obtain statistically significant spectral data. The spectral characteristics of the turbulent fluctuations of the horizontal and vertical wind speed components as well as air temperature fluctuations (spectral densities $\Phi_u(k)$, $\Phi_v(k)$, $\Phi_w(k)$,

IOP Conf. Series: Earth and Environmental Science 1040 (2022) 012013 doi:10.1088/1755-1315/1040/1/012013

$\Phi_T(k)$ and spectra of coherence between wind speed components and air temperature fluctuations $Coh_{uv}(k)$), $Coh_{wT}(k)$) were calculated for horizontal sampling legs at different altitudes based on well-known windowed Fourier transform [10]. All spectral characteristics were presented as dependence on the wave number k ($k = \frac{1}{\lambda}$, where λ was the wave length). Figure 4 demonstrates normalized spectral densities $\frac{\Phi_u(k)}{\sigma_u^2}$, $\frac{\Phi_v(k)}{\sigma_v^2}$, $\frac{\Phi_w(k)}{\sigma_w^2}$, and $\frac{\Phi_T(k)}{\sigma_T^2}$ as well as spectra of coherence $Coh_{uv}(k)$), $Coh_{wT}(k)$ obtained for different flight levels.

The spectra of coherence between air temperature and vertical wind speed fluctuations, shown in figure 4, had maxima with very high amplitudes of about 0.8 at vortex scales of 900 - 1200 m. This value of coherence corresponded to the dip in the normalized spectrum of energy of the vertical wind speed fluctuations. Apparently, the energy of the vertical wind speed fluctuations was suppressed on

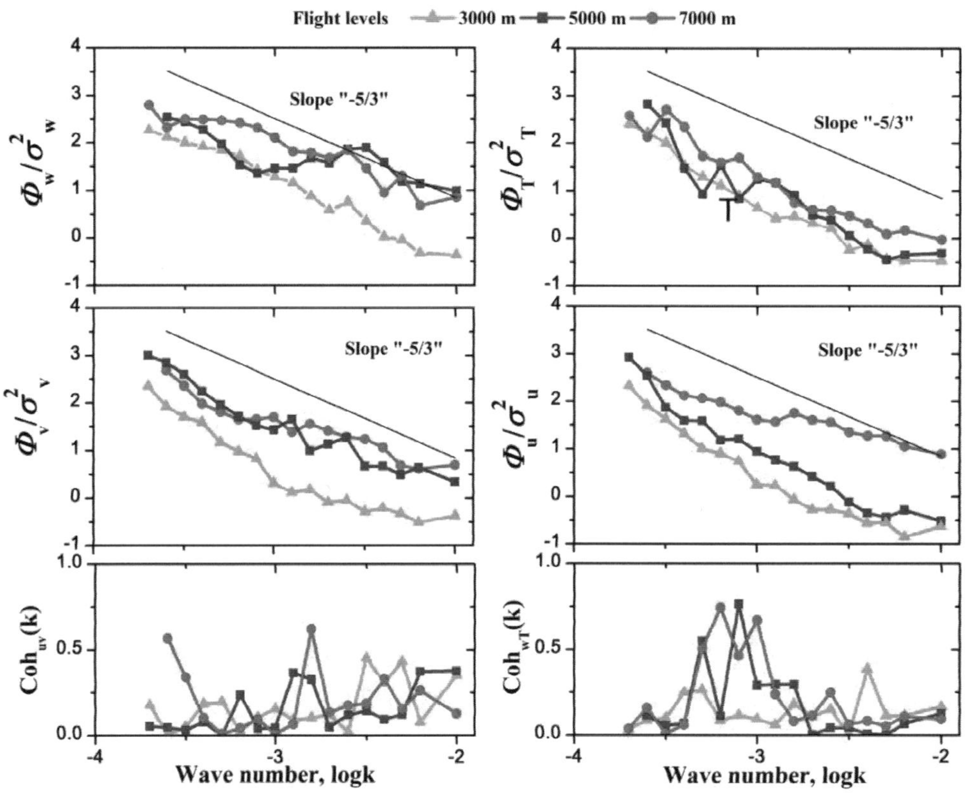

Figure 4. Normalized spectral densities of fluctuations of the vertical wind speed component $\frac{\Phi_w(k)}{\sigma_w^2}$, air temperature $\frac{\Phi_T(k)}{\sigma_T^2}$, the lateral wind speed component $\frac{\Phi_u(k)}{\sigma_u^2}$, the longitudinal wind speed component $\frac{\Phi_u(k)}{\sigma_u^2}$, and spectrum of coherence between the longitudinal and lateral wind speed components $Coh_{uv}(k)$ and between the vertical wind speed component and air temperature $Coh_{wT}(k)$ fluctuations.

these scales by negative buoyancy in a stable air layer. Significant values up to 0.7 were also revealed in the spectrum of coherence between temperature and the vertical wind speed fluctuations, calculated for an altitude of 7000 m. Simultaneously, a maximum of the energy spectrum (spectral densities) of vertical wind speed fluctuations was observed at the same scales. A similar nature of the spectral structure of vertical wind speed fluctuations was characteristic of an unstable air mass in a cloud system at heights of 6 - 8 km. This gives possibility to suggest that the zone of instability in the clouds led to the formation of precipitation. The energy spectrum of vertical wind speed fluctuations for the lower part of the cloud system at an altitude of about 3 km had a slope, which close to the Kolmogorov-Obukhov law for the inertial interval [2].

The horizontal wind speed components were used to determine the values of the turbulent fluctuations of longitudinal and lateral wind speed components (relative to the direction of the mean wind). The spectra of longitudinal and lateral wind speed fluctuations for the lower part of the cloud system (at an altitude of 3 km) had a slope of the curves close to the degree "-2" on scales from 100 to 1000 m. This slope of spectra was typical for the stable layers in the atmosphere [2]. The slope of the energy spectra of longitudinal wind speed fluctuations for the clouds of the upper and middle tier (at heights of 6 and 8 km) was close to the one of Kolmogorov-Obukhov law "-5/3". The spectrum of lateral and longitudinal wind speed fluctuations for an altitude of 7000 m (close to the axis of the atmospheric jet stream) had maxima on a scale of about 600 m. The same scales corresponded to the significant maximum in the spectrum of coherence which was equal to 0.8. According to the generally accepted theory [6], the main flow transfers energy to longitudinal wind speed fluctuations and, apparently, a high value of coherence indicates the existence of a mechanism that effectively transfers this energy to lateral wind speed fluctuations.

7. Conclusion

Aircraft observations of the dynamic structure of a warm front cloud system, including the altitude distributions of vertical and horizontal wind shears, turbulence intensity, the spectral structure of turbulent wind speed and air temperature fluctuations, were fulfilled in the vicinity of Moscow city.

It was shown that the vertical structure of wind shears and the intensity of turbulence depended on the degree of stability in cloudy atmospheric layers.

A relationship between the dynamic characteristics of wind fields and a cloud system of warm atmospheric front consisting of stratus (Ns), altostratus (As) and cirrostratus (Cs) clouds was revealed. Thus it was demonstrated that a solid Cs-As-Ns cloud system could be divided into sub-layers, depending solely on the characteristics of the dynamic air-mass structure.

The spectral densities of wind speed and temperature fluctuations, calculated for both stable and unstable air layers in clouds, were presented.

Significant peaks in the spectra of coherence between vertical wind speed and air temperature fluctuations, as well as in the spectra of coherence between longitudinal and lateral wind speed fluctuations, were revealed. It was suggested that these peaks were associated with the existence of the mechanisms of energy transfer though vortices of a certain size.

References

[1] Bazanin N V, Borisov Y A, Volkov V V, Dmitriev V K, Zhivoglotov D N, Makosko A A, Strunin A M and Strunin M A 2014 Onboard instrumentation of the Yak-42D "Roshydromet" new-generation research aircraft for measuring and registering flight parameters and atmospheric thermodynamic parameters *Russ. Meteorol. Hydrol* 39 768–82

[2] Vinnichenko N K, Pinus N Z, Shmeter S M and Shur G N 1980 *Turbulence in the Free Atmosphere* (New York: Springer Sc.) p 310

[3] Volkov V V, Kolokutin G E, Strunin M A and Bazanin N V 2019 The onboard data-acquisition system of research aircraft for studying atmospheric processes *Instruments and Experimental Techniques* 62 401–7

[4] Strunin A M and Zhivoglotov D N 2014 A method to determine true air temperature

fluctuations in clouds with liquid water fraction and estimate water droplet effect on the calculations of the spectral structure of turbulent heat fluxes in cumulus clouds based on aircraft data *Atmos. Res.* 138 98–111

[5] Zhivoglotov D N and Strunin M A 2015 *Proc. 2nd All-Russia Scientific Conference "Ecology and Space" Named after Academician K. Ya. Kondrat'ev* (St. Petersburg: Mozhaisky Military Space Academy) pp 124–130

[6] Lumley J L and Panofsky H A 1964 *The structure of atmospheric turbulence* (New York: Interscience Publishers) p 239

[7] ICAO 2005 *Manual on low-level wind shear* Doc 9817 AN/449 (International Civil Aviation Organization) p 213

[8] Strunin M A 2013 *Turbulence in the Cloudy Atmosphere (in Clouds and Near-cloud Space). Empirical Model of Turbulence in the Cloudy Atmosphere. Reference Book for Researchers and Engineer-meteorologists* (Dolgoprudny: Fizmatkniga) p 192

[9] Strunin M A 2020 *Investigation of methods of the thermodynamic condition of atmosphere using aircraft-observatory. Handbook for scientists and meteorological engineers, second edition, reviseted and updated* (Moscow: Shans) p 212

[10] Jenkins G and Watts D 1969 *Spectral Analysis and Its Applications* (San Francisco: Holden-Day) p 525

IOP Conf. Series: Earth and Environmental Science 1040 (2022) 012014 doi:10.1088/1755-1315/1040/1/012014

Model estimates for climatic effects of anthropogenic GHG emission scenarios in the 21st century

S Denisov[1], **A Eliseev**[1,2,3] **and I Mokhov**[1,2]

[1] A.M. Obukhov Institute of Atmospheric Physics RAS, Moscow, Russia
[2] Lomonosov Moscow State University, Moscow, Russia
[3] Kazan Federal University, Kazan, Russia

E-mail: denisov@ifaran.ru

Abstract. To estimate the climatic effects of anthropogenic CO_2 and CH_4 emission scenarios we performed numerical experiments using IAP RAS CM global climate model. Emissions pathways of 5 RCP (Representative Concentration Pathways) regions were used as evaluated scenarios. It is shown that the anthropogenic contribution of CO_2 to global surface temperature change starts to decrease in the second half of the century only for RCP 2.6 and only for ASIA, OECD and REF emission scenarios. The rest of the CO_2 emission scenarios contribute to an increase in the global surface temperature of the atmosphere throughout the 21st century. The growth of the compensatory effect of natural CO_2 fluxes generally slows down by the end of the 21st century. Impact of anthropogenic methane emissions on climate stabilizes in the 21st century for all scenarios under RCP 2.6, 4.5 and 6.0. The estimates of GHG climatic cost equal up to 24 mK per PgC for CO_2 and up to 0.9 mK per Tg for CH_4. Methane is shown to have at least 40 times stronger impact on global surface temperature per molecule than CO_2.

1. Introduction

Various indicators can be used to quantify the relative and absolute contribution to climate change of emissions of different greenhouse gases, as well as emissions from different regions, countries or individual sources. The climatic effect of emissions can be estimated for a specific time instant or integrated over a given time interval. The most common indicators are based on the radiative forcing (RF) [1], which is used to compare the contribution to changes in global mean surface temperature of various factors affecting the Earth's radiative budget.

For this purpose, the UN Framework Convention on Climate Change, the Kyoto Protocol, and the Paris Agreement use global warming potentials (GWP) of various GHG's, calculated as the integral of radiative forcing on a defined time horizon divided by the same integral of reference gas, normally CO_2. At the same time, targets for climate policy are usually formulated as some given temperature thresholds to be avoided. Because of inherent inertia of the climate system, such goals are partly incompatible with a metric based on cumulative radiative forcing [2]. The most used of the alternative metrics is the global temperature change potential (GTP) [1,3]. The GTP indicator reflects changes in global temperature at a selected time interval after a pulse release of a selected gas relative to changes resulting from a similar CO_2 release, and thus accounts for climate response along with radiative efficiency and atmospheric lifetime of the gas.

In this paper, the cumulative influence of the anthropogenic CO_2 and CH_4 emission scenarios to the surface air temperature changes in the 21st century is estimated using the cumulative temperature potential CT [4] based on GTP, which was modified to account for changing background conditions.

2. Methods and data

To estimate the climatic effects of anthropogenic CO_2 and CH_4 emission scenarios we performed numerical experiments using IAP RAS CM global climate model [5-7]. Thus, the necessary background parameters for calculating the cumulative temperature potential of CO_2 and CH_4 emissions were obtained. The range of possible anthropogenic GHG emission scenarios from various sources was obtained by rescaling the anthropogenic emission scenarios for different regions of the Earth according to the RCP (Representative Concentration Pathways) data.

2.1. Model and Setup of Numerical Experiments

The IAP RAS CM belongs to the class of the global climate models of intermediate complexity [8,9]. A specific feature of the model is that the large-scale atmospheric and oceanic dynamics (with a scale exceeding the synoptic) are described explicitly, whereas the synoptic processes are parameterized. The latter makes it possible to substantially reduce the time required for model simulations. The model contains modules of the carbon cycle, including partly interactive methane cycle, and a module for calculation of emissions from deforestation and from natural fires [10,11].

Using the IAP RAS CM, we performed numerical experiments for 1765–2100 with scenarios of anthropogenic impacts on climate due to changes in the content of greenhouse gases in the atmosphere, tropospheric and stratospheric volcanic sulfate aerosols, changes in the total solar irradiance, and changes in the area of agricultural lands. For 1700–2005, these forcings were given in accordance with the "Historical Simulations" of the CMIP5 project (http://www.iiasa.ac.at/web-apps/tnt/RcpDb). For 2006–2100, the anthropogenic forcings were given prescribed in accordance with the anthropogenic impact scenarios RCP 2.6, 4.5, 6.0 and 8.5 of the CMIP5 project. The numerical experiments described in this study are similar to those conducted earlier with the IAP RAS CM [4,12].

2.2. Cumulative Temperature Potential

Global Temperature Potential of gas x is the ratio of its absolute potential to absolute potential of CO_2:

$$GTP_x(H) = \frac{P_x^{(a)}}{P_{CO2}^{(a)}}, \quad (1)$$

where the Absolute Global Temperature change Potential ($P^{(a)}$) is the change in global mean surface temperature at time H in response to a 1 kg pulse emission of gas x at time t = 0 [13]. It is often written as a convolution of the radiative forcing with the climatic response kernel R_T:

$$P_x^{(a)}(H) = \int_0^H RF_x(t)R_T(H-t)dt, \quad (2)$$

where RF_x is the radiative forcing due to a pulse emission of a gas x, and R_T is the temporally displaced climate response to a unit forcing.

$P^{(a)}$ is determined for a pulse gas emission under constant background conditions, while the assessment of impact of emission scenarios under changing conditions is required. For changing background conditions $P^{(a)}$ can be written as sum of integrals for each year:

$$P_x^{(a)*}(T_0, T_H) = \sum_{k=T_0+1}^{T_H} \int_{k-1}^{k} RF_{x,k}(t)R_T(T_H - T_0 - t)dt, \quad (3)$$

where T_0 is the year of emission and $T_H = T_0 + H$. $RF_{x,k}$ can be achieved in assumption of all necessary changing parameters being step functions, constant for each particular year k.

For sustained emissions started at time T_0, the cumulative effect of the source of gas x at time T_H can be written as cumulative temperature potential:

$$CT_x(T_0, T_H) = \sum_{t=T_0}^{T_H-1} E_x(t)\, P_x^{(a)*}(t, T_H). \quad (4)$$

It represents the total expected global surface temperature change at time T_H caused by the source emitting since time T_0.

2.3. Evaluated GHG emission scenarios

Cumulative impact on climate was estimated for anthropogenic emissions of CO_2 and CH_4 calculated according to the corresponding RCP scenarios from the ASIA (Eastern Asia), MAF (Middle Asia and Africa), LAM (Latin America), OECD (Organization for Economic Co-operation and Development, Western Europe, North America, Australia, Japan) and REF (former USSR, Eastern Europe) regions (http://www.iiasa.ac.at/web-apps/tnt/RcpDb). These emissions pathways are consistent with certain socio-economic assumptions and represent the wide range of possible changes in future anthropogenic GHG emissions. They were linearly rescaled to 1 TgC for CO_2 and 1 Tg for CH_4 in the year 2000 for each region (figure 1). Further the emissions from the specific region will be referred to by corresponding region name and the scenarios used to drive the IAP RAS model will be referred to by the RCP pathway (e.g., OECD 4.5 designates rescaled RCP 4.5 scenario from OECD region).

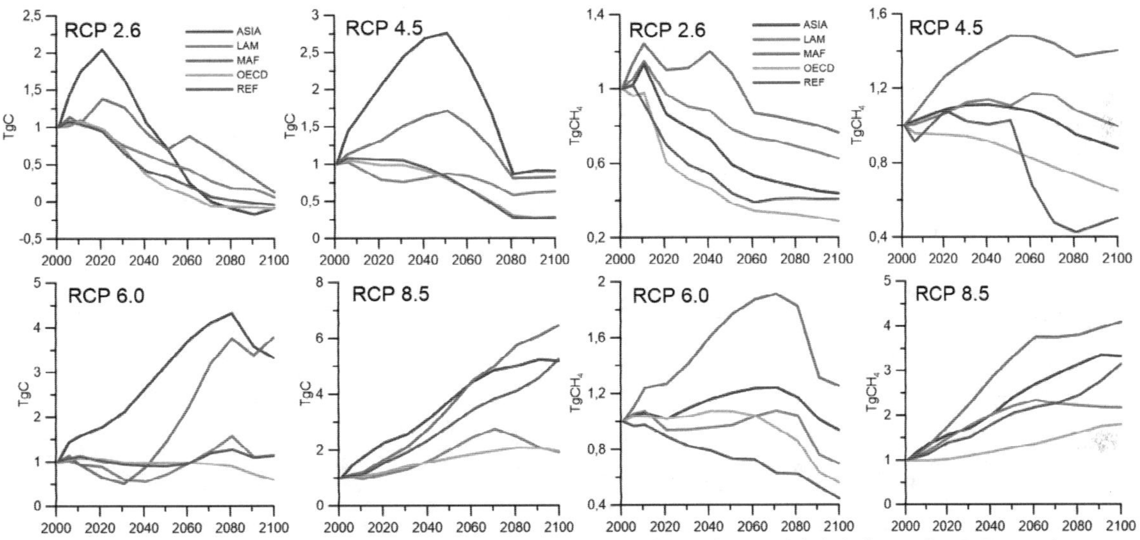

Figure 1. Rescaled anthropogenic emissions of CO_2 (left) and CH_4 (right) from the RCP regions.

For comparison, similar estimates were made for natural GHG emissions from the same regions. For this purpose, the values of carbon dioxide and methane fluxes from terrestrial ecosystems obtained by interactive calculations with the IAP RAS model for the same regions were linearly rescaled using the corresponding values of anthropogenic emissions. Thus, the proportions of natural and anthropogenic scenarios of greenhouse gas fluxes were preserved.

3. Results

Figure 2 shows the values of the cumulative CT potential since the year 2000 separately for carbon dioxide and methane emissions from ASIA, LAM, MAF, OECD and REF regions. It is noted that although CO_2 emissions for RCP 2.6 and 4.5 are declining in the 21st century, the anthropogenic CT of CO_2 starts to decrease in the second half of the century only for ASIA, OECD and REF 2.6 (scenarios with negative emission value in the end of the 21st century). Other regions contribute to an increase in the global surface temperature of the atmosphere, and the OECD region is closest to stabilizing its impact. For the methane emission scenarios, the situation is more encouraging. The contribution to the growth of the surface temperature (as well as the emissions themselves) stabilizes and begins to decrease under all RCPs except 8.5 (and for the LAM scenario even in this case).

For anthropogenic CO_2 emissions from ASIA, it is possible to see the effect of accounting for changes in the background conditions (P(a)* instead of P(a)): although throughout the entire 21st century the ASIA 8.5 CO_2 emissions are higher than ASIA 6.0, due to the higher CO_2 concentration in the atmosphere (and thus a lower forcing) their total impact on the climate by the end of the century is lower.

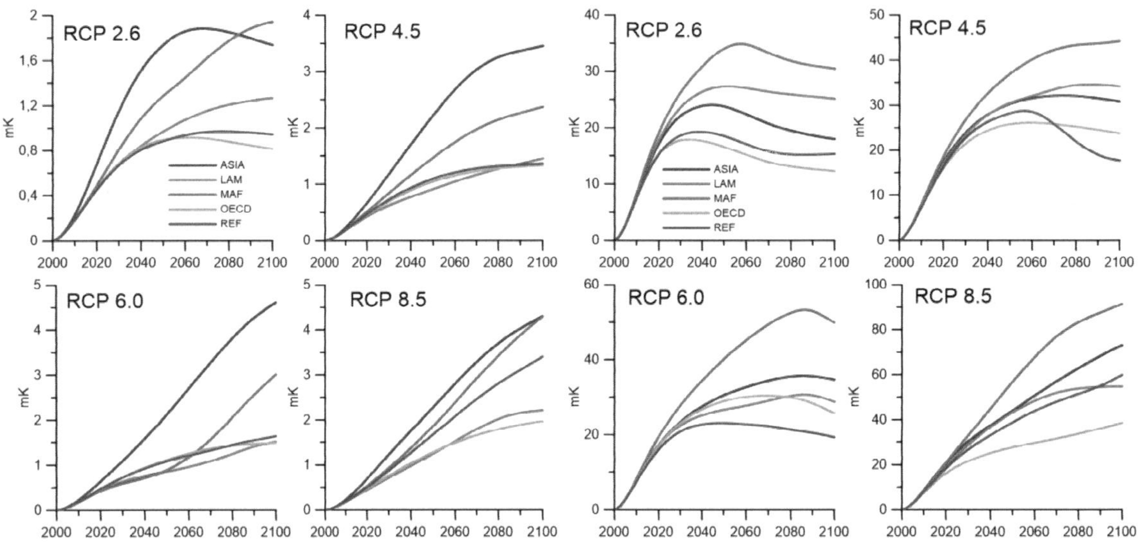

Figure 2. Cumulative potential [mK] of anthropogenic CO_2 (left) and CH_4 (right) emissions.

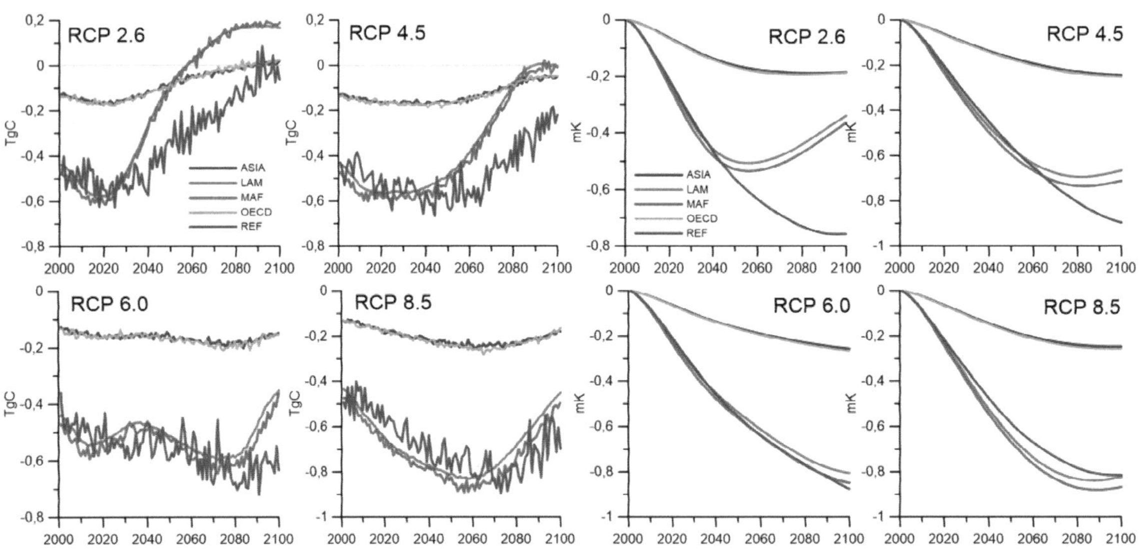

Figure 3. Rescaled natural fluxes of CO_2 (left) from the RCP regions and their cumulative temperature potential [mK] (right).

For comparison, similar estimates were made for rescaled natural emissions of carbon dioxide and methane. CO_2 uptake (figure 3) by terrestrial ecosystems for all scenarios and regions considered increases at the beginning of the 21st century. Further in the first (RCP 2.6 and 4.5) or in the second (RCP 6.0 and 8.5) half of the 21st the maximum absorption is reached after which carbon dioxide uptake begins to decrease, and even switches to emission for LAM and MAF 2.6. Natural CO_2 fluxes range from -60% and -90% to 20% and 10% of anthropogenic emissions, respectively, under scenarios RCP 2.6 and 8.5. Accordingly, they are close to full compensation of anthropogenic emissions only in the second half of the 21st century for the ASIA, OECD and REF regions under RCP 2.6 and for the REF region under RCP 4.5. This can also be observed by the CT value (figure 3) of natural CO_2 fluxes. It is closest to the compensation of anthropogenic emissions effect only for the REF region under the RCP 2.6 scenario. Moreover, due to the reduction of CO_2 uptake, the growth of the temperature potential of its natural fluxes slows down by the end of the 21st century, and for some scenarios is replaced by a reduction.

Natural methane emissions from wetlands at the beginning of the 21st century is estimated to range from 10% for the ASIA region to 100% for the LAM region of corresponding anthropogenic emissions. During the century, they increase due to the growth of the atmospheric temperature. The strongest increase occurs in the OECD and REF regions under the RCP 8.5 scenario where emissions are calculated to double by the end of the century, while in other regions the growth equals 10-20%. This is due to the fact that in boreal latitudes not only temperature increases faster than average, but also the duration of the warm season increases. Cumulative temperature potential of natural methane emissions by the end of the 21st century equals 22-144% of the potential of the corresponding anthropogenic emissions for RCP 2.6 scenario and 4-50% for RCP 8.5.

We can also try to compare the "climatic cost" of different scenarios of greenhouse gas emissions. As can be seen from figure 1, the regional emissions differ significantly. It is possible that approximately the same amount of total greenhouse gas emissions during the 21st century is achieved in different ways (as, for example, in the ASIA and MAF 2.6 scenarios for CO_2 and OECD and REF 4.5 for CH_4). The difference in the calculated response of the surface temperature of the atmosphere can be observed as well (figure 2). Figure 4 shows the value of the "climatic cost" of the emission scenarios obtained by dividing the climate response by the total emissions of the corresponding greenhouse gases by a given moment. The smaller this value, the smaller the temperature increase per unit of emissions. Despite the aforementioned significant differences in both the emission scenarios themselves and their cumulative temperature potential, the resulting climatic cost differs slightly between the scenarios (deviations do not exceed 10%) for both CO_2 and CH_4. Thus, it depends more on total amount of gas emitted over the given period than on specific pathway of such emissions.

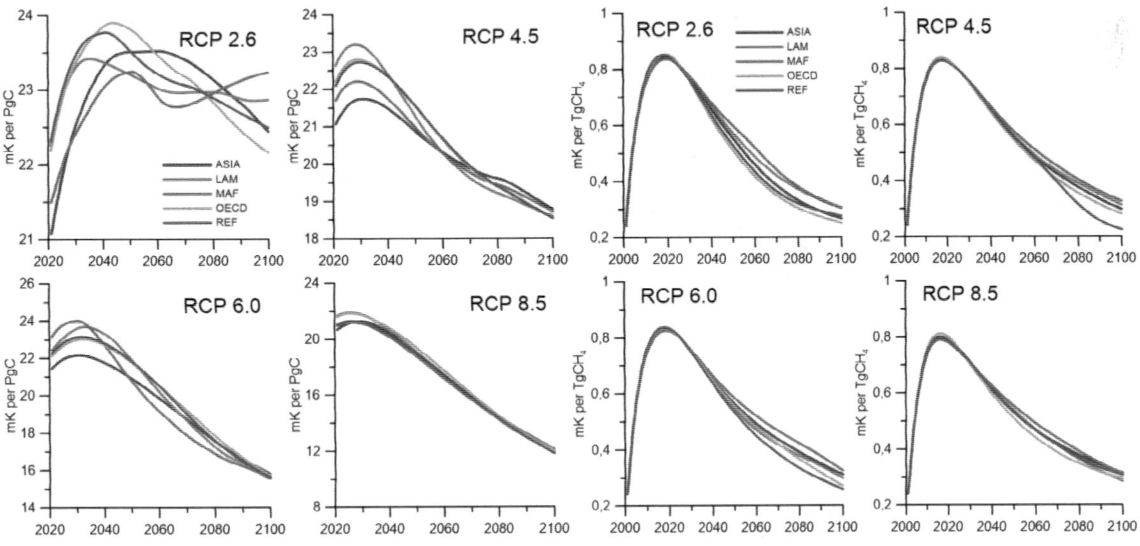

Figure 4. Anthropogenic CO_2 (left) and CH_4 (right) emissions climatic cost.

For both greenhouse gases generally by the end of the century, the emission climatic cost begins to decline, although the reasons for this are different. The decline in CO_2 emission cost is mainly due to an increase in its concentration and a corresponding decrease in RF. Therefore, it falls most rapidly at RCP 8.5, when the concentration rises most rapidly. The decrease in the climatic cost of methane, in turn, is associated with its short (about 10 years) lifetime in the atmosphere. In this case, only recent emissions affect the temperature growth, and the climatic cost decreases as the considered period increases.

In addition, it is also possible to compare the relative contributions of methane and CO_2 to changes in global surface temperature under the considered scenarios. Table 1 shows the ratio of the climatic cost of total methane emissions over different periods beginning in 2000 to the climatic cost of CO_2 emissions (in the same units, for example, kgCO_2 per kgCH_4). The resulting dimensionless value is similar in purpose to GWP and GTP. It is shown that on time horizon of less than a century under expected scenarios of greenhouse gas emissions, methane is at least 40 times more efficient than CO_2. For the most aggressive anthropogenic scenario RCP 8.5 this value exceeds 90 throughout the 21st century. Although this value can be substantially reduced if earlier emissions are taken into account, such an approach reflects the real relationship between the climate impacts of CO_2 and CH_4 emissions on a 50–100-year planning horizon. The 100-year GWP, while conceptually incompatible with current climate policy goals, is much closer in magnitude to the values obtained than the 100-year GTP [13]. Thus, in the case of switching to the use of GTP in a multi-gas policy, it is worth choosing a much shorter time horizon for methane.

Table 1. Efficiency of methane emissions relative to CO_2.

	2020	2040	2060	2080	2100
RCP 2.6	143±3	102±5	73±7	55±5	45±3
RCP 4.5	140±4	108±3	88±3	70±7	57±8
RCP 6.0	138±4	105±4	88±7	80±8	68±7
RCP 8.5	137±3	110±3	98±5	94±5	93±5

4. Conclusions

According to the results obtained, the anthropogenic contribution of CO_2 to global surface temperature change starts to decrease in the second half of the century only for RCP 2.6 and only for ASIA, OECD and REF. The rest of the CO_2 emission scenarios contribute to an increase in the global surface temperature of the atmosphere throughout the 21st century. At the same time, the growth of the compensatory effect of natural CO_2 fluxes slows down by the end of the 21st century, and for some scenarios is replaced by a reduction. Impact of anthropogenic methane emissions on climate stabilizes in the 21st century for all regions under RCP 2.6, 4.5, and 6.0. The results obtained for the REF region can be considered applicable to Russia as well, since it represents a significant share of the region in terms of both area and anthropogenic emissions.

Accounting for changes in climatic conditions can influence the impact indicators of various greenhouse gas emissions on the climate system, especially over large time horizons under the most aggressive scenarios of anthropogenic impact. The more intensive the anthropogenic impact on the climate, the smaller the efficiency of contemporary CO_2 emissions is. At the same time, the relative role of contemporary methane emissions increases.

It is shown, that efficiency of anthropogenic GHG emission scenario impact is weakly dependent on specific emission pathway. The key role is played by the total amount of gas emissions for a given period.

Acknowledgements
This study was supported by the Russian Science Foundation (project 19-17-00240).

IOP Conf. Series: Earth and Environmental Science 1040 (2022) 012014 doi:10.1088/1755-1315/1040/1/012014

References

[1] Shine K P, Fuglestvedt J S, Hailemariam K and Stuber N 2005 *Clim. Change* **68**(3) 281-302
[2] Smith S M, Lowe J A, Bowerman N H A, Gohar L K, Huntingford C and Allen M R 2012 *Nat. Clim. Change* **2** 535-8
[3] Shine K P, Berntsen T K, Fuglestvedt J S, Skeie R B, Stuber N 2007 *Phil. Tran. Roy. Soc.* A. **365** 1903–14
[4] Denisov S N, Eliseev A V, Mokhov I I 2019 *Doklady Earth Sci.* **488**(1) 1066–71
[5] Mokhov I I, Eliseev A V, Demchenko P F, Khon V C, Akperov M G, Arzhanov M M, Karpenko A A, Tikhonov V A, Chernokulsky A V and Sigaeva E V 2005 *Doklady Earth Sci.* **402** 243–7
[6] Eliseev A V and Mokhov I I 2011 *Adv. Atmos. Sci.* **28**(5) 1215–32
[7] Mokhov I I and Eliseev A V 2012 *Doklady Earth Sci.* **443**(2) 532–6
[8] Claussen M *et al.* 2002 *Clim. Dyn.* **18** 579-86
[9] MacDougall A H *et al.* 2020 *Biogeosciences* **17**(11) 2987-3016
[10] Eliseev A V, Mokhov I I, Arzhanov M M, Demchenko P F and Denisov S N 2008 *Izvestiya Atmos. Oceanic Phys.* **44**(2) 139-52
[11] Denisov S N, Eliseev A V and Mokhov I I 2013 *Rus. Meteorol. Hydrol.* **38** 741-9
[12] Denisov S N, Eliseev A V, Mokhov I I and Arzhanov M M 2015 *Izvestiya Atmos. Oceanic Phys.* **51**(5) 482–7
[13] *Climate Change 2013: The Physical Science Basis. Contribution of Working Group I to the Fifth Assessment Report of the Intergovernmental Panel on Climate Change* ed T Stocker et al (New York: Cambridge University Press) pp 659–740

IOP Conf. Series: Earth and Environmental Science 1040 (2022) 012015 doi:10.1088/1755-1315/1040/1/012015

Variability of near-surface aerosol composition in Moscow in the spring of 2020

D P Gubanova[1], A I Skorokhod[1], N F Elansky[1], M A Iordanskii[1], V M Minashkin[2]

[1]A M Obukhov Institute of Atmospheric Physics RAS, Moscow, Russia
[2]JSC "VNIIHT", Moscow, Russia

E-mail: gubanova@ifaran.ru

Abstract. The paper studies variability in mass concentration and elemental composition of near-surface aerosol in Moscow in March-April 2020. During the study period, noticeable fluctuations in concentration of surface aerosol caused by atypical synoptic and meteorological conditions were revealed. Sharp increase in PM_{10} particle concentration (March 25-29, April 13) is associated with anticyclonic activity and advection of air containing combustion aerosols from the areas with biomass fires. In April as a whole, anomalously low values of aerosol particle concentrations were recorded in comparison with the long-term average. The prevailing dry Arctic air masses significantly decreased the atmospheric aerosol pollution. The decrease of anthropogenic load during COVID-19 non-proliferation actions affected on daily variations of the surface aerosol, smoothing out its typical daily maximal concentration values. Results of spring experiment at the IAP RAS showed good agreement with the data of the Obuchi nearest station of State Budgetary Institution "Mosecomonitoring". We analyzed geochemical spectrum of chemical elements in aerosol and its variability under different synoptic and weather conditions in Moscow. Possible sources and sinks of aerosols are discussed taking into account both abnormal weather conditions and decreased anthropogenic load during a lockdown period in the spring of 2020.

1. Introduction

Increased influence of anthropogenic aerosol pollution on chemical composition and optical parameters of atmosphere is important in modeling the climatic effects of atmospheric aerosol and in assessing and predicting the ecosystem state. The spatial-temporal variability of aerosol characteristics is determined by processes of generation, transformation and sinks of aerosol, gas-chemical transformations in the atmosphere, its turbulence, circulation, as well as interaction with the underlying surface [1-3]. Conditions of large industrial metropolis (dense buildings, sealed urban soils, a special wind regime, influence of the "heat island", as well as the presence of numerous local anthropogenic sources) determine specific features and variability of physicochemical properties of near-surface aerosol. Therefore, study of short-period, seasonal and interannual variations of surface aerosol parameters in large cities, taking into account the above factors, is of great interest.

Ehe most significant variability of aerosols, due to their strong dependence on local and regional sources, as well as on long-range transport, is manifested in spring, during unstable transition period of synoptic and meteorological phenomena. The spring of 2020 in the Moscow metropolis, in a series of long-term observations, was characterized by unique conditions, including those associated with a decrease in anthropogenic load during the lockdown caused by COVID-19 pandemic. The first results for Moscow [4,5] and for foreign large cities [6-11] showed improvement of air quality in industrial centers during lockdown, with decreased traffic and activity in industry and economy.

This work studies variations in the microphysical parameters and elemental composition of surface aerosol in Moscow in the spring of 2020, during implementation of measures to prevent the spread of coronavirus infection, taking into account the synoptic and meteorological conditions.

2. Subjects, methods, and tools of experimental studies

The objects of research were submicron and micron aerosol particles in the surface layer of the Moscow atmosphere in spring from March 25 to May 3, 2020. The mass concentration of aerosol particles of sizes smaller than 10 μm (PM_{10}) and 2.5 μm ($PM_{2.5}$) is calculated daily by a numerical method [2] using microparticle number concentrations measured in a continuous automatic regime with the aid of both LAS-P laser and OEAS-05 optical-electrical aerosol spectrometers (with a time resolution of 5 min) developed at the Karpov Institute of Physical Chemistry [12]. These instruments are installed in the courtyard of the IAP located in the Center of Moscow. The observation site is in the vicinity of motorways with medium and light traffic loads, but away from industrial and power enterprises. The main local pollution sources are associated with motor transport. The IAP courtyard is small and is also used as a parking place for up to ten personal vehicles. It is separated from city streets by a fence and auxiliary brick low-rise buildings, which may decrease the emissions of large aerosol particles from the underlying surface due to transport activity.

The atmospheric composition is intensively monitored for 1–1.5 months in each season of a year, when the mass concentration of total aerosol precipitated onto an AFA-type filter with the aid of aspiration samplers for 24 h (from 9:00 to 9:00 next day) is additionally measured using the gravimetrical method. Within the periods of intensive monitoring, aerosol samples are continuously taken onto hydrophobic filters made of the Petryanov cloth and AFA-WP-10 analytical filters with the aid of a six-cascade impactor. These samples are used to study the mean size distribution of aerosol particles and chemical elements contained in them also using both gravimetrical and inductively coupled plasma mass spectrometry methods. Aerosol samples for both gravimetrical and chemical analyses are taken at a height of 2 m above the land surface.

To assess the influence of meteorological and synoptic conditions on the variability of surface aerosol characteristics, we used data on meteorological parameters obtained with a time interval of 3 hours at the Balchug station located 800 meters North-East of the IAP RAS observation point, as well as from open Internet sources [15]. The analysis of synoptic and meteorological conditions of observations was carried out using Internet resources [16,17] and using the method of backward trajectories of air masses calculated by the NOAA HYSPLIT model [18,19] on the website of the NOAA ARL Air Resources Laboratory [20]. The fire situation in the European Territory of Russia (ETR) and in neighboring countries was also taken into account on the basis of the fire maps of SCANEX Group [21].

The data on the mass concentration of submicron and micron particles obtained at the IAP RAS were compared with the results of observations on the ground network of stations of the State Budgetary Institution "Mosecomonitoring" (MEM). For this purpose, a detailed analysis of the variability of the mass concentration of aerosol particles $PM_{2.5}$ and PM_{10} was carried out at the IAP RAS and at the Spiridonovka station of MEM [22] located in the Central Administrative District of Moscow, northwest from of the IAP RAS, at a distance of about 3 km. This is the MEM station nearest in geographic location and urban conditions (building density, traffic intensity, localization of anthropogenic sources) to the IAP RAS observation point.

When analyzing the possible advection of air containing smoke aerosol from the areas with biomass combustion to the Moscow region, we analyzed the concentration fields of black carbon and dust particles, calculated by MERRA-2 reanalysis (The Modern-Era Retrospective Analysis for Research and Applications, version 2) [17,23-25] on the basis of satellite data.

3. Meteorological conditions

Meteorological parameters were analyzed for March and April 2020 and were compared with the same time period for 2007-2019. Figure 1 on the left shows the temporal variability of daily mean values of meteorological parameters (air temperature, atmospheric pressure and relative air humidity) in 2019,

2020 and their averaged time course for 2007-2020, obtained from the data for station Balchug. The picture on the right shows the wind rose in March and April 2019 and 2020 (Fig. 1). As can be seen from Figure 1, the spring of 2020 was characterized by atypical meteorological and synoptic conditions differing from long-term averages. It should be noted that the warm and snowless winter period of 2020 affected the subsequent spring months as well. March 2020 turned out to be one of the warmest over the past 15 years of observations, and its average monthly temperature exceeded the climatic norm by 6 °C.

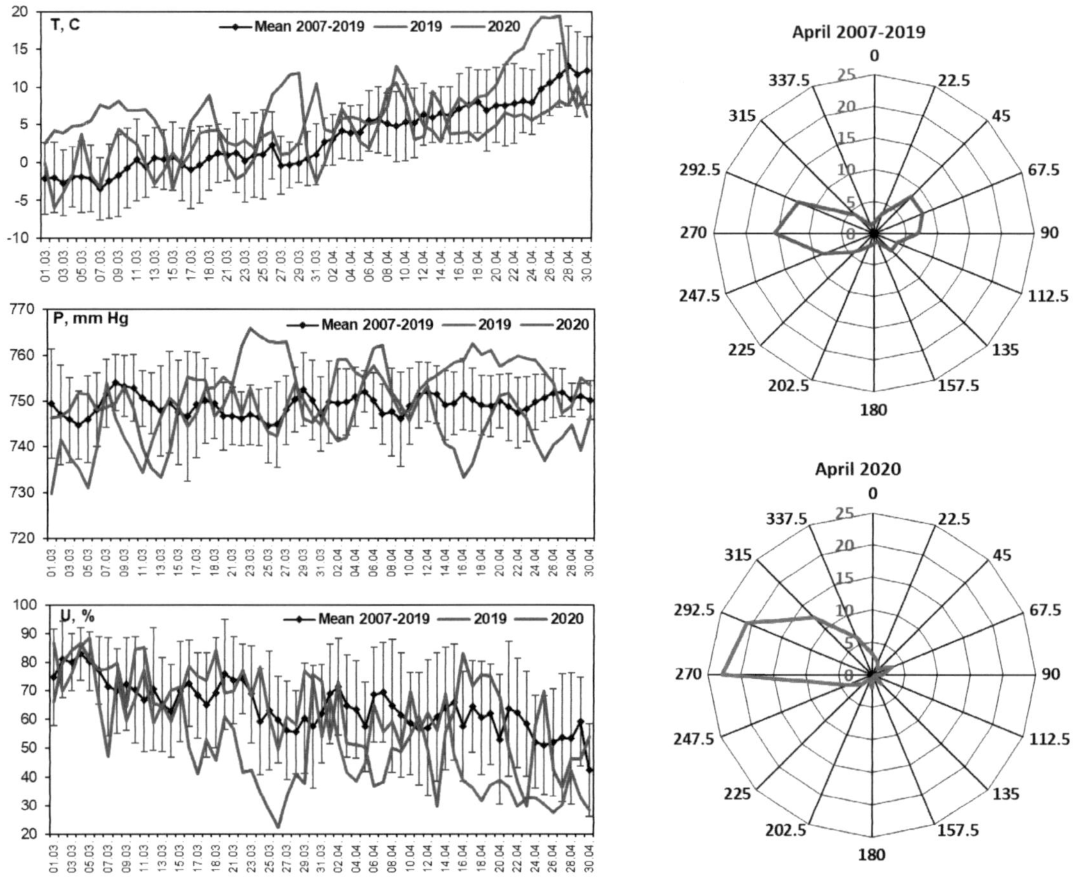

Figure 1. Meteorological conditions in March and April in Moscow: on the left – the time variations in daily averaged temperature, atmospheric pressure and relative humidity (RMS values for means are as vertical dispersion); on the right – wind roses for March and April 2019 and 2020.

April 2020 in Moscow is characterized by anomalies due to two major factors: unfavorable meteorological conditions and lockdown because of the COVID-19 pandemic. April 2020 in Moscow was the coldest over a long-term observation period (see Fig. 1), which was due to the prevalence of the Arctic air masses entering the Moscow region from western, northwestern, and northern directions, and carrying clean air. This is confirmed by results of the analysis of 3-day backward trajectories of air masses, calculated for heights of 100, 250 and 500 m. The wind rose (see Fig. 1 on the right) shows that in April 2020, the western and west-north-west wind directions were predominant.

The second unique phenomenon in April was suspension of vigorous activity of enterprises and institutions, and introduction of mandatory self-isolation of the population of the Moscow region, which led to a significant decrease in anthropogenic load of the metropolis and contributed to purification of the atmosphere from aerosol pollution. On March, 30, a mandatory self-isolation regime was introduced, the activities of most institutions and enterprises were suspended, or a remote mode of operation was established. On April, 13, stricter measures were taken with the introduction of access control. These measures have significantly reduced the activity of local urban sources, including car traffic. The quarantine was cancelled on June 9, 2020.

In general, the presence of unfavorable weather conditions in March-April 2020 in Moscow, coupled with restrictive measures for the non-proliferation of COVID-19, influenced variability of parameters of surface aerosol.

4. Results

4.1. Time variability of the mass concentration of surface aerosols PM_{10}

Figure 2 shows the daily mean values of mass concentration of aerosol particles PM_{10} in the surface atmosphere of Moscow in March and April. The data were obtained from observations at IAP RAS in 2020 and at Spiridonovka (MEM) in 2019, 2020 and the mean values through 2007-2020. The results at both stations in the spring of 2020 correlate very well with each other: the Pearson pairwise correlation coefficients were 0.978 and 0.980 in March and April 2020, respectively. This indirectly confirms the similarity of urban conditions for these two observation points and made it possible to conduct a qualitative study of long-term variability of mass concentration of PM_{10} particles according to the data of previously performed long-term studies at Spiridonovka (MEM).

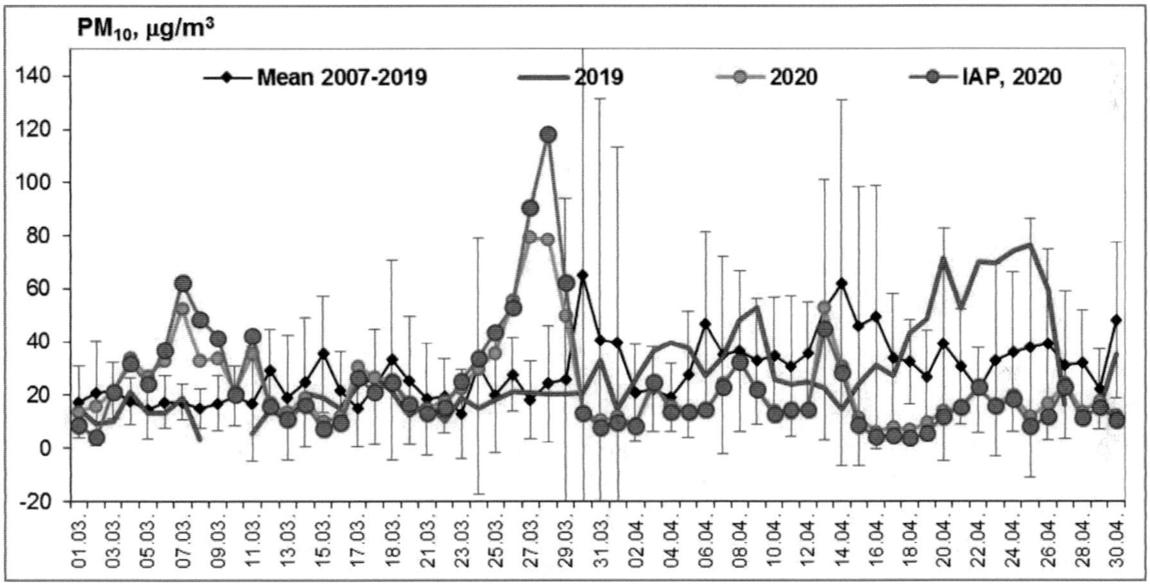

Figure 2. Temporal variability of daily mean mass concentration of surface aerosols PM_{10} in Moscow according to observations at IAP RAS (2020) and st. Spiridonovka MEM (2019, 2020). For comparison, the graph of daily PM_{10} averaged concentration at st. Spiridonovka MEM (black curve with statistical dispersion ± RMS) for 2007-2019 is also shown.

As can be seen from Figure 2, the time curve of daily mass PM_{10} concentration in 2020 differs greatly from both 2019's and (2007-2019)'s ones. The differences were most pronounced in April, due to specifics meteorological conditions discussed above. During the intensive experiment at IAP RAS, two anomalous bursts were observed in aerosol PM_{10} mass concentration in 2020: during 5 days – on March 25-29 (when the average daily concentration of PM_{10} exceeded the daily average MPC = 60 µg/m^3) and on April 13, 2020. Strong aerosol air pollution in Moscow during these periods is associated with the regional transport of combustion aerosols by air masses NW, W, SW, S, SE directions from areas with numerous of biomass combustion spots (Figure 3). The distributions of black carbon concentration over the EPR during these days confirms that fires are the main sources of smoke aerosols (Fig. 3, right column).

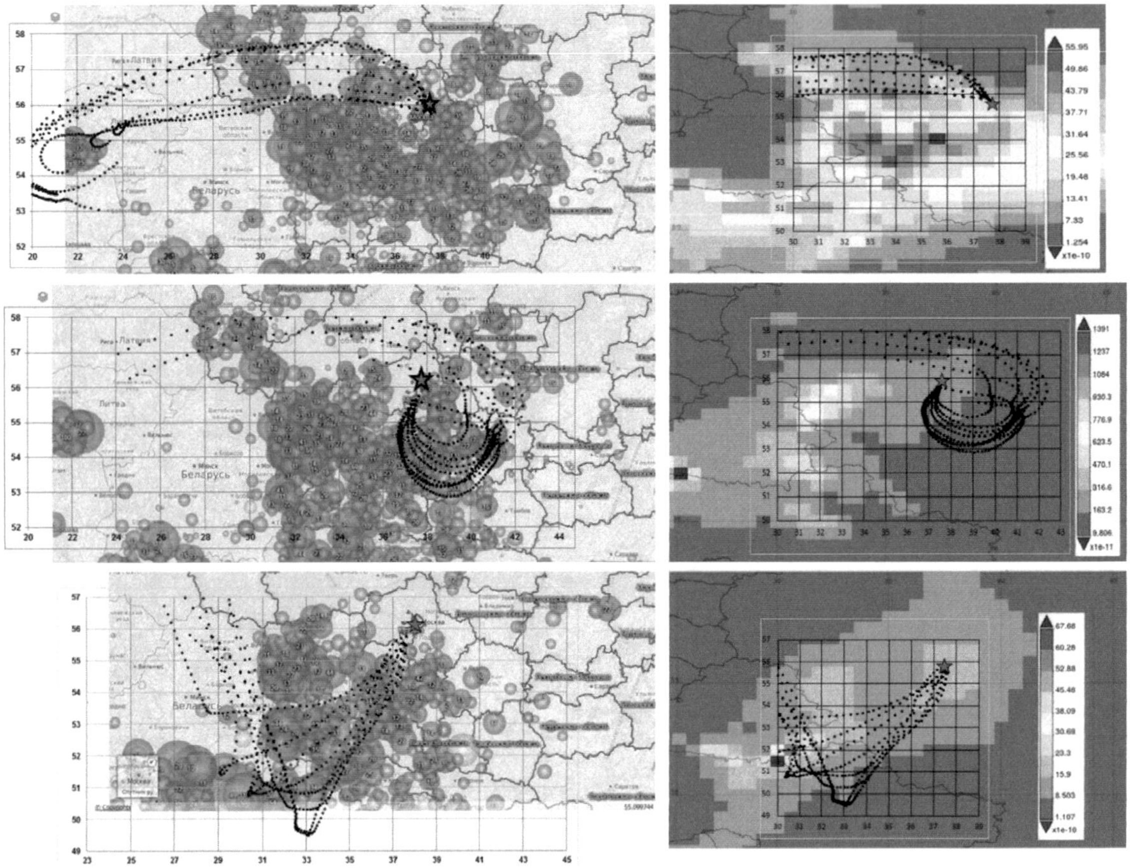

Figure 3. Illustration of fire situation on EPR and the trajectories of long-range air mass transport to Moscow (black dotted lines) during some days of anomalous increase of aerosol PM_{10} concentration in Moscow in March-April 2020. From top to bottom – fire maps according to the data of Scanex service [21] on March 26, 28 and April 13, 2020; the right column – the fields of black carbon near-surface concentration for the same days, obtained from MERRA-2 [17] reanalysis data.

When analyzing anomalous bursts of surface aerosol concentration (Fig. 2), the episode of March 25-29 was conditionally divided into two: March 25-27 and March 28-29. This is associated with the difference in synoptic conditions: in the first case, advection of air containing combustion aerosols occurred with air masses of W-NW directions (regional transfer from Pskov, Tver and Moscow regions), in the second - with air masses of SE-directions (from Ryazan, Tula, Moscow regions). The third episode, April 13, 2020, is associated with the regional transport of combustion aerosols with air masses southwestward (from Kaluga, Smolensk, Bryansk, Kursk, Tula, and Moscow regions and North-West Belarus). These three episodes were characterized by similar synoptic and meteorological conditions: anticyclonic activity, high temperature and low relative humidity. Long-term air advection at the end of March 2020 was mentioned above. On April 13, a short-term invasion of the southern anticyclone was observed between the prolonged dominance of dry Arctic air masses, which clean and cool the air.

During intensive experiment at IAP RAS in Moscow region, lockdown was announced due to the COVID-19 pandemic. This led to reduce in anthropogenic load, mainly associated with decrease in traffic of personal automobile transport and decline in the activity of construction and industrial enterprises, catering and various institutions. In addition, regular washing and disinfection of roads, streets and courtyards of the metropolis began. These measures led to decrease in aerosol pollution of surface air in Moscow and smooth daytime peaks in diurnal variation of the concentration of aerosol particles. However, analysis of short-period variations in surface aerosol is beyond the scope of this work and is the subject of our further studies.

An indirect factor that makes it possible to assess the influence of urban vehicles on the concentration of aerosol particles in surface layer of the atmosphere of Moscow is the so-called Yandex self-isolation indices (SII – from Yandex service [26]) reflecting business activity in the city. The absolute values of the calculated pairwise Pearson correlation coefficients between the daily mean values of the aerosol mass concentration PM_{10} and SII in Moscow were: in April 0.186 (with IAP RAS data), 0.195 (with Spiridonovka MEM data); in May – 0.419 and 0.418, respectively. The obtained values do not allow us to make a conclusion about the significance and suitability of this indicator yet.

In addition, it should be noted that under abnormal weather conditions in April 2020, which significantly influenced the parameters of the surface aerosol in Moscow, it is rather difficult to identify the contribution of anthropogenic load decrease into cleaning the city atmosphere.

4.2. Elemental composition of surface aerosols

Information on the elemental composition of aerosol particles is extremely important, since it can indirectly show the local and remote sources of atmospheric pollution [28]. During the intensive experiment at IAP RAS in the spring of 2020, aerosol samples were taken for elemental analysis every day. In total, 65 chemical elements (from Li to U) were determined in the samples. For visual presentation of results, 25 elements are considered, conventionally divided into groups depending on their origin and degree of distribution [14, 27].

To determine the origin of the element (terrigenous or anthropogenic), enrichment coefficients were calculated in relation to the composition of the Earth's crust [14] (the reference element of terrigenous origin is La, data on the average composition of the Earth's crust are taken from [28]). Enrichment factor of the element EF_{el} is calculated from the concentrations C for the element and La as follows: $EF_{el} = (C_{el}/C_{La})^{sample}/(C_{el}/C_{La})^{Earth\ crust}$. In urban aerosol elements with EF>10 are usually classified as anthropogenic (not terrigenous). Figure 4 at the top shows the geochemical spectrum of mass concentration of selected elements contained in Moscow surface aerosol and their EFs. So, the elements W, Sn, Cu, Mo, Bi, Sb, Zn, S, Pb, Cd, Se do not have terrigenous origin. There are anthropogenic elements (Sn, Pb, Cd, etc.) and elements of widespread distribution (for example, S, Se). The others are of previously terrigenous origin.

The elemental composition of aerosol can give us possibility to differ the composition of aerosols coming from different regions or from different sources in the city. We combined the samples collected in the days with winds of NW and N directions, when air was come from the Arctic. These days, as a rule, were characterized by low air temperature, lack of precipitation and low concentrations of aerosol particles. The mean element composition for these samples was preliminarily called "suggested background composition" or simply BG. Further, for mass concentration of each element, the excess value relative to BG was calculated as follows:

$$Exc_{el}=(C_{el} - BG_{el})/BG_{el}.$$

Figure 4 at the bottom shows three profiles of elements in relative values to BG for three anomalous episodes associated with regional transport of combustion aerosols (see Fig. 3). Note, that along with combustion aerosols due to convective transfer and advection, particles of mineral origin also entered the surface layer of Moscow atmosphere, which led to increase in concentration of larger, micron-sized aerosol particles in Moscow air. These profiles of elemental concentrations differ significantly both for terrigenous elements (Al, Na, K, Mg, Fe, Ca, La, etc.) and for anthropogenic elements (S, Cr, Mn, Ni, Cu, Zn, Cd, Sn, etc.). This is due to a number of factors, main of which is the difference between anthropogenic sources and sources of biomass combustion, structure and properties of underlying surface (soils and ground). The most variable are heavy metals, including anthropogenic elements in Moscow. The variability of soil elements mainly depends on properties of underlying surface.

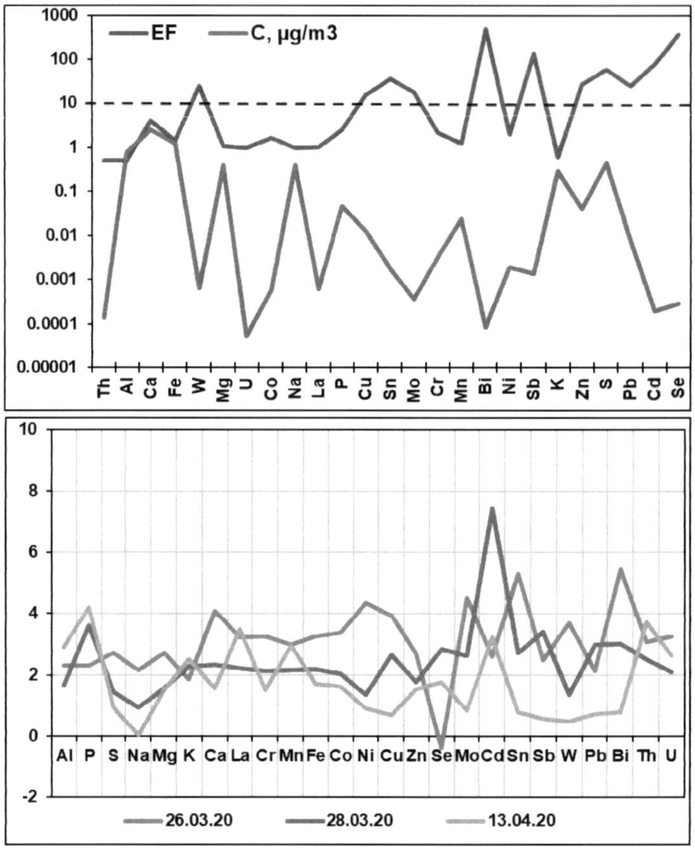

Figure 4. Geochemical spectra of elements in surface aerosol of Moscow: top – element concentrations and enrichment factors averaged over all spring 2020 period of observation; bottom – excess (Exc) of elemental mass concentrations relative to BG level for three samples under regional atmospheric transport of combustion aerosols.

However, for detailed classification and deeper study of possible sources of aerosols transported to Moscow with air masses of various origins, we must take into account the dispersed composition and morphology of aerosol particles, physicochemical properties and structure of soil and vegetation cover in regions surrounding the Moscow metropolis, as well as the classification of large regional anthropogenic sources of aerosol emissions.

5. Conclusions

During the spring intensive observations in 2020, two episodes were detected (March 25-29, April 13) associated with strong increase in mass concentration of aerosol particles PM_{10} (including excess of the average daily MPC). These episodes were mainly due to synoptic processes and meteorological conditions. Anticyclonic activity was accompanied by advection of air masses containing combustion aerosols and mineral particles from the areas with multiple biomass fires. The results of elemental analysis of aerosol particles composition confirm this assumption. Significant variability of geochemical profiles of aerosol chemical elements in the samples obtained under various weather and synoptic conditions was revealed.

The rest days of the observation period, lower mass concentration of near-surface aerosols was recorded in comparison with the averaged one. This was significantly influenced by unfavorable meteorological conditions in April, 2020 in Moscow (dominance of dry Arctic air masses with low temperatures). A certain contribution to reduction of aerosol pollution in Moscow during the lockdown was made by decrease in anthropogenic load. The measures for non-proliferation of COVID-19 (decrease in vehicle traffic, decline in the activity of industrial enterprises, regular washing and disinfection of city streets) helped to smooth out the daily high concentration of PM_{10} particles in a typical daily cycle. However, the study of short-term variations in Moscow surface aerosol connected with restrictive measures for non-proliferation of COVID-19 needs more reliable information about them and further research.

Acknowledgments
The reported study was funded by RFBR, projects ## 19-05-00352 and 19-05-50088.

References

[1] Kondrat'ev K Ya, Ivlev L S and Krapivin V F 2007 *Atmosfernye aerozoli: Svoistva, protsessy obrazovaniya i vozdeistviya. Ot nano- do global'nykh masshtabov (Atmospheric Aerosol: Properties, Formation, and Impacts. Nano- to-Global Scale Processes)*, St. Petersburg: VVM, p 860 (In Russian)
[2] Seinfeld J H and Pandis S N 2006 *Atmospheric chemistry and physics: from air pollution to climate change* 2nd Edit (New York: Wiley, USA) p 1232
[3] Ivlev L S 2011 *Regional Ecology* **32**(3-4) 83–93
[4] Ginzburg A S, Semenov V A, Aleshina M A, Semutnikova E G, Zakharova P V and Lezina E A 2020 *Doklady Earth Sciences* **495** (1) 862–866 DOI: 10.31857/S2686739720110067
[5] Popovicheva O B, Chichaeva M A and Kasimov N S 2021 *Vestnik Rossiiskoy Academii nauk* **91**(4) 350–360 (In Russian)
[6] Sannino A, D'Emilio M, Castellano P, Amoruso S, Boselli A 2021 *Aerosol Air Qual. Res.* **21** (2) 200381 DOI: https://doi.org/10.4209/aaqr.2020.07.0381
[7] Li L et al. 2020 *Science of the Total Environment* **732** 139282 DOI:10.1016/j.scitotenv.2020.139282.
[8] Mahato S, Pal S and Ghosh K G 2020 *Science of the Total Environment* **730** 139086 DOI: 10.1016/j.scitotenv.2020.139086.
[9] Krecl P, Targino A C, Oukawa G Y and Cassino Junior R R 2020 *Environmental Pollution* **265** part B 114883 DOI: 0.1016/j.envpol.2020.114883
[10] Sharma S, Zhang M, Gao A J, Zhang H and Kota S H 2020 *Science of The Total Environment* **728** 138878 DOI: 10.1016/j.scitotenv.2020.138878
[11] Han B-S, Park K, Kwak K-H, Park S-B, Jin H-G, Moon S, Kim J-W and Baik J-J 2020 *Int. J. Environ. Res. Public Health* **17** 6208 DOI: 10.3390/ijerph17176208
[12] Gubanova D P, Kuderina T M, Chkhetiani O G, Iordanskii M A, Obvintsev Yu I and Artamonova M S 2018 *Izvestiya, Atmospheric and Oceanic Physics* **54** (10) 1430–1448 DOI: 10.1134/S0001433818100043
[13] Gubanova D P, Elansky N F, Skorokhod A I, Kuderina T M, Iordansky M A, Sadovskaya N V and Anikin P P 2020 *IOP Conf. Ser.: Earth Environ. Sci.* 606 012019 DOI: https://doi.org/10.1088/1755-1315/606/1/012019
[14] Gubanova D P, Vinogradova A A, Iordanskii M A and Skorokhod A I 2021 *Izvestiya, Atmospheric and Oceanic Physics* **57** (3) 297–309 DOI: 10.1134/S0001433821030051
[15] https://rp5.ru
[16] https://www.windy.com/ru
[17] https://giovanni.gsfc.nasa.gov/giovanni/
[18] Stein A F, Draxler R R, Rolph G D, Stunder B J B, Cohen M D and Ngan F 2015 *Bull. Amer. Meteor. Soc.* **96** 2059–2077 DOI: http://dx.doi.org/10.1175/BAMS-D-14-00110.1
[19] Rolph G, Stein A and Stunder B 2017 *Environmental Modelling & Software* **95** 210–228 DOI: https://doi.org/10.1016/j.envsoft.2017.06.025

[20] www.arl.noaa.gov
[21] https://fires.ru
[22] https://mosecom.mos.ru/spiridonovka/
[23] Gelaro R et al. 2017 *J. Climate.* **30** (14) 5419–5454 DOI:10.1175/JCLI-D-16-0758.1.
[24] Zhuravleva T B, Artyushina A V, Vinogradova A A and Voronina Yu V 2020 *Atmospheric and Oceanic Optics* **33** (6) 591–601 DOI: 10.1134/S1024856020060251
[25] Randles C A et al. 2017 *J. Climate* **30** (17) 6823–6850 DOI:10.1175/JCLI-D-16-0609.1
[26] https://yandex.ru/company/researches/2020/podomam
[27] Rasmussen P E 1998 *Environ. Geology* **33** (2/3) 96–108
[28] Dobrovol'skii V V 2009 *Izbrannye Trudy / Biogeokhimiya mirovoi sushi (Selected Works / Biogeochemistry of World Land)* vol. 3 (Moscow, Nauch. Mir) p 440 (In Russian)

IOP Conf. Series: Earth and Environmental Science 1040 (2022) 012016 doi:10.1088/1755-1315/1040/1/012016

Changes of the sea ice and snow cover extent associated with temperature changes in the Northern and Southern Hemispheres in recent decades

I I Mokhov[1, 2], M R Parfenova[1]

[1]A.M. Obukhov Institute of Atmospheric Physics RAS, 119017, Moscow, Russia
[2]M.V. Lomonosov Moscow State University, 119991, Moscow, Russia

Abstract. Changes in snow cover and sea ice extents associated with temperature changes in the Northern and Southern Hemispheres for the period 1979-2020 are analysed using monthly-mean satellite and reanalysis data. Quantitative estimates of the relationship between the Antarctic and Arctic sea ice and changes in the surface air temperature were obtained. Overall increase of the Antarctic sea ice extent is associated with the regional manifestation of natural multidecadal climate modes with periods of up to several decades (against the background of global warming and a rapid decrease in the extent of sea ice in the Arctic).The results of correlation and cross-wavelet analyses show significant coherence and negative correlation of the surface air temperature in both Arctic and Antarctic with the respective sea ice extent in recent decades. Seasonal and regional features of the snow cover sensitivity to changes in the temperature regime in the Northern Hemisphere for the past four decades are noted. The features of snow cover variability in Eurasia and North America are presented.

1. Introduction

The snow-ice cover plays a crucial role in the formation of regional and global climatic regimes [1]. On time scales from seasons to millions of years, the most noticeable and extensive changes in the Earth's surface properties have been associated with changes in the snow-ice cover extent [2]. A decrease in snow-ice cover area (and a corresponding decrease in surface albedo) under the temperature rise contributes to an increased absorption of solar radiation and thereby to a further temperature growth. This process is associated with the strongest positive climatic feedback, which increases the sensitivity of the temperature regimes to external influences – both natural and anthropogenic ones.

The most rapid climatic changes of recent decades have been observed in Arctic latitudes – clearly manifested in a sharp decrease in the total extent of Arctic sea ice [1,3,4]. At the same time, the total Antarctic sea ice extent has been statistically insignificantly growing since the late 1970s, according to satellite data. By 2014, its average annual area reached a record high of about 13 million km^2, and since 2016 there has been a sharp decrease with reaching the record low since the end of the 1970s.

Multiple studies have been devoted to understanding the Arctic and Antarctic sea ice extent variability [1,3-28]. The peculiarities of Antarctic sea ice variability are attributed to various causes, such as oceanic processes and atmospheric circulation features in the Southern Hemisphere. In particular, [8] the influence of different climatic variability modes, including El Niño and the Antarctic Oscillation, on the Antarctic sea ice extent has been analyzed. In [12], the features of climatic changes in the Antarctic latitudes are associated with the features of oceanic circulation in the Southern Ocean,

IOP Conf. Series: Earth and Environmental Science 1040 (2022) 012016 doi:10.1088/1755-1315/1040/1/012016

more precisely, in the Antarctic latitudes. The connection between the Antarctic sea ice extent and the interannual Pacific Ocean oscillation is revealed in [13]. In [20], features of the relationship between Antarctic sea ice extent and the position of the Antarctic polar front were noted.

According to satellite data since the late 1970s, the maximum of sea ice extent in the Arctic reached 16 million km^2, in the Antarctic – about 20 million km^2. Variations in the annual extent of snow cover are substantially greater – in the Northern Hemisphere their range reaches 50 million km^2. As for the changes in snow cover, both increases and decreases are manifested at the regional level during the warming. Increased precipitation (including solid precipitation during the cold months) due to atmospheric moisture capacity increase contributes to the growth in snow cover extent with warming. Many studies have been dedicated to estimating the current trends of snow cover changes for various regions [1,29-48].

This paper presents the results of analysis of the Arctic and Antarctic sea ice extent changes, as well as Northern Hemisphere snow cover changes (including Eurasia and North America), in relation to the surface air temperature variations for recent decades.

2. Data and methods

Changes in the extent of snow cover and sea ice associated with temperature changes in the Northern and Southern Hemispheres are analyzed by using monthly-mean NSIDC (https://www.ncdc.noaa.gov/) and GSL (https://climate.rutgers.edu/) data, as well as ERA5 reanalysis data (https://www.ecmwf.int/en/forecasts/datasets/reanalysis-datasets/era5) for the period 1979-2020 [49].

3. Results and discussion

Figure 1 (a, b) presents the dependences of the total extent of sea ice S_I in the Arctic (a) and Antarctic (b) (according to satellite data) on the surface air temperature T_H in the Northern (a) and Southern (b) Hemispheres (according to the ERA5 reanalysis data) for the months of the period 1979-2020. Considering strong variations in the surface air temperature and the sea ice extent in annual cycle, the corresponding variations in the interannual variability are noted for each month. Figure 1a shows negative correlation of surface air temperature variations and the sea ice extent S_I in the Arctic for each month more clearly.

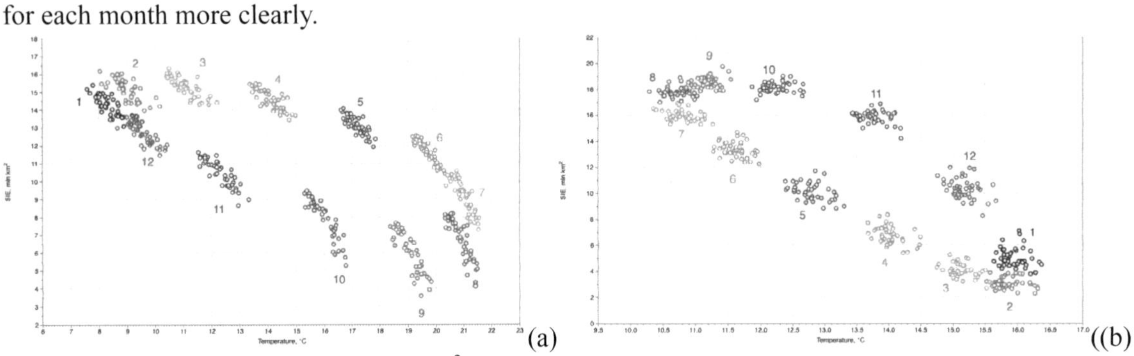

Figure 1. The extent of sea ice (mln km^2) in the Arctic (a) and Antarctic (b) for different months (1–12) depending on the surface temperature (K) in the Northern (a) and Southern (b) hemispheres according to data for the period 1979-2020.

Table 1 presents quantitative estimates of the sensitivity of the total extent of Arctic and Antarctic sea ice S_I [mln km^2] to the changes in the surface air temperature T_P [K] in the Arctic and Antarctic for different months in interannual variability for the period 1980-2019. The parameters of sensitivity, estimated by the coefficients of the corresponding linear regressions, are significant within two standard deviations, as well as the corresponding correlation coefficients highlighted. In the Northern

Hemisphere, the estimates are statistically significant for all months. In the Southern Hemisphere, significant estimates (within two standard deviation) were obtained for most of the months.

Table 1. Estimates of the sensitivity of the total extent S_I of the Arctic and Antarctic sea ice to changes in the surface temperature T_P in the Arctic and Antarctic for different months by monthly-mean data in interannual variability (1980–2019).

Months	dS_I/dT, mln km^2 / K 1980–2019 Arctic	Antarctic
January	**-0.28(±0.05)**	**-0.54(±0.27)**
February	**-0.25(±0.06)**	-0.19(±0.14)
March	**-0.28(±0.06)**	**-0.69(±0.12)**
April	**-0.31(±0.05)**	**-0.57(±0.12)**
May	**-0.47(±0.08)**	**-0.34(±0.10)**
June	**-0.90(±0.09)**	**-0.34(±0.11)**
July	**-1.58(±0.22)**	**-0.23(±0.06)**
August	**-1.63(±0.17)**	-0.03(±0.08)
September	**-1.17(±0.11)**	0.08(±0.10)
October	**-0.77(±0.05)**	-0.08(±0.12)
November	**-0.41(±0.04)**	**-0.45(±0.14)**
December	**-0.37(±0.05)**	**-0.80(±0.24)**

The highest (absolute) values of the sensitivity parameter dS_I/dT_P for Arctic sea ice were estimated for the summer months (July and August) – up to -1.6 million km^2 with 1 K warming. The highest (absolute) value of the sensitivity parameter dS_I/dT_P for Antarctic sea ice was also obtained for the summer season: -0.8 million km^2 with 1 K warming in December.

According to the data for recent decades, the negative correlation of the total Antarctic sea ice extent with the Antarctic surface air temperature is more significantly manifested, rather than with the surface air temperature of the Southern Hemisphere as a whole [27]. At the same time, the relationship between the interannual variations of sea ice extent and the surface air temperature for Antarctic region is more statistically significant, than for the whole Southern Hemisphere air temperature. Simultaneously, the relation of the interannual variations in Arctic sea ice extent to the hemispheric surface temperature is stronger than its relation to the regional temperature for most months of the year.

Figure 2 shows the dependence of the total extent of snow cover S_S in the Northern Hemisphere for different months based on satellite data for the period 1979-2020 on the corresponding surface air temperature T_H in the whole Northern Hemisphere based on the ERA5 reanalysis data. Considering stronger variations in the surface air temperature and the snow cover extent within the annual cycle, the corresponding interannual variations also appear in each different month. The sensitivity of the Northern Hemisphere snow cover S_S to the changes in the Northern Hemisphere surface air temperature T_H is determined by the coefficient of the corresponding linear regression. It is characterized by the straight line in Figure 2, and is estimated to be –3.6 million km^2 / K.

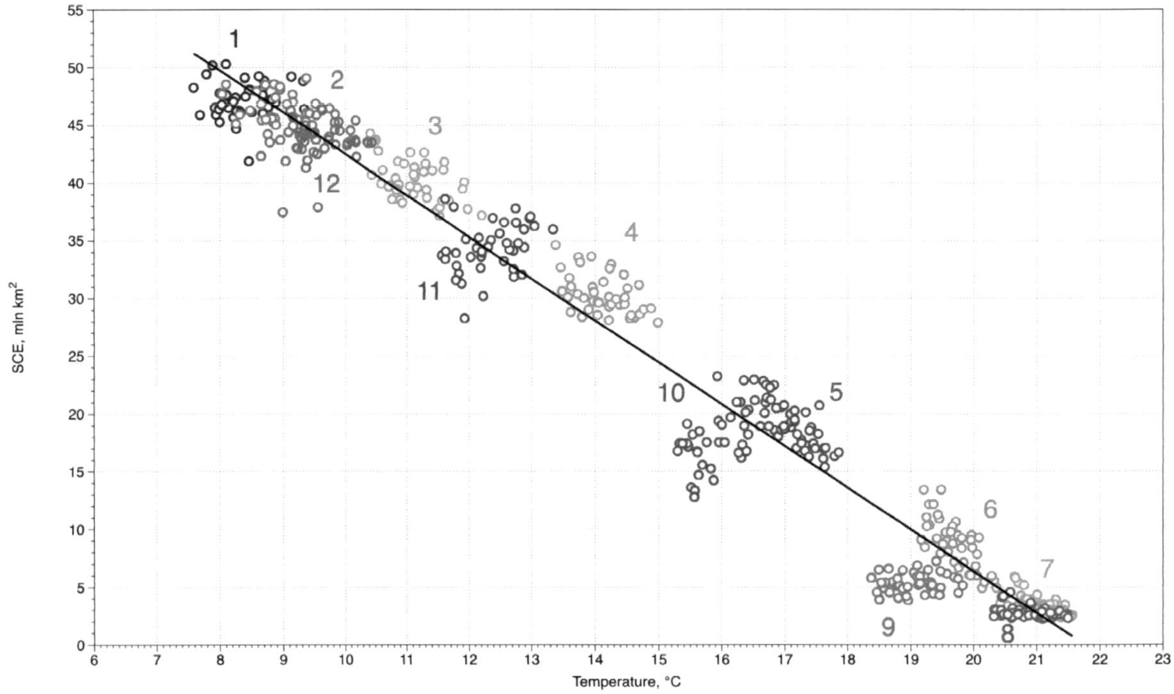

Figure 2. Dependance of the monthly mean total snow extent (S_S) in the Northern Hemisphere on the hemispheric-mean surface temperature (T) for the time interval 1979–2020.

Table 2 presents quantitative estimates of the sensitivity of the total snow cover extent S_S [mln km^2] in Eurasia and North America to changes in the surface air temperature T_C [K], respectively, in Eurasia and North America for different months in interannual variability for the period 1980–2019. The sensitivity parameters are estimated by the coefficients of the corresponding linear regressions that are significant within two standard deviations. The corresponding correlation coefficients are highlighted. A significant negative correlation between S_S and T_C is present for most of the month for Eurasia and North America.

Table 2. Estimates of the sensitivity of the total snow cover extent S_S in Eurasia and North America to the corresponding changes of the surface temperature T_C in Eurasia and North America by monthly-mean data for different months, 1980–2019.

Months	dS_S/dT_C, mln km^2 / K 1980–2019	
	Eurasia	North America
January	**-0.52 (±0.21)**	**-0.14 (±0.07)**
February	**-0.84 (±0.13)**	**-0.25 (±0.08)**
March	**-0.95 (±0.15)**	**-0.34 (±0.09)**
April	**-0.82 (±0.23)**	**-0.83 (±0.11)**
May	**-1.66 (±0.24)**	**-0.67 (±0.15)**
June	**-1.56 (±0.27)**	**-1.07 (±0.18)**
July	**-0.64 (±0.12)**	**-0.40 (±0.13)**
August	**-0.19 (±0.07)**	-0.08 (±0.10)

September	-0.24 (±0.18)	-0.13 (±0.09)
October	0.76 (±0.46)	0.03 (±0.16)
November	-0.41 (±0.25)	**-0.39 (±0.12)**
December	**-0.41 (±0.19)**	-0.15 (±0.09)

The highest absolute value of the sensitivity parameter dS_S/dT_C for Eurasia was obtained for May, which is -1.7 million km^2 for the increase in the surface air temperature in Eurasia by 1 K. The highest absolute value of the sensitivity parameter dS_S/dT_C for North America was obtained for June, which is about -1.1 million km^2 for the increase in the surface air temperature in North America by 1 K.

In [50], based on the data on the annual variation for Northern Hemisphere, the sensitivity of the total extent of the snow-ice cover to changes in the surface air temperature was estimated in the range of 3.5÷5.1 million km^2 / K. In this paper, the sensitivity of the total snow-ice cover extent of the Northern Hemisphere to changes in the hemispheric surface air temperature based on monthly average data for the same period is estimated equal to –4.1 million km^2/ K. At the same time, the sensitivity of the snow cover extent of the Northern Hemisphere to changes in the hemispheric near-surface temperature is estimated as –3.6 million km^2 / K for the period 1980–2019 (–3.5 million km^2 / K for the period 1980–1999 and –3.7 million km^2 / K for the period 2000–2019). The sensitivity of the extent of the Arctic sea ice to changes in the hemispheric surface air temperature was calculated at about –0.7 million km^2 / K for the period 1979–2020. This estimate was obtained under the condition of maximum correlation between the extent of the Arctic sea ice and hemispheric near-surface temperature variations. The lag of 1 month of the ice regime in the annual cycle relative to the temperature regime has been considered.

4. Conclusions

Interannual variations in the total extent of sea ice in the Arctic are statistically significantly associated with interannual variations in surface air temperatures both in the Arctic region and for the whole Northern Hemisphere in recent decades. At the same time, for most months of the year, the connection with hemispheric air temperature is manifested even more significant than a connection with regional temperature. Simultaneously, interannual variations in the total extent of sea ice in Antarctica are more statistically significantly associated with interannual variations in surface air temperature for Antarctic region than for the whole Southern Hemisphere.

For variations in the of snow cover extent in the Northern Hemisphere in recent decades, including Eurasia and North America, for most months, a significant negative correlation was noted with the corresponding interannual variations in surface air temperature. The negative correlation of the total snow cover extent on the continents of the Northern Hemisphere for most months is generally more significant with the continental surface temperature (for Eurasia and North America) than with the surface temperature for the Arctic latitudes and the whole Northern Hemisphere. At the same time, a more significant relationship between the total extent of the snow cover and the temperature in the corresponding Arctic sector is noted by the late spring – early summer. The weakening of statistically insignificant large-scale correlation between the snow cover and the temperature regime is manifested during the transition to the autumn season. At the same time, a statistically significant negative correlation of snow cover and the temperature in October-November was observed for the last two decades in the autumn months, particularly for Eurasia.

Acknowledgements

The work was carried out within the framework of project 19-17-00240 of the Russian Science Foundation using results of the satellite data analysis, obtained in accordance with the grant agreement No. 075-15-2020-776 with the Ministry of Science and Higher Education of the Russian Federation.

Regional features of snow variability were carried out within the framework of the Russian Foundation for Basic Research, project no. 19-35-90118.

References

[1] *Climate Change 2013: The Physical Science Basis. Contribution of Working Group I to the Fifth Assessment Report of the Intergovernmental Panel on Climate Change* ed T Stocker *et al* (New York: Cambridge University Press) pp 1535
[2] Mokhov I I 1993 *Diagnostics of the Climatic System Structure* (Saint Petersburg: Gidrometeoizdat) 271 pp (In Russian)
[3] Mokhov I I 2015 Contemporary climate changes in the Arctic *Herald Russ. Acad. Sci.* **85** 265–271
[4] Alekseev G V, Aleksandrov E I, Glok N I, Ivanov N E, Smolyanitsky V M, Kharlanenkova N E and Yulin A V 2015 Arctic sea ice cover in connection with climate change *Izvestiya, Atmos. Oceanic Phys.* **51** (9) 889–902
[5] Mokhov I I 2019 Contemporary climate changes: Anomalies and trends *IOP Conf. Series: Earth and Environ. Sci.* **231** 012037
[6] Liu J, Curry J A, Martinson D G 2004 Interpretation of recent Antarctic sea ice variability *Geophys. Res. Lett.* **31** L02205
[7] Stammerjohn S E, Martinson D G, Smith R C, Yuan X and Rind D 2008 Trends in Antarctic annual sea ice retreat and advance and their relation to El Niño–Southern Oscillation and southern annular mode variability *J. Geophys. Res.* **113** C03S90
[8] Yuan X, Li C 2008 Climatic modes in southern high latitudes and their impacts on Antarctic sea ice *J. Geophys. Res.* **113** C06S91. 8
[9] Parkinson C L, Cavalieri D J 2012 Antarctic sea ice variability and trends, 1979-2010 *The Cryosphere* **6** 871-880.
[10] Stammerjohn S, Massom R, Rind D and Martinson D 2012 Regions of rapid sea ice change: An inter-hemispheric seasonal comparison *Geophys. Res. Lett.* **39** L06501.
[11] Simmonds I 2015 Comparing and contrasting the behavior of Arctic and Antarcticsea ice over the 35-year period 1979-2013 *Ann. Glaciol.* **56** 18-28
[12] Armour K C, Marshall J, Scott J R, Donohoe A and Newsom E R 2016 Southern Ocean warming delayed by circumpolar upwelling and equatorward transport *Nat. Geosci.* **9** 549-554
[13] Meehl G A, Arblaster J M, Bitz C M, Chung C T Y and Teng H 2016 Antarctic sea-ice expansion between 2000 and 2014 driven by tropical Pacific decadal climate variability *Nature Geosci.* **9** 590-595
[14] Doddridge E W and Marshall J 2017 Modulation of the seasonal cycle of Antarctic sea ice extent related to the Southern Annual Mode *Geophys. Res. Lett.* **44** 9761-9768
[15] Schroeter S, Hobbs W and Bindoff N L 2017 Interactions between Antarctic sea ice and large-scale atmospheric modes in CMIP5 models *The Cryosphere* **11** 789-803
[16] Stuecker M F, Bitz C M and Armour K C 2017 Conditions leading to the unprecedented low Antarctic sea ice extent during the 2016 austral spring season *Geophys. Res. Lett.* **44** 9008-9019
[17] Turner J, Phillips T, Marshall G J, Hosking J S, Pope J O, Bracegirdle T J and Deb P 2017 Unprecedented springtime retreat of Antarctic sea ice in 2016 *Geophys. Res. Lett.* **44** 6868-6875
[18] Schlosser E, Haumann F A and Raphael M N 2018 Atmospheric influences of the anomalous 2016 Antarctic sea ice decay *The Cryosphere* **12** 1103-1119
[19] Screen J A, Bracegirdle T J and Simmonds I 2018 Polar climate change as manifest in atmospheric circulation *Curr. Clim. Change Rep.* **4** 383-395

[20] Alekseev G V, Glok N I, Vyasilova A E, Ivanov N E, Kharlanenkova N E and Smirnov A V 2019 Influence of sea surface temperature in the tropics on the Antarctic sea ice under global warming *Ice and Snow* **59** (2) 213-221 (In Russian)

[21] Meehl G A, Arblaster J M, Chung C T Y, Holland M M, DuVivier A, Thompson L A, Yang D and Bitz C M 2019 Sustained ocean changes contributed to sudden Antarctic sea ice retreat in late 2016 *Nature Comm.* **10** Art. Number. 14.

[22] Parkinson C L 2019 A 40-y record reveals gradual Antarctic sea ice increases followed by decreases at rates far exceeding the rates seen in the Arctic *PNAS* **116** (29) 14414-14423

[23] Wang G, Hendon H H, Arblaster J M, Lim E-P, Abhik S and van Rensch P 2019 Compounding tropical and stratospheric forcing of record low Antarctic sea-ice in 2016 // Nature Comm. **10** (13) 9 p.

[24] Wang Z, Turner J, Wu Y and Liu C 2019 Rapid decline of total Antarctic sea ice extent during 2014-16 controlled by wind-driven sea ice drift *J. Clim.* **32** 5381-5395

[25] Mokhov I I 2020 Rapid climate change in the Arctic and their regional and large-scale impacts *Arctic and Antarctic Reseach* **66** (3) 446–462 (In Russian)

[26] Mokhov I I 2020 Russian Climate Research in 2015–2018 *Izvestiya, Atmos. Oceanic Phys.* **56** (4) 325–343

[27] Mokhov I I, Parfenova M P 2020 Features of variability of the Antarctic and Arctic sea ice in recent decades against the background of global and regional climatic changes *Problems of Geography* **150** (Exploration of Antarctica) 304–319 (In Russian)

[28] Mokhov I I, Parfenova M R 2021 Relationship of the Extent of Antarctic and Arctic Ice with Temperature Changes, 1979–2020 *Doklady Earth Sci.* **496** (1) 66–71

[29] Foster J, Owe M and Rango A 1983 Snow cover and temperature relationships in North America and Eurasia // J. Clim/ Appl. Meteorol. **22** 460-469

[30] Robinson D A, Dewey K F and Heim R R Jr 1993 Global snow cover monitoring: an update Bull. Am. Meteorol. Soc. **74** 1689–1696

[31] Brown R D and Mote P W 2009 The response of Northern Hemisphere snow cover to a changing climate *J. Clim.* **22** 2124-2145

[32] Bulygina O N, Groisman P Ya, Razuvaev V N and Korshunova N N 2011 Changes in snow cover characteristics over Northern Eurasia since 1966 Environ. Res. Lett. **6** 045204 (10 pp)

[33] Callaghan T V, Johansson M, Brown R D, Groisman P Ya, Labba N and Radionov V and Contributors 2011 *Snow, Water, Ice and Permafrost in the Arctic (SWIPA), AMAP Report to the Arctic Council* chapter 4 (Changing snow cover and its impacts) 59 pp (http://amap.no/swipa/)

[34] Cohen J L, Furtado J, Barlow M A, Alexeev V A and Cherry J E 2012 Arctic warming, increasing snow cover and widespread boreal winter cooling *Environ. Res. Lett.* **7** 014007

[35] Robinson D A, Estilow T W and NOAA CDR Program 2012 NOAA Climate Data Record (CDR) of Northern Hemisphere (NH) Snow Cover Extent (SCE), Version 1. [indicate subset used]. NOAA Nat. Centers for Environ. Inform. doi: 10.7289/V5N014G9

[36] Brown R D and Derksen C 2013 Is Eurasian October snow cover extent increasing? *Environ. Res. Lett.* **8** 024006 (7 pp)

[37] Mudruk L R, Kushner P J and Derksen C 2014 Interpreting observed northern hemisphere snow trends with large ensembles of climate simulations *Clim. Dyn.* **43** 345-359

[38] Estilow T W, Young A H and Robinson D A 2015 A long-term Northern Hemisphere snow cover extent data record for climate studies and monitoring *Earth Syst. Sci. Data* **7** 137–142

[39] Hernandez-Henriquez M A, Dery S J and Derksen C 2015 Polar amplification and elevation-dependence in trends of Northern Hemisphere snow cover extent, 1971-2014 *Environ. Res. Lett.* **10** 044010

[40] Mankin J S, Diffenbaugh N S 2015 Influence of temperature and precipitation variability on near-term snow trends *Clim. Dyn.* **45** (3) 1099–1116

IOP Conf. Series: Earth and Environmental Science 1040 (2022) 012016 doi:10.1088/1755-1315/1040/1/012016

[41] Najafi M R, Zwiers F W, Gillett N P 2016 Attribution of the spring snow cover extent decline in the Northern Hemisphere, Eurasia and North America to anthropogenic influence *Clim. Change* **136** 571-586
[42] Lin Y and Jiang M 2017 Maximum temperature drove snow cover expansion from the Arctic, 2000-2008 *Sci. Rep.* **7** 15090
[43] Mudruk L R, Kushner P J, Derksen C and Thackeray C 2017 Snow cover response to temperature in observational and climate model ensembles *Geophys. Res. Lett.* **44** 919-926
[44] Popova V V, Shiryaeva A V and Morozova P A 2018 Changes in the snow depth characteristics in the territory of Russia in 1950-2013: The regional features and connection with the global warming *Kriosfera Zemli* **XXII** (4) 58–67
[45] Connolly R, Connolly M, Soon W, Legates D R, Cionco R G and Herrera V M V 2019 Northern Hemisphere snow-cover trends (1967-2018): A comparison between climate models and observations *Geosci.* **9** 135 (23 pp)
[46] Thackeray C W, Derksen C, Fletcher C G and Hall A 2019 Snow and climate: Feedbacks, drivers, and indices of change *Curr. Cli. Change Rep.* **5** 322-333
[47] Mudruk L R, Santolaria-Otin M, Krinner G, Menegoz M, Derksen C, Brutel-Vuilmet C, Brady M and Essery R 2020 Historical Northern Hemisphere snow cover trends and projected changes in the CMIP6 multi-model ensemble *The Cryosphere* **14** 2495-2514
[48] Santolaria-Otin M and Zolina O 2020 Evaluation of snow cover and snow water equivalent in the continental Arctic in CMIP5 models *Clim. Dyn.* **55** 2993-3016
[49] Hersbach H, Bell B, Berrisford P et al. 2020 The ERA5 global reanalysis. *Q J R Meteorol Soc.* 146 1999– 2049
[50] Mokhov I I 1984 The temperature sensitivity of cryosphere area of Northern Hemisphere *Izvestiya, Atmos. Oceanic Phys.* **20** (2)

IOP Conf. Series: Earth and Environmental Science 1040 (2022) 012017 doi:10.1088/1755-1315/1040/1/012017

Estimates of natural methane emissions into the atmosphere in the regions of Western Siberia by model simulations

M M Arzhanov[1,2], S N Denisov[1], I I Mokhov[1, 2] and M R Parfenova[1]

[1]*A.M. Obukhov Institute of Atmospheric Physics RAS, 119017, Moscow, Russia*
[2]*M.V. Lomonosov Moscow State University, 119991, Moscow, Russia*

E-mail: arzhanov@ifaran.ru

Abstract. Model estimates of changes in natural methane emissions into the atmosphere were obtained for the regions of Western Siberia, including the estimates of wetland ecosystems and decomposition of relict methane hydrates contributions to these changes in the high latitude cryolithozone. Regional model estimates of methane fluxes into the atmosphere were obtained using the results of calculations with the ensemble of global climate models CMIP6 under different scenarios of anthropogenic forcing for the 21st century. According to the model estimates, the increase in regional natural methane emissions in the 21st century under the SSP585 scenario with the most aggressive anthropogenic impacts is more than 2 times greater than under the SSP126 scenario with the least aggressive anthropogenic impacts. Dissociation of relic methane hydrates in the permafrost resulting in methane emissions into the atmosphere in the north of Western Siberia are estimated to be considerably less significant compared to the regional emissions of wetland ecosystems under current climatic changes.

1. Introduction

According to observational data and modeling results, the spatial structure of the response of permafrost temperature to the observed general increase in surface air temperature and Arctic amplification [1, 2] reveals regional maxima, particularly in the north of Western Siberia [3, 4]. An increase in the permafrost temperature contributes to increase in the depth of the active layer [5, 6] and to thawing of organic matter formed during interglacial periods and preserved in permafrost. The global carbon deposits in the upper 1 meter of soil are estimated at 1500–1600 PgC, which is approximately twice as large as those in the atmosphere [7]. At the same time, the reserves of carbon substrate the result of permafrost thawing by the end of the 21st century) that can participate in the global biogeochemical cycle, are estimated from 50-100 PgC [8] to more than 170-300 PgC [9, 10].
Soil temperature increase contributes to degradation of permafrost, affects the rate of decomposition of organic matter and the emission of greenhouse gases into the atmosphere from wetland ecosystems. Most of the soil carbon is found in wetland ecosystems, although they only cover 6-7% of the Earth's surface area. The specific regional hydrological soil regime is manifested in the permafrost conditions. About half of the bogs are located in the high latitudes of the Northern Hemisphere. In [11], it was noted that with the observed warming, the increase in methane emissions from wetland ecosystems of extratropical latitudes significantly exceeds the increase in global emissions. At the same time, a significant part of methane emissions is associated with bogs located in a relatively narrow strip from

50 N to 70 N latitude [12]. The total emissions of methane into the atmosphere from wetland ecosystems of Western Siberia, according to various estimates, range from 1 to 20 $TgCH_4$ yr^{-1} [13-15]. An increase in soil temperature of the continental permafrost zone regions reduces the durability of permafrost and contributes to the dissociation of relict methane hydrates (formed during ice ages), which are located above the modern border of the hydrate stability zone [16] and are sensitive to temperature changes. The formation of methane hydrates in the upper horizons of the permafrost zone in the past can be explained by the influence of the overlying ice sheet and marine transgression. The results of calculations with the dynamic model of thermophysical processes in the glacier-soil system showed that in the north of Western Siberia under the ice sheet [17, 18], the upper boundary of the zone of stability of methane hydrates could reach the surface during the periods of glaciation at about 95-80 and 70-60 thousand years ago [19, 20]. Modern warming, especially in high latitudes, accompanied by permafrost melting, increases the risk of gas emissions during the dissociation of shallow methane hydrates [20] with the formation of craters with traces of discarded rock [21, 22], similar to those recorded in 2014–2020 in the permafrost zone in the north of Western Siberia.

Changes in the thermal and hydrological regimes of permafrost and wetland ecosystems and associated changes in methane fluxes into the atmosphere have significant regional and global consequences [23]. The purpose of this work is to assess the contribution of regional natural methane emissions to the carbon cycle using ensemble model calculations under various scenarios of anthropogenic forcing.

2. Methods and data

The approach to determining the response of natural methane emissions to climatic changes is similar to that used in [11, 24]. In particular, when calculating methane emissions for wetland ecosystems, the soil layers were considered saturated with moisture, which overestimates the flux values. It should be noted that the characteristic area of the bogs is much less than the area of the cells of the model grids. In this regard, when determining the average soil moisture index for the model cell, the fluxes are significantly underestimated. Methane fluxes were summed up for thawed soil layers up to a maximum depth of 1 m. For frozen soil layers, methane emissions were absent. The total flow of methane from the soil to the atmosphere linearly depended on the intensity of methane production per unit volume of the thawed soil layer and exponentially depended on the soil temperature.

Regional calculations based on the model algorithm [11, 24] were performed using the results of calculations with an ensemble of global climate models, in particular, the calculations of wetland ecosystems proportion in model cells and soil temperature profiles. The monthly average values of the corresponding parameters were used. In particular, regional numerical calculations were carried out for years 2015 to 2100 using the ensemble of global climate models of the international project CMIP6 [25]: ACCESS-CM2, CanESM5, NorESM2-LM, NorESM2-MM, UKESM1-0-LL. Two scenarios of anthropogenic forcing were used: the least aggressive SSP126 and the most aggressive SSP585. The selected models differ in the parametrizations of physical processes and spatial resolution. Due to the fact that the results of various global model calculations for the vertical distribution of soil temperature have different spatial resolution, the model soil temperature profiles were interpolated to a regular grid with a vertical step of 0.2 m. In addition, the emissions of methane into the atmosphere as a result of violation of thermobaric conditions of metastable relict methane hydrates were obtained. According to [26, 27], the volume of methane released as a result of the decomposition of gas hydrates in the pore space is proportional to the moisture content of the rock, hydrate saturation, the percent of water in a hydrate (hydration coefficient), density and volume of the rock. When calculating the volume, it was assumed that the gas craters (GEC) have a cylindrical shape. The diameter and depth were set in accordance with the observational data [21]. The values of hydrate saturation, the proportion of moisture transferred to the hydrate, and the moisture content of the soil were set in accordance with

[28]. This paper presents the results of the analysis of model calculations for the regions of Western Siberia (55-75 N, 65-85 E).

3. Results and discussion

According to model estimates for the region (55-75 N, 65-85 E) of Western Siberia by the end of the 21st century, an increase in the average annual soil temperature was revealed, which in the upper soil layer 1 m thick on average for the ensemble of global climate models is 1.4° C under the SSP126 scenario and 6.1° C under the SSP585 scenario. For the modern period, the estimates of the average annual soil temperature have a significant scatter. For various models of the ensemble, the spread in soil temperature in some years reaches 7° C. Under the SSP585 scenario, a linear increase in the average annual ground temperature is observed in the analyzed area for all models of the ensemble. Under the least aggressive scenario SSP126, the increase in ground temperature continues until the middle of the 21st century, then the rate of temperature increase decreases. The greatest increase in ground temperature for the two scenarios for the whole ensemble of models was revealed at the highest latitudes, including the Yamal Peninsula (figure 1). The estimates of the rise in ground temperature in Western Siberia under different scenarios differ greatly. The difference in regional estimates of trends for different models is noted. Under the SSP126 scenario, the highest values of the trends in the mean annual ground temperature at high latitudes reach and exceed 0.02° C yr^{-1}. Under the most aggressive scenario SSP585, the trend values at high latitudes reach and exceed 0.09°C yr^{-1}.

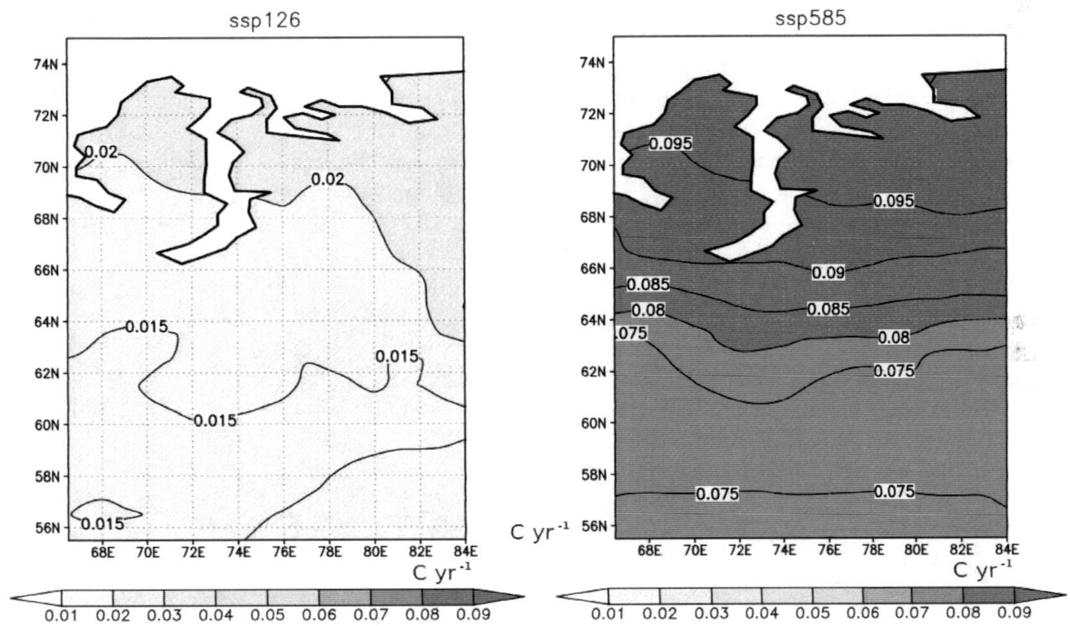

Figure 1. Linear trends of the mean annual ground temperature for the ensemble of climatic models (up to a depth of 1 m) for the period 2015-2100 under scenarios SSP126 and SSP585 for the region (55-75 N, 65-85 E) of Western Siberia.

The share of wetland ecosystems in the studied part of Western Siberia under current climatic conditions is about 11% for the ensemble of global climate models on average, and the spread in the

IOP Conf. Series: Earth and Environmental Science 1040 (2022) 012017 doi:10.1088/1755-1315/1040/1/012017

values of the wetland ecosystems areas among different models reaches 10%. According to model estimates for the SSP126 scenario in the 21st century, the area of wetland ecosystems in the considered part of Western Siberia does not change significantly, and under the SSP585 scenario it increases slightly and reaches 14% by the end of the 21st century. (figure 2). At the same time, for most of the analyzed models, no significant increase in the proportion of wetland ecosystems was revealed. The largest increase in both scenarios is obtained for the CanESM5 model. This model also shows the highest interannual variability in the proportion of wetland ecosystems. For the UKESM1-0-LL model, the proportion of wetland ecosystems was obtained the same for both scenarios and no noticeable changes were oserved in the 21st century. The sensitivity of methane emissions to changes in the hydrological regime of the atmosphere and soil decreases for the prescribed area of wetland ecosystems [11].
The results of calculations of methane emissions into the atmosphere by wetland ecosystems for the region (55-75 N, 65-85 E) of Western Siberia are presented in figure 3. According to model estimates, methane emissions from wetland ecosystems increase on average 1.5 times (from 4.6 to 7.1 $TgCH_4\ yr^{-1}$) under the SSP126 scenario and 3 times (from 5.3 to 16.4 TgCH4 yr^{-1}) under the SSP585 scenario by the end of the 21st century. There is a significant inter-model spread of methane emissions, increasing from the SSP126 to SSP585 scenario. The existing estimates of methane emissions into the atmosphere from wetland ecosystems of Western Siberia also differ significantly. According to [29], methane emissions for this region are about 1.7 $TgCH_4\ yr^{-1}$. According to the results of observations [30], the total methane emissions from bogs in the middle and southern taiga of Western Siberia are 3.2 $TgCH_4\ yr^{-1}$. According to these data, the corresponding emissions for the region are about 3 $TgCH_4\ yr^{-1}$. Model values of regional methane emissions at the beginning of the 21st century are estimated from 9.9 [31] to 22.2 $TgCH_4\ yr^{-1}$ [32]. It was noted in [31] that, due to the insufficiently high spatial model resolution, the sources of methane emission from wetland ecosystems are more localized than according to the observational data, with an overestimation of their intensity. According to the results of calculations using the regional model, methane emissions for this area vary from 6.2 $TgCH_4\ yr^{-1}$ at the end of the 20th century up to 9.8 $TgCH_4\ yr^{-1}$ at the end of the 21st century. [33].

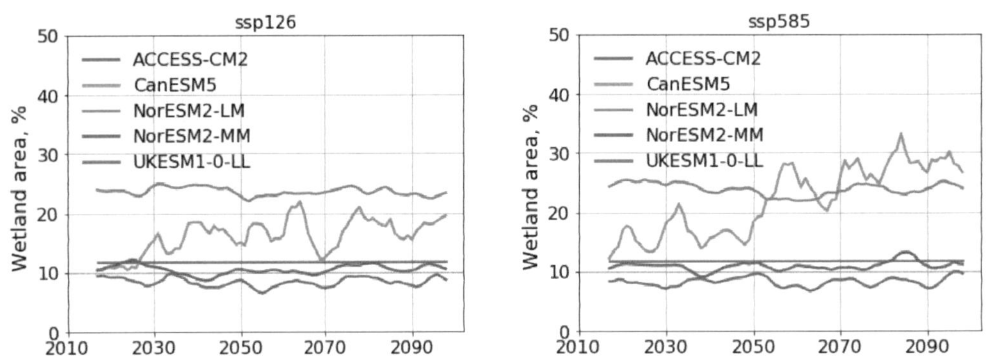

Figure 2. Five year moving average of wetland ecosystems area fractions for the region (55-75 N, 65-85 E) of Western Siberia according to calculations with various global climate models under the SSP126 and SSP585 scenarios.

According to the calculations, the greatest natural increase in methane emissions into the atmosphere, as well as the greatest interannual variability of emission values and the proportions of wetland ecosystems under different scenarios, were obtained for the CanESM5 model. For the same model, the highest average annual soil temperature was revealed in comparison with the rest of the ensemble models. The proportion of wetland ecosystems for this model exceeds the values obtained for the rest of the ensemble models only in the second half of the 21st century under the scenario SSP585. High values of methane emissions into the atmosphere were obtained for models with higher ground temperature, even with a relatively small proportion of bogs. According to the results obtained, regional methane emissions into the atmosphere are most dependent on soil temperature. The spread in methane emissions calculated with an ensemble of global climate models is also determined by the spread in soil temperature values. Regional trends in methane flux changes also depend on regional changes in the hydrological regime. An increase in precipitation with an increase in the moisture content of the soil enhances the thermal effect, while a decrease in the moisture content of the soil and a reduction in the area of bogs contributes to the weakening of methane emissions into the atmosphere by wetland ecosystems.

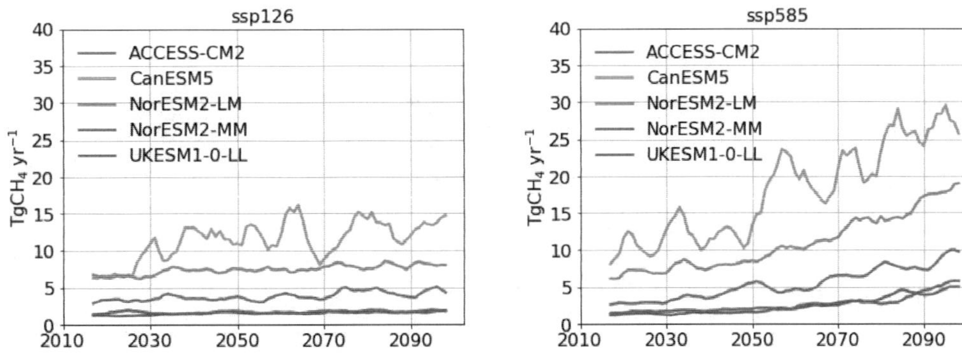

Figure 3. Five-year rolling average of methane emissions into the atmosphere from wetland ecosystems (TgCH$_4$ yr^{-1}) for the Western Siberia region according to calculations with different global climate models under the SSP126 and SSP585 scenarios.

An increase in soil temperature at high latitudes, including the Yamal and Gydansky Peninsulas, contributes to a decrease in the stability of frozen rocks and an increase in the risk of dissociation of relict methane hydrates occurring in the upper horizons of the permafrost zone above the modern boundary of the stability zone [19]. Considering that the time of temperature fluctuations propagation from the surface to a depth of 60-80 m is 20-25 years [34] with an amplitude of 10-20 degrees, the warming that began at the end of the 20th century can cause the dissociation of relict methane hydrates occurring in a given depth interval and under nonequilibrium conditions. Since 2014, in the regions of the Yamal and Gydansky peninsulas, 17 craters with traces of discarded rock have been identified, which are naturally associated with gas emissions [19]. Based on the data on craters dimensions [21], on rocks features, including density and porosity, calculations of the volume of methane that could be released during the dissociation of shallow relict methane hydrates were performed. In these calculations, the values of the moisture content of the soil and the percent of water in a hydrate were defined according to [26, 28]. The hydrate saturation of the rocks was taken equal to 37% [28]. It was assumed that methane emissions can occur in the regions of Western Siberia, which were under the ice sheet about 70-59 thousand years ago [17, 18]. The values of methane fluxes into the atmosphere as a result of the dissociation of shallow relict methane hydrates in the north of Western Siberia under current climatic conditions are estimated at 240 MgCH$_4$ yr^{-1}. The contribution of these emissions to the total natural emission of methane into the atmosphere is estimated to be small not only at the regional, but also at the local level at present time.

4. Conclusions

The estimates of changes in natural methane emissions into the atmosphere in the regions of Western Siberia (55-75 N, 65-85 E), including the contribution of wetland ecosystems and decomposition of relict methane hydrates in the cryolithozone of high-latitude regions, were obtained using model simulations. At the same time, a regional analysis of the distributions of soil temperature and the proportion of wetland ecosystems was carried out according to calculations with the ensemble of global climate models CMIP6 in the 21st century under the least aggressive (SSP126) and most aggressive (SSP585) scenarios of anthropogenic forcing. By the end of the 21st century, the regional increase in the average annual temperature of the upper soil layer on average for the ensemble of models reaches 1.4° C under the SSP126 scenario and 6.1° C under the SSP585 scenario. For the share of wetland ecosystems, no significant change was noted by the end of the 21st century for most of the analyzed models. Estimates of changes in regional methane emissions into the atmosphere from wetland ecosystems of Western Siberia and the contribution of methane fluxes as a result of decomposition of relict methane hydrates in the permafrost zone have been obtained. A general tendency of an increase in methane fluxes from wetland ecosystems into the atmosphere was noted under different scenarios. According to the calculation results, regional methane emissions into the atmosphere increase on average for the ensemble of models by 2.5 $TgCH_4$ yr^{-1} under the SSP126 scenario and by 11 $TgCH_4$ yr^{-1} under the SSP585 scenario. At the same time, there was a significant spread in the values of fluxes for different models. Methane emissions into the atmosphere as a result of the dissociation of shallow relict methane hydrates in the north of Western Siberia under current climatic conditions are estimated at 240 $MgCH_4$ yr^{-1}. This is much less than the estimates of regional methane emissions into the atmosphere from wetland ecosystems. It is important to emphasize that the described changes, accompanied by formation of a new geophysical phenomena – craters in the permafrost zone in high-latitude regions, are of particular significance and serve as an important indicator of modern climate change.

Acknowledgments

This study was supported by the Russian Foundation for Basic Research (project no. 19-05-00409, project no. 19-35-90118). Changes in the hydrate characteristics were estimated with the support of the Russian Science Foundation (project no. 19-17-00240). Permafrost response to regional climatic changes were estimated with the support of the Russian Science Foundation (project no. 21-17-00012).

References

[1] *Climate Change 2013: The Physical Science Basis. Contribution of Working Group I to the Fifth Assessment Report of the Intergovernmental Panel on Climate Change* ed T Stocker *et al* (New York: Cambridge University Press) p 1535
[2] Mokhov I I 2015 *Herald Russ. Acad. Sci.* **85** 265
[3] Pavlov A V 2008 *Permafrost monitoring (*Novosibirsk: Geo) p 230 (In Russian)
[4] Arzhanov M M and Mokhov I I 2013 *Dokl. Earth Sci.* **449** 319
[5] Slater A G and Lawrence D M 2013 *J. Climate* **26** 5608
[6] Kitover D C, Van Balen R T, Vandenberghe J, Roche D M and Renssen H 2016 *Permafr. Periglac. Process.* **27** 31
[7] Hugelius G *et al* 2013 *Earth System Sci. Data* **5** 393
[8] Jobbagy E N and Jackson R B 2000 *Ecol. Appl.* **10** 423

[9] Schuur E *et al* 2015 *Nature* **520** 171
[10] Kleinen T and Brovkin V 2018 *Environ. Res. Lett.* **13** 094001
[11] Denisov S N, Eliseev A V, Mokhov I I and Arzhanov M M 2015 *Izv. Atmos. Ocean. Phys.* **51** 482
[12] Matthews E and Fung I 1987 *Glob. Biogeochem. Cycles* **1** 61
[13] Andronova N G and Karol I L 1993 *Chemosphere* **26** 111
[14] Kim H-S, Maksyutov S, Glagolev M V, Machida T, Patra P K, Sudo K and Inoue G 2011 *Environ. Res. Lett.* **6** 035201
[15] Morimoto S *et al* 2021 *Polar Science* **27** 100553
[16] Chuvilin E M, Yakushev V S and Perlova E V 1998 *Polarforschung* **68** 215
[17] Ingolfsson O, Moller P and Lokrantz H 2008 *Polar Res.* **27** 152
[18] Kleman J, Fastook J, Ebert K, Nilsson J and Caballero R 2013 *Clim. Past.* **9** 2365
[19] Arzhanov M M, Mokhov I I and Denisov S N 2016 *Dokl. Earth Sci.* **468** 572
[20] Arzhanov M M, Malakhova V V and Mokhov I I 2020 *Permafr. Periglac. Process.* **31** 487
[21] Kizyakov A, Zimin M, Sonyushkin A, Dvornikov Yu, Khomutov A and Leibman M 2017 *Remote Sens.* **9** 1023
[22] Bogoyavlensky V, Bogoyavlensky I, Nikonov R, Kargina T, Chuvilin E, Bukhanov B and Umnikov A 2021 *Geosciences* **11** 71
[23] Denisov S N, Eliseev A V and Mokhov I I 2019 *Dokl. Earth Sci.* **488** 1066
[24] Denisov S N, Arzhanov M M, Eliseev A V and Mokhov I I 2011 *Atmos. Ocean. Opt.* **24** 319
[25] Eyring V, Bony S, Meehl G A, Senior C A, Stevens B, Stouffer R J and Taylor K E2016 *Geosci. Model Dev.* **9** 1937
[26] Fedoseev S M 2014 *Nauka i Obrazovanie* **1** 40
[27] Makogon Yu F 2010 *Geologiya i poleznye iskopaemye Mirovogo okeana* **22** 5 (In Russian)
[28] Davletshina D A and Chuvilin E M 2020 *Kriosfera Zemli* **XXIV** 25 (In Russian)
[29] Matthews E and Fung I 1987 *Glob. Biogeochem. Cycles* **1** 61
[30] Glagolev M V and Shnyrev N A 2008 *Mosc. Univ. Soil Sci. Bull.* **63** 67
[31] Eliseev A V, Mokhov I I, Arzhanov M M, Demchenko P F and Denisov S N 2008 *Izv. Atmos. Ocean. Phys.* **44** 139
[32] Anisimov O A and Reneva S A 2006 *Ambio* **35** 169
[33] Denisov S N, Eliseev A V and Mokhov I I 2010 *Russ. Meteorol. Hydrol.* **2** 55
[34] Melnikov P I and Tolstikhin O N 1974 *General geocryology* (Novosibirsk: The USSR Academy of Sciences, Siberian Branch) p 291 (In Russian)

IOP Conf. Series: Earth and Environmental Science 1040 (2022) 012018 doi:10.1088/1755-1315/1040/1/012018

Influence of aerosols on the meteorological regime of Northern Eurasia according to COSMO-Ru Model

A A Poliukhov[1,2] and D V Blinov[2]

[1]Lomonosov Moscow State University, Leninskie Gory 1, Moscow, 119991, Russia
[2]Hydrometcenter of Russia, Bolshoy Predtechensky per., Moscow, 123242, Russia

email: aeromsu@gmail.com

Abstract. We provided the estimation of the aerosols effect at the surface and at the top of the atmosphere, as well as on the sensible and latent heat flux for the Northern Eurasia using the MACv2 aerosol climatology. The numerical experiments were performed with the COSMO-Ru model for the January and July in 2017. It is found that the aerosol radiation effect on the surface can reach -50 W/m^2. At the same time, it is shown that the net shortwave radiation increase does not unequivocally leads to the air temperature at two meters increase. We revealed that the temperature at 2 meters decreases on -0.5±0.1 degrees per 100 W/m^2 in the clear sky condition for Mediterranean, Red, Black and Caspian seas areas. The temperature sensitivity is positive and equals 0.9±0.3 degrees per 100 W/m^2 for Middle East and southern Europe and 1.7±0.7 degrees per 100 W/m^2 for northern Africa.

1. Introduction

Aerosols affect the climate system through several physical mechanisms. The direct aerosol effect associated with the absorption and scattering of solar radiation has the most significant influence [1]. The aerosol radiation effect is defined as a change in the net radiation caused by the combined scattering and absorption of radiation by anthropogenic and natural aerosols relative to the non-aerosol atmosphere at the top of the atmosphere.

The first studies of the aerosol radiation effect have been started since the 60s -70s of the XX century. The main goal was to evaluate the hypothesis that aerosols reduce solar radiation on a global scale [2, 3]. For some cases, the contribution of the aerosol to the downward shortwave radiation fluxes may exceed 100 W/m^2. For example, such assessments were obtained during the aircraft measurements in a mineral dust plume over the ocean [4].

The quantitative assessment of the aerosols direct effect is often complicated due to aerosol high spatial and temporal variability. However, ground- and satellite-based observations as well as the theoretical modelling developments have helped to reduce estimation uncertainty significantly. Furthermore, the set of complex chemical transport models have been developed to reproduce the entire life cycle of various aerosol types from their formation to deposition [5, 6]. Various aerosol climatologies have been created using the output from these models [7-10]. Usage of aerosol climatologies is necessary for evaluation of the aerosol impact in numerical weather prediction (NWP) models. In recent decades, the approach to describing aerosols' properties for NWP models has changed significantly from semi-empirical formulas to ensemble calculations using chemical transport models and correction from observational data.

The inclusion of new aerosol climatologies in weather forecasting models inevitably modify the calculation processes of radiation fluxes, heat balance, and, as a result, the surface air temperature. Previously, the positive experience of replacing the aerosol climatology Tanré [11] with Tegen [10] in the ALADIN-HIRLAM model [12] was noted. A significant aerosols effect on surface air temperature simulation quality was also shown [13].

Up-to-date MACv2 aerosol climatology is taken in to account in COSMO-Ru model. Significant improvement of the surface temperature forecast was obtained for the simulations using MACv2 climatology compared to using Tanre's one [14].

This work aims to evaluate the effect of MACv2 aerosol climatology on the surface heat balance and the sensitivity of the surface air temperature to changes in the surface net shortwave radiation in the COSMO-Ru model.

2. Data and methods

2.1. COSMO-Ru Model

We performed the simulations with the COSMO-Ru mesoscale nonhydrostatic model [15]. The model radiation scheme [16] is based on the two-stream approximation. The spectrum was divided into three intervals in the solar range and five intervals in the far infrared range. The AFGL 1982 spectroscopic database was used to calculate the radiation flux density extinction by different gases [17]. The radiation absorption by water vapor, methane, carbon dioxide, and nitrogen oxide was considered.

Previous studies have evaluated the estimation of the calculation error for the total radiation in this algorithm compared to ground-based measurements and the two-stream CLIRAD(FC05)-SW algorithm [18]. It was shown the error does not exceed 5 % for different climatic regions [19,20].

2.2. Experimental setup

The COSMO-Ru model experiments were performed for the territory of Northern Eurasia with adjacent regions. We used the horizontal grid resolution of about 13 km, and 40 vertical levels for 24 hours experiment with the integration time step of 120 seconds. The forecast was calculated for January and July 2017. The ICON global model data were used as lateral boundary and initial conditions [21]. The calculations were carried out on the Roshydromet supercomputer CRAY-XC40.

Two experiments were performed, including the control experiment with AOT=0 and with MACv2 aerosol climatology.

The MACv2 climatology [7, 8] has a spatial resolution of 1°x1°. Monthly mean values of aerosol optical thickness, single scattering albedo, and asymmetry factor were considered for the following wavelengths 0.29, 0.32, 0.55, 0.87, 1.47, 2.0, 2.6 microns. That values are correspond to 2005 year. The results of calculations of 14 chemo-transport models of the AeroCom project are used [6]. Ground station measurements of AERONET [22] from 1995 to 2015 are used to correct model data for climatology.

3. Results

The average monthly values of the net shortwave radiation at the top of the atmosphere (TOA) and the surface (figure 1), the sensible and latent heat fluxes (figure 2), and the surface air temperature (figure 3) were calculated for each month. Furthermore, the sample included not only daytime as well as nighttime and the cases with clouds.

According to the COSMO-Ru model, the MACv2 aerosol climatology results in the average radiation effect of aerosols is 0.4 W/m^2 in January and -2.1 W/m^2 in July at the TOA for the selected area. At the same time, the positive radiation effect (up to 10 W/m^2) is observed over Central Asia in January. That effect could be because a large mineral aerosol concentration over a bright surface causes to decrease in the surface albedo due to the radiation absorption [23].

IOP Conf. Series: Earth and Environmental Science 1040 (2022) 012018 doi:10.1088/1755-1315/1040/1/012018

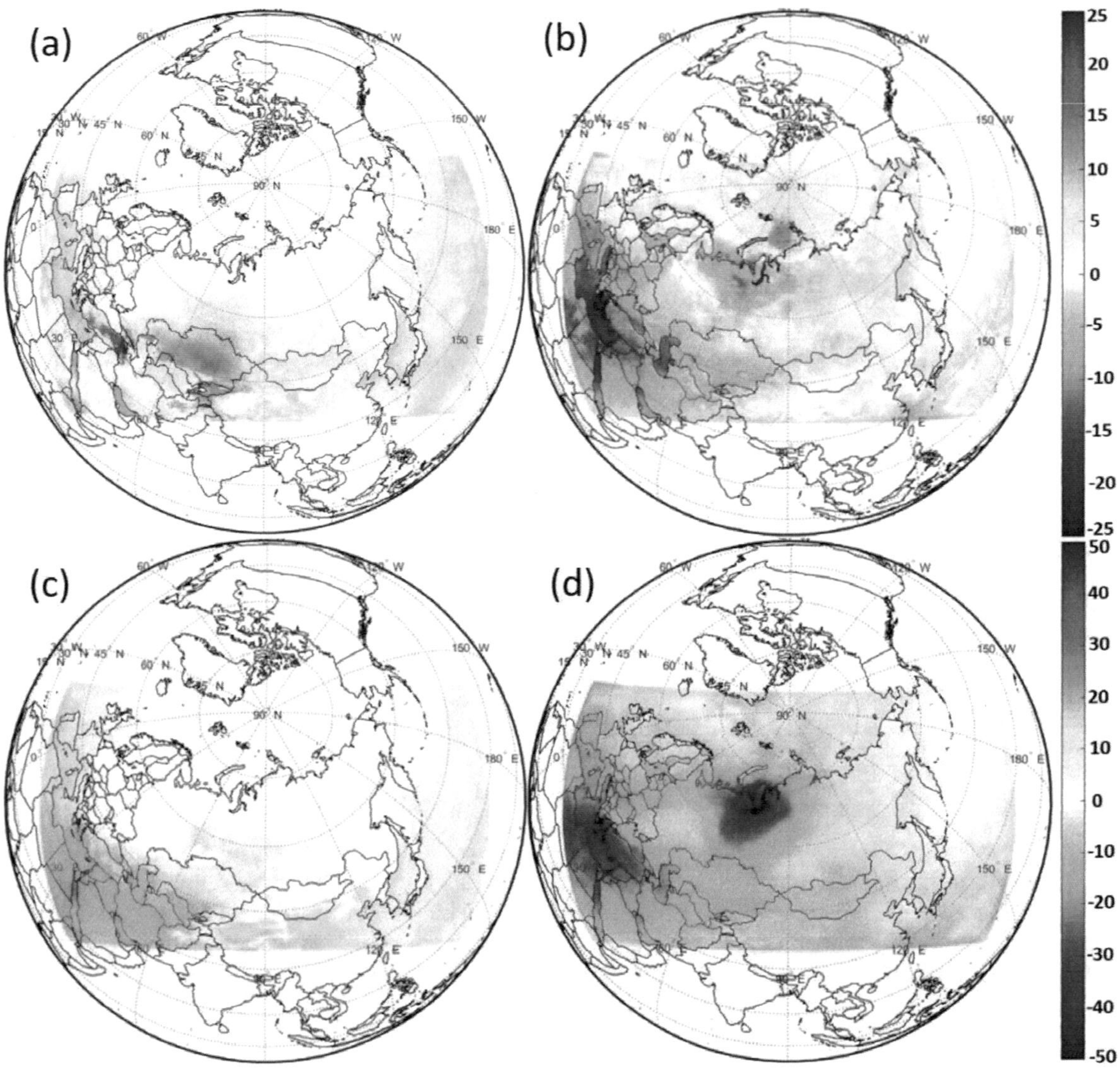

Figure 1. Differences between MACv2 and AOT= 0 experiments for the net shortwave radiation on the TOA (a,b) and at the surface (c,d) in January (a,c), and in July (b,d).

The difference in the net shortwave radiation is on average -3.2 W/m^2 in January and -9.1 W/m^2 in July at the surface. The maximum effect (up to -50 W/m^2) is observed for the territory of Southern Europe and the Mediterranean Sea. The same decrease in the net shortwave radiation is obtained for the territory of the Gulf of Ob, probably due to an increase in the cloud cover due to the semidirect effect.

Thus, the radiation effect of the aerosol at the surface are less than those at the TOA due to the absorption of radiation by aerosols in the atmosphere.

IOP Conf. Series: Earth and Environmental Science	1040 (2022) 012018	doi:10.1088/1755-1315/1040/1/012018

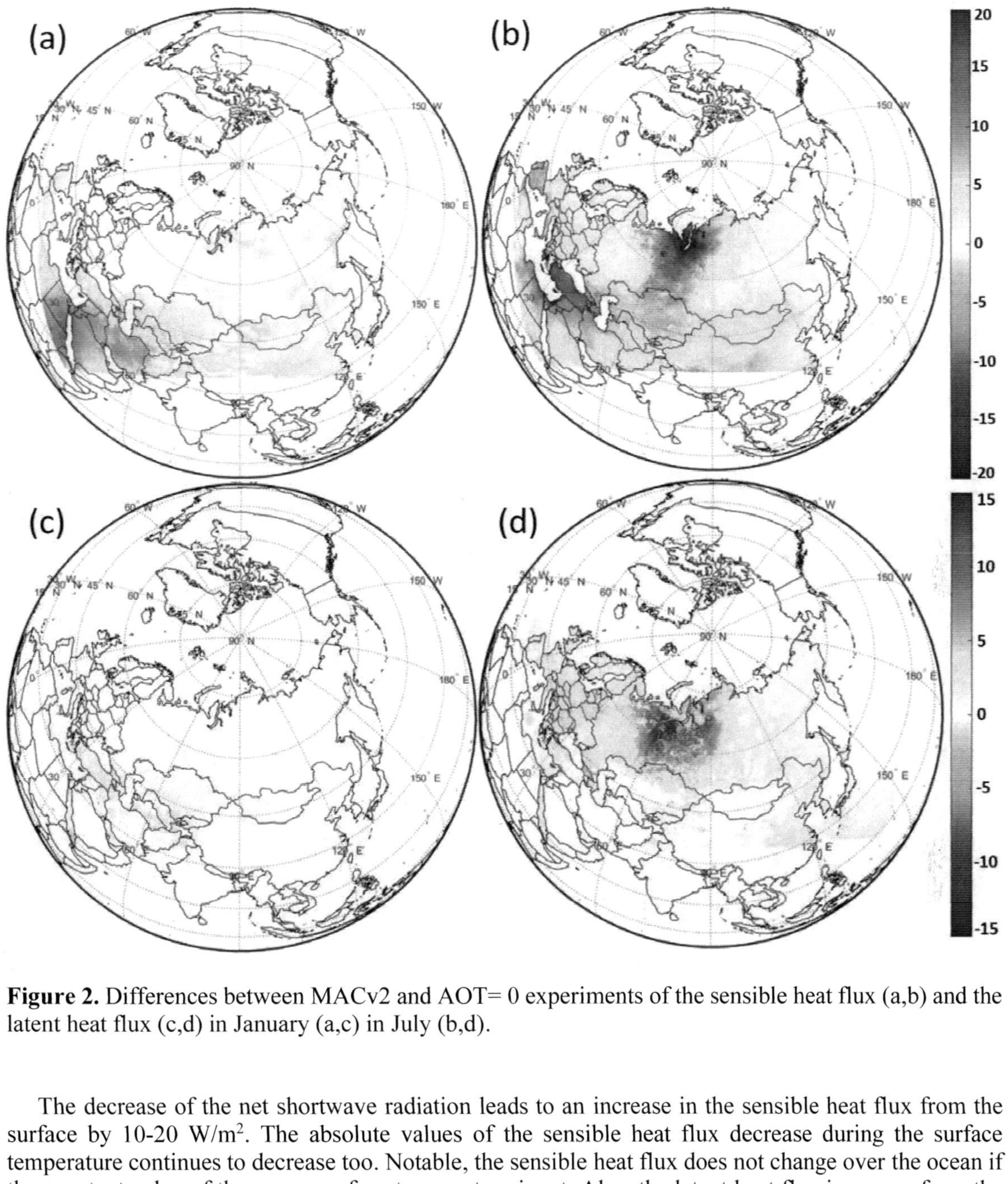

Figure 2. Differences between MACv2 and AOT= 0 experiments of the sensible heat flux (a,b) and the latent heat flux (c,d) in January (a,c) in July (b,d).

The decrease of the net shortwave radiation leads to an increase in the sensible heat flux from the surface by 10-20 W/m^2. The absolute values of the sensible heat flux decrease during the surface temperature continues to decrease too. Notable, the sensible heat flux does not change over the ocean if the constant value of the ocean surface temperature is set. Also, the latent heat flux increases from the surface by 5-15 W/m^2. The sensible and latent heat fluxes increase for the territory of the Gulf of Ob. So, this feature requires an additional research.

The aerosol effect on the surface air temperature has the opposite response sign over land and over the sea (figure 3). The surface air temperature over the land decreases due to the decrease of the net shortwave radiation. The most significant changes are observed in July in Africa and the Middle East (up to 1 degree). The temperature decrease reaches 0.3 degrees in January. However, the temperature increase is combined with the sensible heat flux decrease in Mongolia and Eastern Siberia.

The surface air temperature over the oceans depends primarily on the ocean surface temperature, which is derived from the global model data and assumed to be constant during forecasting period. However, the temperature at the lower model level could increase due to increased radiation absorption. Therefore, the surface air temperature should be permitted to increase even with a decrease in the net shortwave radiation. For example, this effect reaches 0.2 degrees in the Red and Caspian Seas and the western part of the Mediterranean in January.

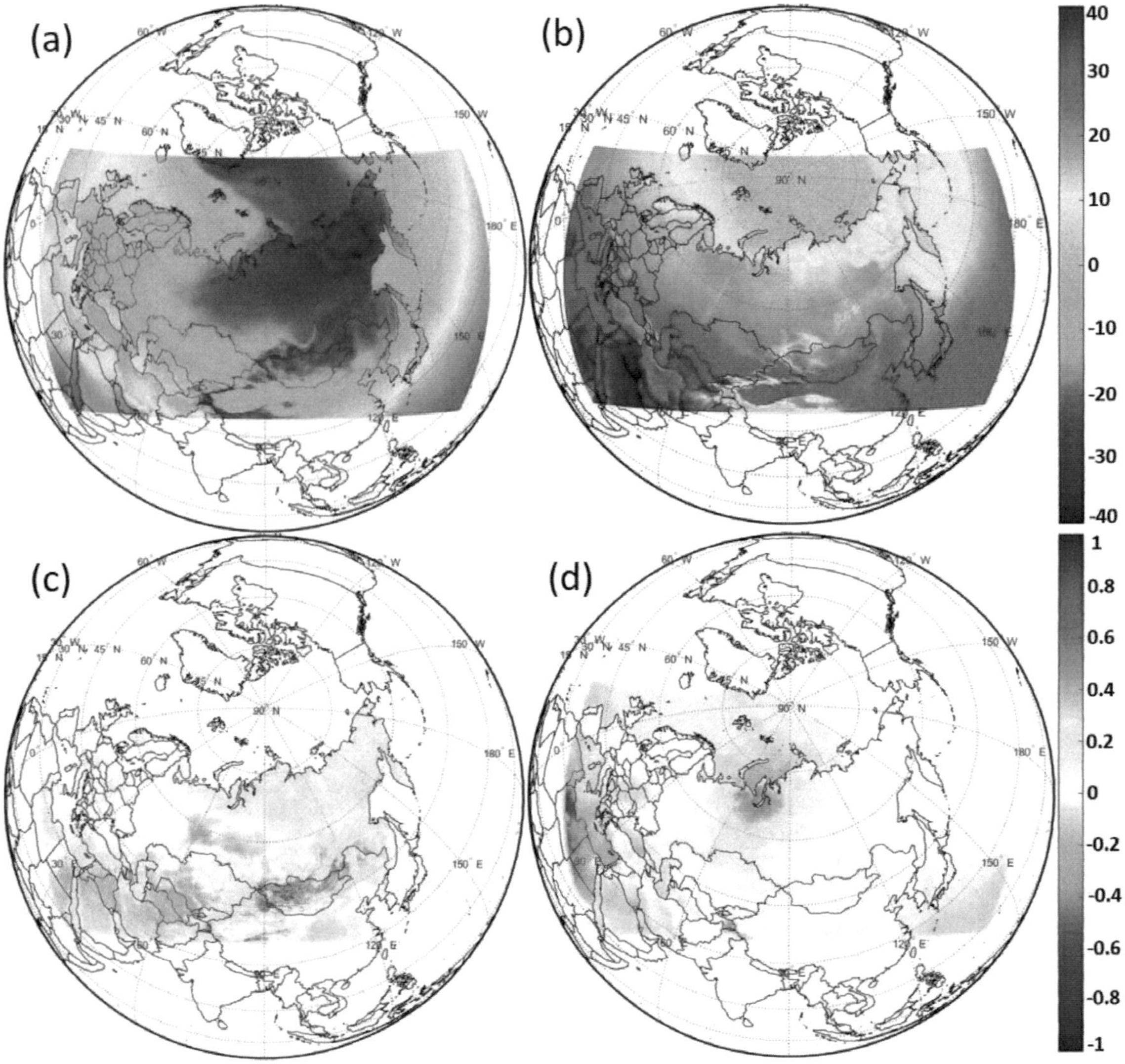

Figure 3. The monthly average surface air temperature in AOT= 0 experiment (a,b) and temperature differences between MACv2 and AOT= 0 experiments (c,d) in January (a,c), and in July (b,d).

The surface air temperature is one of the most important characteristics of the weather forecast. The surface temperature computation procedure in NWP models primarily depends on the radiation balance of the surface, advection, parametrization of heat transfer within the surface layer, and the soil layer parametrizations.

IOP Conf. Series: Earth and Environmental Science 1040 (2022) 012018 doi:10.1088/1755-1315/1040/1/012018

Previously, the changes in the surface air temperature for the cloudless conditions were estimated due to the aerosol radiation effect at the 4 stations located in different climatic zones. The sensitivity is amounted to 0.9±0.2 degrees per 100 W/m^2 [24].

We selected clear sky conditions in the area (defined as cloud cover is less than 1%), and with the net shortwave radiation of more than 100 W/m^2. Thus, the night and morning hours were removed. Also, we calculated linear approximation coefficients and estimated their statistical significance.

Because of the small number of clear sky condition cases in our study area, a statistically significant relationship (at the level of 0.05) was obtained only for Northern Africa, the Mediterranean Sea, southern Europe and the Middle East (figure 4).

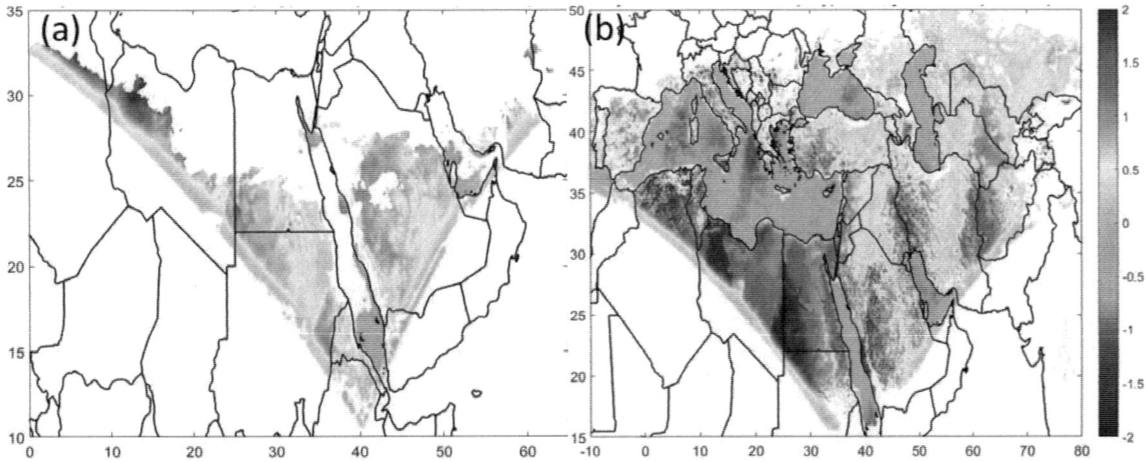

Figure 4. The sensitivity of the surface air temperature to changes in the net shortwave radiation in January (a) and July (b) (degree/100 W/m^2). The white color inside the calculation boundaries indicates the insignificant sensitivity.

In clear sky conditions, the effect of multidirectional changes in surface air temperature over the ocean and land area is most pronounced. The temperature sensitivity is -0.5±0.1 degrees per 100 W/m^2 over the ocean in January and July. The sensitivity has a larger variability over the land. For Middle East, Southern Europe and Central Asia, the temperature sensitivity is 0.9±0.3 degrees per 100 W/m^2, which confirms the previous work conclusions [24]. However, the temperature sensitivity can exceed 2 degrees per 100 W/m^2, and on average is 1.7±0.7 degrees per 100 W/m^2 in July over the North Africa. This may be due to the extremely low moisture content in the region. In January, the temperature sensitivity decreases to 1.2±0.2 degrees per 100 W/m^2 for the territory of North Africa.

4. Conclusions

In this paper we evaluated the impact of the MACv2 aerosol climatology on the calculation of radiation fluxes and sensible and latent heat fluxes for the territory of Northern Eurasia by the COSMO-Ru model.

Experiments have shown the aerosol radiation effect at TOA of about -2.1 W/m^2 in July. The aerosol radiation effect at the surface is greater (-9.1 W/m^2), and in some grid points reaches -50 W/m^2. Changes in the net shortwave radiation lead to an increase in the sensible and latent heat fluxes by 5-15 W/m^2. It was also found that the air temperature at 2 meters decreases by 0.3 degrees in January and by 0.8-1 degrees in July in the North Africa and the Middle East. However, the air temperature at 2 meters above the ocean could increase by 0.1-0.2 degrees due to the constant surface temperature during forecasting period.

We estimated the temperature change due to the difference in the net shortwave radiation in clear sky conditions. The temperature sensitivity is 0.9±0.3 degrees per 100 W/m^2 for Middle East and southern Europe, and 1.7±0.7 degrees per 100 W/m^2 for northern Africa. The changes are opposite and about -0.5±0.1 degrees per 100 W/m^2 for Mediterranean, Red, Black and Caspian seas.

Acknowledgements
The model simulation was funded by the Roshydromet project AAAA-A20-120021490079-3. The analysis of aerosol effects was funded by RFBR, project number 19-35-90129.

References

[1] Haywood J and Boucher O 2000 Estimates of the direct and indirect radiative forcing due to tropospheric aerosols: A review *Reviews of geophysics* **38** 4 513-543

[2] McCormick R A and Ludwig J H 1967 Climate modification by atmospheric aerosols *Science* **156** 3780 1358-1359

[3] Mitchell Jr J M 1971 The effect of atmospheric aerosols on climate with special reference to temperature near the earth's surface *Journal of Applied Meteorology and Climatology* **10** 4 703-714

[4] Haywood J, Francis P, Osborne S, Glew M, Loeb N, Highwood E and Hirst E 2003 Radiative properties and direct radiative effect of Saharan dust measured by the C-130 aircraft during SHADE: 1 Solar spectrum *Journal of Geophysical Research: Atmospheres* **108** D18

[5] Baklanov A, Brunner D, Carmichael G, Flemming J, Freitas S, Gauss M and Vogel B 2017 Key issues for seamless integrated chemistry–meteorology modeling *Bulletin of the American Meteorological Society* **98** 11 2285-2292

[6] Kinne S, Schulz M, Textor C, Guibert S, Balkanski Y, Bauer S E and Tie X 2006 An AeroCom initial assessment–optical properties in aerosol component modules of global models *Atmospheric Chemistry and Physics* **6** 7 1815-1834

[7] Kinne S, O'Donnel D, Stier P, Kloster S, Zhang K, Schmidt H, and Stevens B 2013 MAC-v1: A new global aerosol climatology for climate studies *Journal of Advances in Modeling Earth Systems* **5** 4 704-740

[8] Kinne S 2019 The MACv2 aerosol climatology *Tellus B: Chemical and Physical Meteorology* **71** 1 1-21

[9] Koepke P, Hess M, Schult I, and Shettle E P 1997 Global aerosol data set *Max-Plank-Inst fur Meteorol Rep* **243** 44 p

[10] Tegen I, Hollrig P, Chin M, Fung I, Jacob D, and Penner J 1997 Contribution of different aerosol species to the global aerosol extinction optical thickness: Estimates from model results *Journal of Geophysical Research: Atmospheres* **102** D20 23895-23915

[11] Tanre D, Geleyn J F and Slingo J 1984 First results of the introduction of an advanced aerosol-radiation interaction in the ECMWF low resolution global model *Aerosols and their climatic effects*, **133**, 177.

[12] Toll V, Gleeson E, Nielsen K P, Männik A, Mašek J, Rontu L and Post P 2016 Impacts of the direct radiative effect of aerosols in numerical weather prediction over Europe using the ALADIN-HIRLAM NWP system *Atmospheric Research* **172** 163-173

[13] Tompkins A M, Cardinali C, Morcrette J J and Rodwell M 2005 Influence of aerosol climatology on forecasts of the African Easterly Jet *Geophysical research letters* **32** 10

[14] Poliukhov A A and Blinov D V 2021 Aerosol Effects on the Temperature Forecast in the COSMO-Ru Model *Russian Meteorology and Hydrology* **46** 1 29-39

[15] Doms G, Foerstner J, Heise E, Herzog H-J, Mironov D, Raschendorfer M, Reinhardt T, Ritter B, Schrodin R, Schulz J-P and Vogel G A 2011 Description of the Nonhydrostatic Regional COSMO-Model Part II: Physical Parameterization Consortium for Small-Scale Modelling *Deutscher Wetterdienst* DWD

[16] Ritter B and Geleyn J F 1992 A comprehensive radiation scheme for numerical weather prediction

models with potential applications in climate simulations *Monthly weather review* **120** 2 303-325

[17] Rothman L S, Gamache R R, Barbe A, Goldman A, Gillis J R, Brown L R and Camy-Peyret C 1983 AFGL atmospheric absorption line parameters compilation: 1982 edition *Applied Optics* **22** 15 2247-2256

[18] Tarasova T A and Fomin B A 2007 The use of new parameterizations for gaseous absorption in the CLIRAD-SW solar radiation code for models. *Journal of Atmospheric and Oceanic Technology* **24** 6 1157-1162.

[19] Chubarova N, Poliukhov A, Shatunova M, Rivin G, Becker R and Kinne S 2018 Clear-sky radiative and temperature effects of different aerosol climatologies in the COSMO model *Geography Environment Sustainability* **11** 1 74-84

[20] Poliukhov A, Chubarova N, Kinne S, Rivin G, Shatunova M and Tarasova T 2017 February Comparison between calculations of shortwave radiation with different aerosol datasets and measured data at the MSU MO Russia *In AIP Conference Proceedings* **1810** 1 p 100006

[21] Zängl G, Reinert D, Rípodas P and Baldauf M 2015 The ICON ICOsahedral Non-hydrostatic modelling framework of DWD and MPI-M: Description of the non-hydrostatic dynamical core *Quarterly Journal of the Royal Meteorological Society* **141** 687 563-579

[22] Holben B N, Eck T F, Slutsker I A, Tanre D, Buis J P, Setzer A and Smirnov A 1998 AERONET—A federated instrument network and data archive for aerosol characterization *Remote sensing of environment* **66** 1 1-16

[23] Kinne S 2019 Aerosol radiative effects with MACv2 *Atmospheric Chemistry and Physics* **19** 16 10919-10959

[24] Poliukhov A A, Chubarova N E, Blinov D V, Tarasova T A, Makshtas A P and Muskatel H 2019 Radiation Effects of Different Types of Aerosol in Eurasia According to Observations and Model Calculations *Russian Meteorology and Hydrology* **44** 9 579-587

IOP Conf. Series: Earth and Environmental Science 1040 (2022) 012019 doi:10.1088/1755-1315/1040/1/012019

Variability of thoron distribution in the surface atmosphere at Borok Geophysical Observatory

S V Anisimov, E M Dmitriev, K V Aphinogenov, A S Kozmina

Borok Geophysical Observatory of Schmidt Institute of Physics of the Earth RAS, Borok, Yaroslavl region, RF

Corresponding author e-mail: eldar@borok.yar.ru

Abstract. The variability of radon-220 (thoron) distribution in the surface atmosphere is investigated by mathematical modeling of thoron turbulent transport, based on the data of continuous field natural observations at Borok Geophysical Observatory (Central Russia). The thoron flux from the surface was set according to the observations. Variations of turbulent diffusion coefficient altitude profiles were calculated by the observations of wind velocity pulsations at two altitudes made synchronously on the same location. The altitude profiles of thoron volumetric activity typical for the atmospheric surface layer are estimated. It is shown that the thoron volumetric activities at any altitudes vary concurrently, but the largest thoron volumetric activity, as well as its gradient, occurs below 0.5 m at nighttime.

1. Introduction

Radon isotopes ^{222}Rn and ^{220}Rn (thoron) make the main contribution to the surface atmosphere ionization over the land under natural conditions [1–3]. Natural radon isotopes are noble radioactive gases escaped from minerals into pore space of rocks and soil and then transported to the atmosphere by diffusion and advection [4, 5]. The soil radon is actively investigated, first of all due to increasing of its volumetric activity is considered to be earthquake precursors [6]. Note that the volumetric activities of the both natural radon isotopes ^{222}Rn and ^{220}Rn are of interest to be observed [7, 8]. Studies of atmospheric radon are important because of its radiation hazard to of the people’s health [9].

The short lifetime of ^{220}Rn leads to the fact that all ^{220}Rn entering the atmosphere decays in its near-surface layer, in contrast to the relatively long-lived ^{222}Rn, which has time to spread within the entire atmospheric boundary layer. Thus, the contribution of ^{220}Rn to ionization of the near-surface atmosphere can be quite significant, even despite its low efflux from the soil surface [10]. At the same time, the short lifetime of ^{220}Rn requires a special approach to measure its volumetric activity. The authors have already proposed a method to estimate ultra-low background volumetric activities of ^{222}Rn and ^{220}Rn in soil air and in the surface atmosphere [11]. In this paper the early experience is expanded by continuous natural observations of atmospheric surface layer turbulent parameters and mathematical modeling of thoron turbulent transport. The paper presents the variations of thoron volumetric activity altitude profiles, based on the data of continuous observations at the experimental site of the Borok Geophysical Observatory of Schmidt Institute of Physics of the Earth of the Russian Academy of Sciences (BGO IPE RAS).

2. Atmosphere thoron transport modeling

Thoron transport in the atmosphere, taking into account its diffusion and radioactive decay, is described by the equation

$$\frac{\partial A}{\partial t}+\frac{\partial}{\partial z}\left(K(z,t)\frac{\partial A}{\partial z}\right)-\frac{1}{\tau}A=0, \qquad (1)$$

where A is a thoron volumetric activity in the atmosphere, τ is a thoron lifetime (80 s), $K(z, t)$ is an effective diffusion coefficient dependent on altitude z and time t.

A mean surface thoron flux density J_0 and a zero value of thoron volumetric activity on infinity altitude can be given as a boundary conditions

$$-K(0,t)\left(\frac{\partial A}{\partial z}\right)\Big|_{z=0}=\frac{J_0}{\tau},\quad A|_{z\to\infty}=0. \qquad (2)$$

The diffusion coefficient $K(z, t)$ is assumed to be

$$K(z,t)=K_{\mathrm{M}}+K_{\mathrm{T}}(z,t),\quad K_{\mathrm{T}}(z,t)=\begin{cases}K_0(t)(z/z_0)^{m(t)}, & z>z_0,\\ 0, & z<z_0,\end{cases} \qquad (3)$$

where K_{M} is a molecular diffusion coefficient, $K_{\mathrm{T}}(z, t)$ is a turbulent diffusion coefficient and z_0 is a surface roughness parameter. The dependence of the turbulent diffusion coefficient on altitude above the surface roughness parameter z_0 is assumed to be approximated by the power law [12] with a factor $K_0(t)$ and an exponent $m(t)$. These approximation parameters can be determined from the observations.

Through discontinuity of a function $K(z, t)$ in $z = z_0$, the equation (1) must be replaced in $z = z_0$ with a continuity condition of thoron flux density

$$-K_{\mathrm{M}}\left(\frac{\partial A}{\partial z}\right)\Big|_{z-z_0\to-0}=-\left(K_{\mathrm{M}}+K_0(t)\right)\left(\frac{\partial A}{\partial z}\right)\Big|_{z-z_0\to+0}. \qquad (4)$$

The equations (1–4) can be greatly simplified. Firstly, we can eliminate a term $\partial A/\partial t$ from (1), because a thoron lifetime is much less than a diffusion time scale. So a time t can be considered as a parameter in a boundary problem (1–4). Secondly, we can eliminate K_{M} from the equations in region $z > z_0$, because a molecular diffusion coefficient is much less than a turbulent one. Finally, we can assume that an exponent $m(t)$ takes only two possible values: $m = 1$ under a stable atmosphere stratification and $m = 4/3$ under unstable one. The equation (1) can be solved analytically after these simplifications. The solution is given by

$$A(z)=\begin{cases}A_0F_m\left(\dfrac{z}{h_m}\right), & z\geq z_0,\\[2ex] A_+\exp\left(\dfrac{z}{h_{\mathrm{M}}}\right)+A_-\exp\left(-\dfrac{z}{h_{\mathrm{M}}}\right), & z\leq z_0,\end{cases} \qquad (5)$$

where $h_{\mathrm{M}} = (\tau K_{\mathrm{M}})^{1/2}$, $h_1 = \tau K_0$, $h_{4/3} = (\tau K_0)^{3/2}/z_0^2$, $F_1(x) = \mathrm{K}_0(2x^{1/2})$, $F_{4/3}(x) = x^{-1/3}\exp(-x^{1/3})$, K_0 is modified Bessel function of the second kind of zero order. The factors A_0, A_+ and A_- can be calculated from equations (2) and (4).

3. Measurement equipment and data

The continuous monitoring of thoron volumetric activity as well as a turbulent regime of the atmospheric surface layer was conducted at the experimental site of the BGO IPE RAS during the summer season of 2016.

A seismic radon station SRS-05 [13] installed near the surface in the weather shelter was used as a sensor of the thoron volumetric activity. The SRS-05 is widely used for monitoring the volumetric activity of radon, primarily in order to detect its increase as one of the precursors of earthquakes. Low sensitivity of the sensor does not allow real-time registration of background values of thoron volumetric activity. At the same time background values of thoron volumetric activity can be estimated by the longtime measurement using the seismic radon station [11].

Two ultrasonic stations METEO-2H [14] installed at heights of 2 and 10 m recorded the main meteorological parameters (air temperature, horizontal components of wind speed and wind direction, vertical wind speed, relative air humidity and air pressure) with a 10 Hz sampling frequency. The software of these stations enables also evaluation of the vertical velocity variance, total energy of turbulent pulsations, friction velocity and the Monin-Obukhov length scale. It makes also possible to estimate the turbulent diffusion coefficient at the corresponding heights. A configuration of the measuring instruments is presented in figure 1.

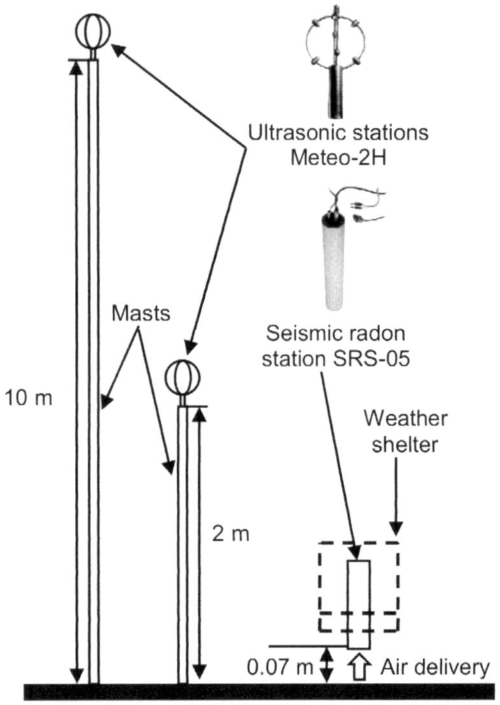

Figure 1. A configuration of the measuring instruments, installed on the experimental site of the BGO IPE RAS during June – July, 2016.

The obtained data in the so-called fair weather days were chosen to be analysed. Fair weather implies the following conditions: (1) thunderstorms, precipitation, fog, mist and haze were absent; (2) wind speed did not exceed 6 m/s; (3) no low clouds were present, and the total cloudiness did not exceed three.

4. Estimation of model parameters by the observational results

To estimate the surface thoron flux density J_0, the data observed under stable atmosphere stratification were chosen, because maximum of thoron volumetric activity is observed precisely under stable atmosphere stratification. The stable atmosphere stratification has been identified by the down-directed vertical turbulent temperature flux, estimated by the software of ultrasonic stations. The example of diurnal variations of vertical turbulent temperature flux on altitudes 2 and 10 m are presented on figure 2 (a). It is obvious that stable atmosphere stratification existed at night (00:00–02:30 UT and 15:30–00:00 UT). The diurnal variations of turbulent diffusion coefficient on same altitudes are presented on figure 2 (b). It is apparent that a turbulent diffusion coefficient can be well approximated indeed by the power law (3) with $m = 1$ for these time intervals.

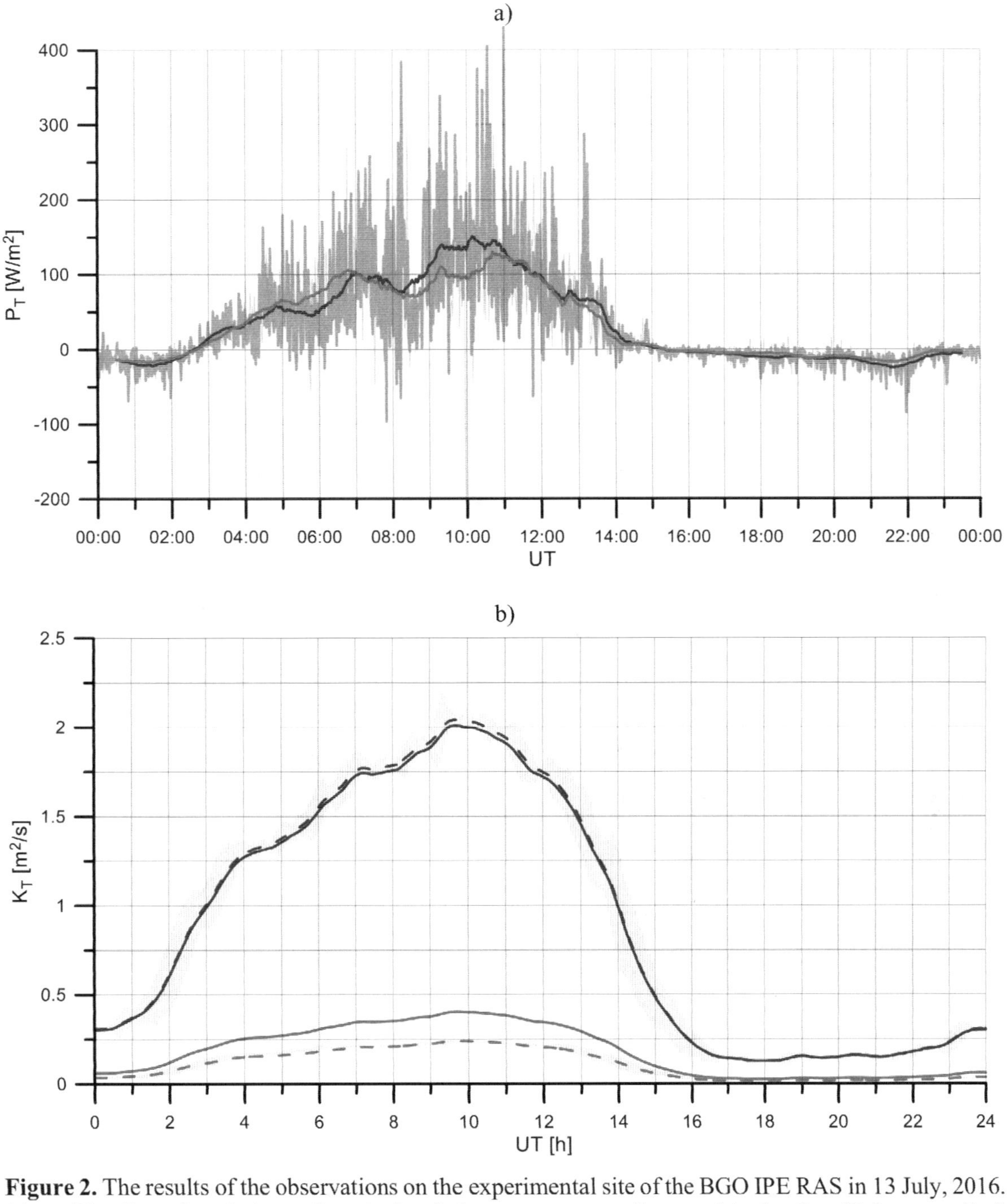

Figure 2. The results of the observations on the experimental site of the BGO IPE RAS in 13 July, 2016. (a) — The diurnal variations of vertical turbulent temperature flux on altitudes 2 m (light green line) and 10 m (light blue line) with one hour running averaging (green and blue lines). (b) — The approximation of diurnal variations of turbulent diffusion coefficient on altitudes 2 m (green lines) and 10 m (blue line) for $m = 1$ (solid lines) and $m = 4/3$ (dashed lines), vertical bars denote standard deviations for measured turbulent diffusion coefficients at hour time interval centered on the current minute.

The mean thoron volumetric activity observed by the seismic radon station at time intervals with stable atmosphere stratification (14 days totally) was $A(7\ \text{cm}) = (4.27 \pm 3.67)\ \text{Bq/m}^3$. The mean value of the parameter K_0 estimated from the ultrasonic stations data at the same time intervals was $K_0 = (0.032 \pm 0.016)$ m/s. The analytical solution (5) of stationary boundary problem (1–4), corresponding to these parameters, allows to estimate the surface thoron flux density $J_0 = (1.5 \pm 1.3)\ \text{atoms/m}^2\text{/s}$. The corresponding estimations of mean altitude profiles of thoron volumetric activity and thoron flux density with the surface roughness parameter $z_0 = 0.1$ m, according to grassland and low crops [15], are presented on figure 3. Taking into account that influence of radon turbulent transport in the atmosphere on surface radon flux density is negligible [16], one can use the estimation of J_0 as a stable model parameter.

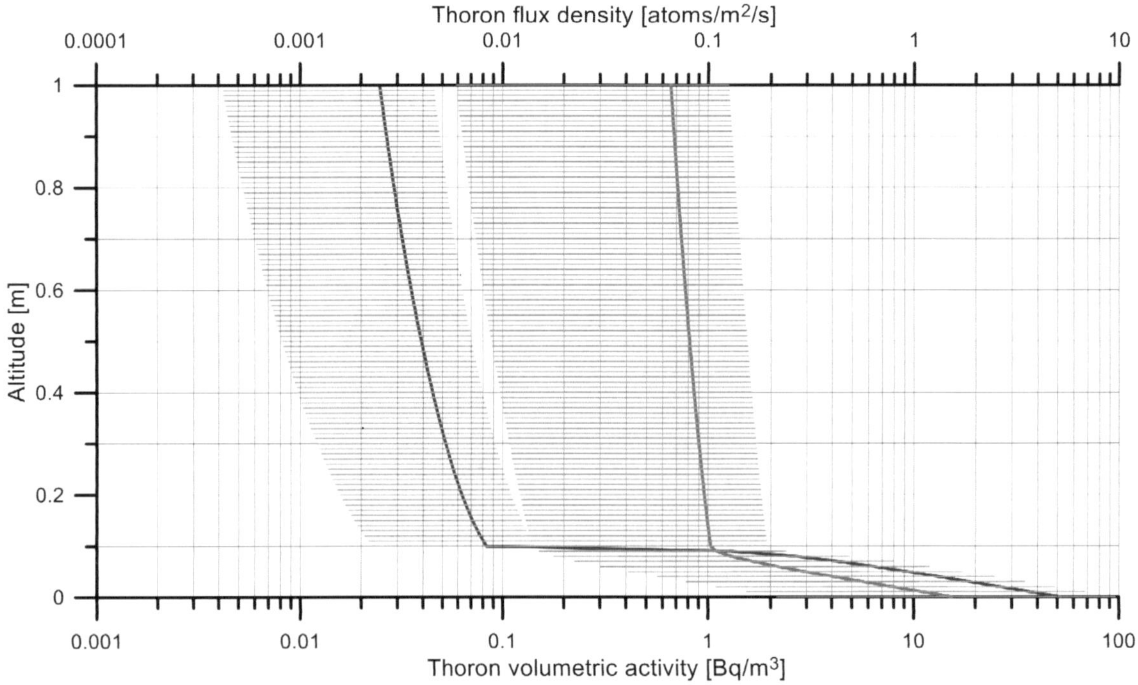

Figure 3. The model estimations of mean altitude profiles of thoron volumetric activity (blue line) and thoron vertical flux density (green line), calculated from BGO IPE RAS observations in summer of 2016 under fair-weather conditions and stable atmosphere stratification. Horizontal bars denote standard deviations for estimated values according standard deviations of measured thoron volumetric activity at 0.07 m altitude.

5. Results and conclusions

The typical approximation of $K(z,t)$ diurnal variation, calculated from (3) with $z_0 = 0.1$ m, $m = 4/3$ and $K_0(t)$, estimated by results of the observations on the experimental site of the BGO IPE RAS in 13 July, 2016 (see figure 2), is presented in figure 4 (a). The corresponding diurnal variation of thoron volumetric activity altitude profile obtained through analytical solution (5) of the boundary problem (1, 2, 4) with $J_0 = 2.5\ \text{atoms/m}^2\text{/s}$ is presented in figure 4 (b).

It is easy to see that the thoron volumetric activity at any altitude across the atmospheric surface layer varies synchronously with the near-surface thoron volumetric activity. The thoron volumetric activity is 2–5 times greater at midnight than at midday. At the same time the gradient of thoron volumetric activity is also much larger at nighttime then daytime. The largest gradient of thoron volumetric activity occurs below 0.5 m at night. So the thoron volumetric activity already above 4 m is always very small and does not depend on time of day in contrast to the volumetric activity of surface ^{222}Rn, which is considerably varied over the day, especially below 2 m [17].

a)

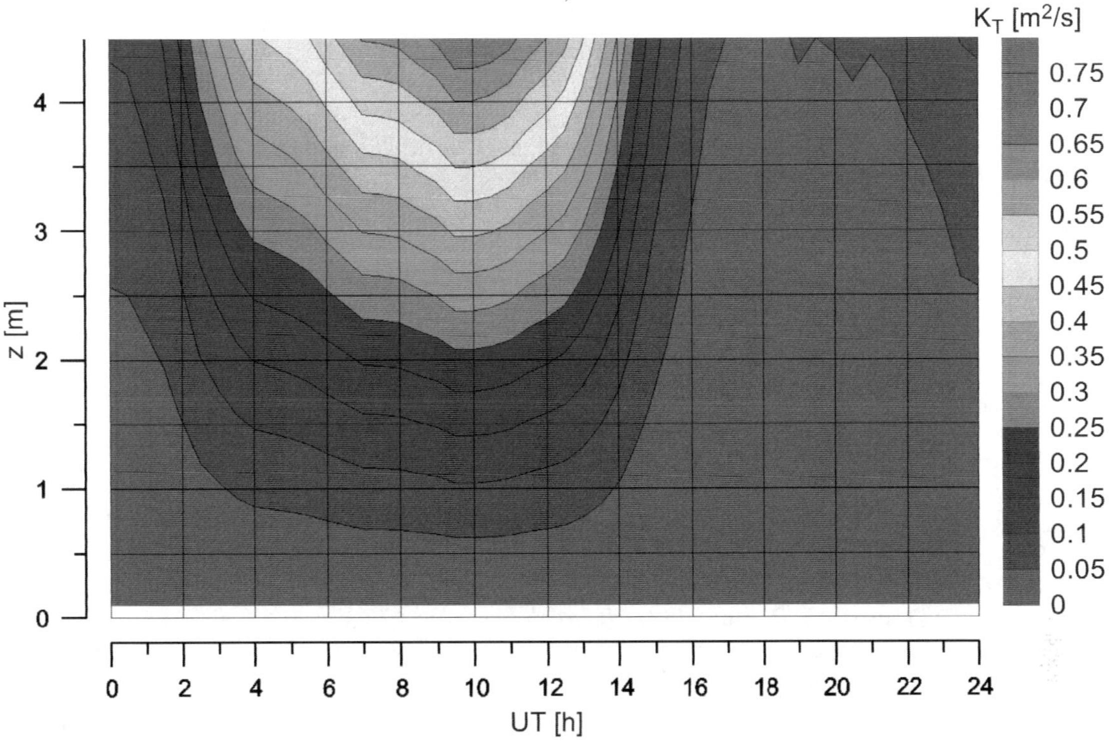

b)

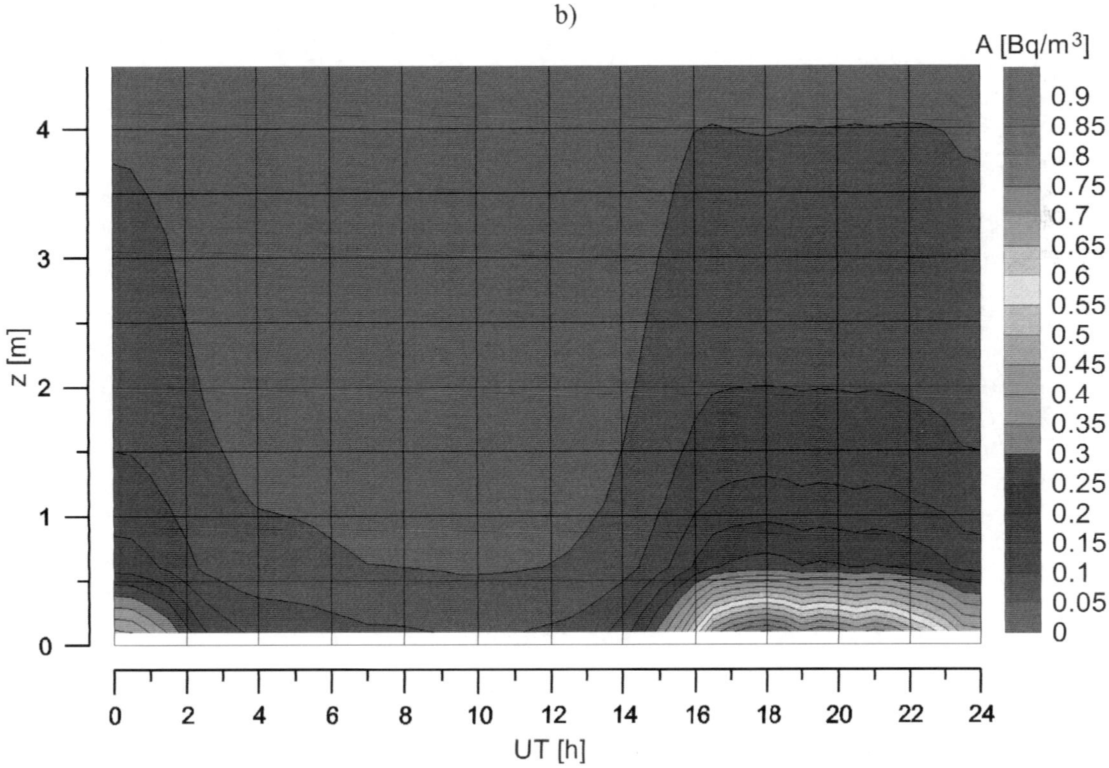

Figure 4. The diurnal variations of altitude profile of turbulent diffusion coefficient (a) and thoron volumetric activity (b) calculated from the results of the observations on the experimental site of the BGO IPE RAS in 13 July 2016.

The presented data agree with the results obtained by direct near-surface measurements of ^{222}Rn and ^{220}Rn activities [18]. It is highly likely that diurnal variations of thoron volumetric activity altitude profiles are typical for the atmospheric surface layer over the land. The proposed technique makes most exact and complete estimates of atmospheric ^{220}Rn variations in the surface layer, based on the continuous monitoring of surface atmosphere turbulent regime. So, the including of ^{220}Rn into variability of integral natural radioactivity in the surface atmosphere can be estimated by the proposed technique in many locations over the world.

Acknowledgments
This work was supported by government program No. 0144-2014-00115.

References

[1] Smirnov V V 1992 *Ionization in the troposphere* (St. Petersburg: Gidrometeoizdat) pp 312
[2] Morozov V N 2011 *Mathematical modeling of atmospheric and electric processes taking into account influence of aerosol particles and radioactive substances* (St. Petersburg: RSHU) pp 53
[3] Anisimov S V, Galichenko S V, Aphinogenov K V, Makrushin A P, Shikhova N M 2017 *Izvestiya. Physics of the solid earth* **53** 147–161
[4] Tanner A B 1964 *The Natural Radiation Environment* ed J A S Adams and W M Lowder (University of Chicago Press) 161–190
[5] Nazaroff W W 1992 *Rev. Geophys.* **30** 137–160
[6] Cicerone R D, Ebel J E, Beitton J 2009 *Tectonophysics* no 476 371–396
[7] Yang T F, Walia V, Chyi L L, Fu C C, Chen C H, Liu T K, Song S R, Lee C Y, Lee M 2005 *Radiation Measurements* **40** 496–502
[8] Perez N M, Hernandez P A, Padron E, Melian G, Marrero R, Padilla R, Barrancos J, Nolasco D 2007 *Pure and Applied Geophysics* **164** 2431–2448
[9] Al-Zoughool M, Krewski D 2009 *Int. J. Radiat. Biol.* **85** 57–69
[10] Anisimov S V, Galichenko S V, Aphinogenov K V, Prokhorchuk A A 2017 *Boundary-Layer Meteorol.* **127** 327–348
[11] Anisimov S V, Dmitriev E M, Kozmina A S 2020 *Seismic Instruments* **56** 245–253
[12] Wyngaard J C 2010 *Turbulence in the atmosphere* (New York: Cambridge University Press.) pp 393
[13] SRS-05 radon-measuring seismic station, a manual. http://www.ntm.ru/UserFiles/File/product/ION/srs-05/rukovodstvo_po_ekspluatazii.doc. Accessed April 1, 2021.
[14] Gladkikh V A, Makienko A E 2009 *Pribory* **109** 21–25
[15] Wieringa J 1981 *Air Pollution Modeling and Its Application 1* ed C De Wispelaere (Plenum Press, New York) 279–295
[16] Dmitriev E M 2018 *Izvestiya. Physics of the solid Earth* **54** 775–781
[17] Anisimov S V, Dmitriev E M, Aphinogenov K V, Guriev A V, Kozmina A S 2019 *IOP Conf. Ser.: Earth Environ. Sci.* **231** 012006
[18] Modibo, O.B., Tamakuma, Y., Suzuki, T., Yamada, R., Zhuo, W., Kranrod, C., Iwaoka, K., Akata, N., Hosoda, M. and Tokonami, S. 2021 *Int. J. Environ. Res. Public Health* **18** 1489

IOP Conf. Series: Earth and Environmental Science 1040 (2022) 012020 doi:10.1088/1755-1315/1040/1/012020

Unmanned aerial vehicle "Tsimlyanin" for studying turbulent structure of atmospheric boundary layer

D G Chechin[1], A Yu Artamonov[1], N YE Bodunkov[2], M Yu Kalyagin[2], A A Kunashuk[2], A A Shestakova[1], A M Shevchenko[3] and D N Zhivoglotov[4]

[1] Obukhov Institute of Atmospheric Physics of the Russian Academy of Sciences, 119017, Pyzevsky per., 3, Moscow, Russian Federation
[2] Moscow Aviation Institute (National Research University), 4 Volokolamskoye Shosse, 125993, Moscow, Russian Federation
[3] Khristianovich Institute of Theoretical and Applied Mechanics of the Siberian Branch of the Russian Academy of Sciences, 4/1 Institutskaya Street, 630090, Novosibirsk, Russian Federation
[4] Central Aerological Observatory, Ulitsa Pervomayskaya, 3, 141700, Dolgoprudny, Moscow Oblast, Russian Federation

E-mail: chechin@ifaran.ru

Abstract. A fixed-wing unmanned aerial vehicle "Tsimlyanin" having a hybrid vertical take-off and landing (VTOL) scheme with a compact meteorological payload is designed for an application to the atmospheric boundary layer (ABL) studies. The main purpose of such an observational platform is to supplement traditional mast observations by sampling horizontal and vertical inhomogeneities of mean variables and turbulent statistics. Such observations are needed to advance turbulence theory especially for the ABL over inhomogeneous landscapes and also in non-stationary conditions. The VTOL scheme is chosen as it provides a safer take-off and landing and is less demanding to the landing conditions. The designed turbulence payload consists of a seven-hole probe, inertial navigation system and a fast-response platinum-wire temperature sensor. It is complemented with the static pressure sensor, a slow-response temperature and humidity sensor and an infrared surface temperature sensor. The data acquisition and logging is performed by an on-board computer. The designed system has been successfully tested in the summer campaign in 2020 in Tsimlyansk, Russia.

1. Introduction

A theoretical description of the atmospheric turbulence is well established for rather idealized, horizontally-homogeneous and stationary conditions. In nature, however, such conditions are rare. Most of the natural surfaces are heterogeneous with respect to roughness, temperature and humidity and are characterized by a multiscale character of heterogeneities. The effect of such inhomogeneities on the atmospheric boundary layer remains an open challenge [1]. Moreover, even over a flat surface, the atmospheric boundary-layer (ABL) turbulence experiences the effect of non-stationarities of various scales, such as rapid transitions during the diurnal cycle, effects of clouds and mesoscale motions [2].

From the observational point of view, traditional eddy-covariance systems installed on a mast are not sufficient to resolve the turbulent structure of the ABL over inhomogeneities [1]. They need to be supplemented with aircraft and multiplatform observations. UAVs offer a much more affordable and flexible solution as compared to the piloted flying laboratories.

Meteorological application of the unmanned-aerial vehicles (UAV) has been experiencing a rapid growth. This is associated with the fact that the components of the UAVs are becoming lighter and cheaper and are commercially available. An increased number of observational campaigns involve the usage of meteorological UAVs [3, 4].

UAVs are also more frequently applied for turbulence research. Typically they are using the fixed-wing scheme [5]. It has several advantages over the multi-rotor platforms, such as a longer endurance and a smaller distortion of the air flow. Recently, the vertical take-off and landing (VTOL) scheme have started to be used on such UAVs [6, 7]. It allows for a safer landing and provides more universality with respect to take-off and landing conditions.

This paper presents the VTOL UAV "Tsimlyanin" and its turbulence payload developed for the atmospheric boundary layer studies. The results of the first test flights are demonstrated.

2. Unmanned aerial vehicle "Tsimlyanin"

2.1. The UAV scheme

The UAV "Tsimlyanin" is shown in Figure 1a. It was designed in the Moscow Aviation Institute specially for the application to the atmospheric boundary-layer studies. The chosen scheme meets the following requirements: payload mass up to 1.2 kg, mission endurance up to 60-90 min, cruising altitude up to 2000 m, the near-surface wind speed limit not less than 8 ms^{-1}, the possibility to locate sensors in the undisturbed flow and safe landing.

The two schemes were tested and compared: the VTOL fixed-wing scheme and the flying-wing scheme. The latter was based on the commercially available Skywalker-X8 model. During test flights it was found that the steadiness and the level of control of the flying-wing scheme strongly depends on the location of the payload which limits the flexibility of the whole system design. The VTOL fixed-wing scheme proved to provide wider limits for the payload placement. Thus, although both schemes meet the listed requirements, the VTOL fixed-wing scheme was chosen as optimal.

The developed aircraft has the frame design in which the central section and the twin-boom tail are tied. On the latter, the propeller-engine systems of the vertical take-off and landing are located. The truss structure fuselage is attached to the frame. The sensors of the payload are located in the nose of the fuselage. Inside the fuselage there is also an automatic control system and a power battery. The UAV is using the electric propulsion engine.

2.2. Meteorological payload

The turbulence payload (Figure 1b) consists of the seven-hole probe designed in the Khristianovich Institute of Theoretical and Applied Mechanics SB RAS (Novosibirsk, Russia) [8], a fast-response platinum-wire temperature sensor designed in the Central Aerological Observatory (Dolgoprudniy, Russia) and of the inertial navigation system (INS) Ellipse-2N (SBG Systems). The meteorological payload also includes the slow-response temperature and humidity sensor HMP110 (Vaisala), the infrared surface temperature sensor (Melexis) and the relative humidity sensor P14 Rapid (IST AG) which has a relatively fast response (about 1 s in still air according to the manufacturer). The data acquisition and logging is performed by a Linux-based single-board computer RaspberryPi4 at 100 Hz sampling rate.

The seven-hole probe is using a hemispheric tip with a diameter of 10-mm. At the central hole, the full pressure is measured, while at the six holes located at 60 degree angles relative to the central axis the difference between the local pressure and the pressure at the central

Figure 1. (a): the VTOL UAV "Tsimlyanin" during the Tsimlyansk 2020 campaign. (b): sensors at the nose of the UAV: 1) seven-hole pressure probe, 2) fast-response platinum-wire temperature sensor; 3) slow-response temperature and relative humidity Vaisala HMP110 sensor; 4) P14 Rapid IST AG relative humidity sensor.

hole is measured. Four holes at the sides are connected to a chamber where the static pressure is measured. Honeywell differential and absolute pressure sensors from The TruStability High Accuracy Silicon Ceramic (HSC) Series are used. The pressure transducers were calibrated in a laboratory in order to test and increase their accuracy. The whole probe was further calibrated in a wind tunnel. The seven-hole probe is superior to the frequently used five-hole probes due to an increased accuracy and redundancy of the measured quantities which minimizes potential data loss. The resulting accuracy of the observed wind components using similar systems is in the range 0.1-0.2 $\mathrm{m\,s^{-1}}$ [9, 10].

The fast-response temperature sensor is a resistance thermometer with an open platinum wire placed directly into the air flow. Around the sensor the protective housing is used which does not prevent the airflow through the sensor. The platinum wire has the diameter of 30 μm and the nominal resistance of 100 Ohm. The small diameter of the wire and a rather intense airflow at a cruising speed of about 20 $\mathrm{m\,s^{-1}}$ ensure a small response time. The sensor was first calibrated in a climate chamber where the resistance-temperature calibration curves were obtained. The wind-tunnel calibration showed that the recovery factor is close to unity and that its dependency on speed and direction of the air flow is small, when the latter vary in a range typical for this type of a UAV. The absolute accuracy of air temperature measurements is of order of 0.1 K, while the accuracy of measuring turbulent fluctuations is better and is reported to be 1.8 mK for similar sensors [11].

The general design of the meteorological payload is, thus, similar to the ones developed earlier for UAV applications, such as the one used on MASC-3 [5].

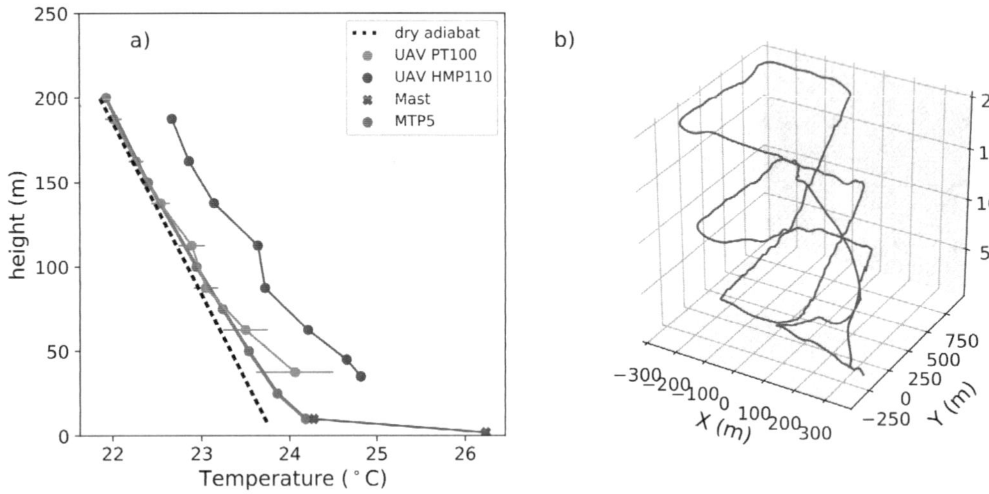

Figure 2. a): vertical temperature profiles obtained from the two sensors onboard the UAV, from the MTP5 profiler and from the Aanderaa automatic weather station on a mast; black dashed line shows the dry adiabat. b): flight track on August 13, 2020.

3. Test flight results

Test flights were carried out over a flat steppe field during the campaign in Tsimlyansk in August 2020 at the fild station of the Obukhov Institute of Atmospheric Physics of the Russian Academy of Sciences (IAP RAS). During the campaign, multiplatform observations were carried out providing the opportunity to verify the observations obtained by the "Tsimlyanin" UAV. Three sonic anemometers (Gill Instruments) were installed on a mast at 2, 10 and 30 m heights. Gradient observations of wind speed and direction, air temperature and humidity were carried out using the Aanderaa automatic weather station with sensors at 2 and 10 m heights. The Scintec mini-sodar provided wind speed and direction profiles up to 100 m height producing 5-min averaged wind profiles. The vertical profiles of air temperature were obtained using the microwave temperature profiler MTP-5 (RPO Attex).

Figure 2a shows vertical temperature profiles obtained during the flight pattern shown in Figure 2b performed by "Tsimlyanin" between 12:10 and 12:20 UTC on August 13, 2020. The flight pattern consisted of elongated rectangular figures (200 × 1000 m) performed at various heights, namely at 50, 100 and 200 m. A well-mixed convective boundary layer developed by the time of the flight due to a strong heating from the surface. The observed temperature structure of the lower ABL was typical for such conditions and consisted of a quasi-logarithmic profile close to the surface transforming into a dry adiabat at higher altitudes. Clearly, the fast-response sensor temperature, the MTP5 data and the mast observations are in close agreement with each other. Air temperature observed by the HMP110 sensor is also closely following the dry adiabat but has a systematic shift towards higher values. It is likely that this is a calibration issue and it will be investigated in more detail the future.

The observed wind speed and direction profiles are presented in Figure 3. The three wind components were calculated using the seven-hole probe and the INS measurements using a procedure described in [12]. Observations from sodar and UAV demonstrate a good agreement with respect to wind direction. The latter showed little variation with height being about 320°. The Scintec sodar observed wind speed in close agreement with the sonic anemometers on the mast. The wind speed observed by "Tsimlyanin" also agrees well with that from sonic

IOP Conf. Series: Earth and Environmental Science 1040 (2022) 012020 doi:10.1088/1755-1315/1040/1/012020

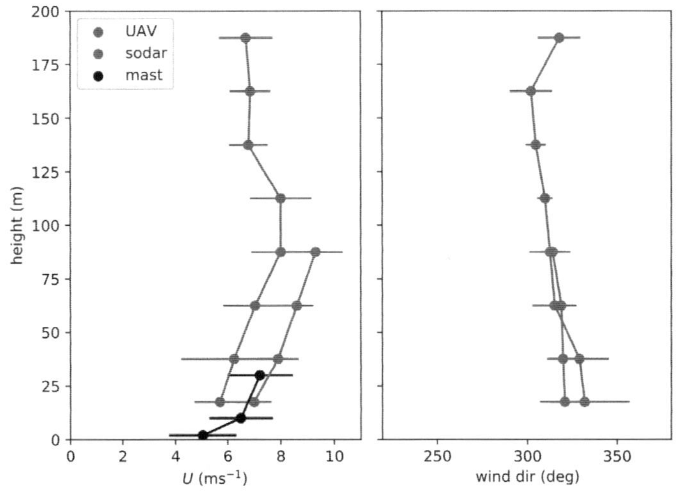

Figure 3. Wind speed (a) and direction (b) profiles obtained by the UAV "Tsimlyanin", the Scintec sodar and three sonic anemometers located at 2, 10 and 30 m heights on a mast.

anemometers and sodar, but is about 1-1.5 $\mathrm{m\,s^{-1}}$ weaker. One should keep in mind that wind speed shows large spatio-temporal variations in convective boundary layer and at least part of the difference between the observations from different platforms could be attributed to this.

Only the observations of the mean ABL structure are considered in this study. Further research will be focusing on the derivation of turbulence statistics from the UAV data.

4. Conclusions

The UAV "Tsimlyanin" and its meteorological payload were designed for the application to the atmospheric boundary layer research. The vertical take-off and landing fixed-wing scheme was chosen as it fullfills best the requirements with respect to the mission endurance, flexibility of the payload design, safety and simplicity of operation.

Several sensors and acquisition devices as part of the meteorological payload were custom designed and calibrated. These include the seven-hole probe and the fast-response platinum-wire temperature sensor.

Test flights performed during the summer campaign in Tsimlyansk demonstrated a robust functionality of the designed system. Comparison of the obtained observations of the vertical structure of convective boundary layer with measurements from sonic anemometers, sodar and temperature profiler showed a good agreement between different platforms.

Further research will be focused on deriving turbulence statistics for the UAV data and on their comparison with those from the ground-based instruments.

Acknowledgments

The the design of the UAV "Tsimlyanin", of the meteorological payload, organization and performance of the test campaigns were funded by the Russian Science Foundation grant 18-77-10072. The development of the algorithms for the UAV data processing was funded by the Russian Science Foundation grant 21-17-00249.

References

[1] Bou-Zeid E, Anderson W, Katul G G and Mahrt L 2020 *Boundary-Layer Meteorology* **177** 227 ISSN 1573-1472 URL https://doi.org/10.1007/s10546-020-00551-8

[2] Mahrt L and Bou-Zeid E 2020 *Boundary-Layer Meteorology* **177** 189 ISSN 1573-1472 URL https://doi.org/10.1007/s10546-020-00533-w

[3] Barbieri L, Kral S T, Bailey S C C, Frazier A E, Jacob J D, Reuder J, Brus D, Chilson P B, Crick C, Detweiler C, Doddi A, Elston J, Foroutan H, González-Rocha J, Greene B R, Guzman M I, Houston A L, Islam A, Kemppinen O, Lawrence D, Pillar-Little E A, Ross S D, Sama M P, Schmale D G, Schuyler T J, Shankar A, Smith S W, Waugh S, Dixon C, Borenstein S and de Boer G 2019 *Sensors* **19** 2179 ISSN 1424-8220 URL https://www.mdpi.com/1424-8220/19/9/2179

[4] Kral S T, Reuder J, Vihma T, Suomi I, Haualand K F, Urbancic G H, Greene B R, Steeneveld G J, Lorenz T, Maronga B, Jonassen M O, Ajosenpaa H, Baserud L, Chilson P B, Holtslag A A M, Jenkins A D, Kouznetsov R, Mayer S, Pillar-Little E A, Rautenberg A, Schwenkel J, Seidl A W and Wrenger B 01 Feb 2021 *Bulletin of the American Meteorological Society* **102** E218 URL https://journals.ametsoc.org/view/journals/bams/102/2/BAMS-D-19-0212.1.xml

[5] Rautenberg A, Schön M, zum Berge K, Mauz M, Manz P, Platis A, van Kesteren B, Suomi I, Kral S T and Bange J 2019 *Sensors* **19** 2292 ISSN 1424-8220 URL https://www.mdpi.com/1424-8220/19/10/2292

[6] Zappa C J, Brown S M, Laxague N J M, Dhakal T, Harris R A, Farber A M and Subramaniam A 2020 *Frontiers in Marine Science* **6** 777 ISSN 2296-7745 URL https://www.frontiersin.org/article/10.3389/fmars.2019.00777

[7] Sun Y, Ma J, Sude B, Lin X, Shang H, Geng B, Diao Z, Du J and Quan Z 2021 *Sensors* **21** 403 ISSN 1424-8220 URL https://www.mdpi.com/1424-8220/21/2/403

[8] Shevchenko A, Afanasiev L, Chechin D and Shmakov A 2021 *AIP Conference Proceedings* **2351** 030042 URL https://aip.scitation.org/doi/abs/10.1063/5.0052070

[9] Bärfuss K, Pätzold F, Altstädter B, Kathe E, Nowak S, Bretschneider L, Bestmann U and Lampert A 2018 *Atmosphere* **9** 28 ISSN 2073-4433 URL https://www.mdpi.com/2073-4433/9/1/28

[10] Witte B M, Singler R F and Bailey S C C 2017 *Atmosphere* **8** 195 ISSN 2073-4433 URL https://www.mdpi.com/2073-4433/8/10/195

[11] Wildmann N, Mauz M and Bange J 2013 *Atmospheric Measurement Techniques* **6** 2101 URL https://amt.copernicus.org/articles/6/2101/2013/

[12] Rautenberg A, Graf M S, Wildmann N, Platis A and Bange J 2018 *Atmosphere* **9** 422 ISSN 2073-4433 URL https://www.mdpi.com/2073-4433/9/11/422

IOP Conf. Series: Earth and Environmental Science 1040 (2022) 012021 doi:10.1088/1755-1315/1040/1/012021

On the reaction of planetary altitudinal frontal zone to climatic changes

S V Morozova[1], **K E Denisov**[2], **K S Kondakov**[3], **E A Polyanskaya**[1], **E I Ormeli**[1], **N K Kononova**[4, 5]

[1]Saratov State University, Saratov, Russia
[2]Saratov State Agrarian University named after N.I. Vavilov, Saratov, Russia
[3]Russian Scientific Research and Design Technological Institute for Sorghum and Corn, Saratov, Russia
[4]Institute of Geography, Russian Academy of Sciences, Moscow, Russia
[5] A.O. Kovalevsky Institute of Biology of the Southern Seas of RAS, Sevastopol, Russia

e-mail: swetwl@yandex.ru

Abstract. This article examines the dynamics of the planetary high-altitude frontal zone in two natural climatic periods of the state of the earth's climate system - the stabilization period and the second wave of global warming. A change in the area and intensity of the altitudinal frontal zone in different climatic periods was found. The planetary altitudinal frontal zone area expands and its intensity decreases from the stabilization period to the second wave of global warming. The climatic change in the area of planetary altitudinal frontal zone is opposite to the seasonal dynamics of this characteristic. Climate change in intensity is similar to seasonal change. The mechanism of action of negative feedback (air temperature - dynamics of the planetary altitudinal frontal zone) is discovered and described. For the first time, negative feedback has been applied to predict climate variability at finite time scales.

1. Introduction

The current state of the earth climatic system is regarded as global warming. The rise in surface air temperature for the entire globe is estimated at 0.075 °C per 100 years. Most experts believe that the observed increase in temperature is anthropogenic and explain this effect by the influence of carbon dioxide [9]. The results of model experiments, regularly published in the IPCC Assessment Reports, show a further steady increase in air temperature [3, 24]. Scenarios of upcoming climatic changes are considered to be rather mild and extremely radical as well [3, 24].

However, it should be noted that progressive changes in the earth climatic system (ECS) are extremely heterogeneous due to complex nonlinear interactions between its components. Assessment of nonlinear interactions in hydrodynamic models of the atmosphere is the most difficult task. Physical and statistical modeling gives an opportunity to assess the dynamics of nonlinear interactions and use obtained results for the development of a future scenario climate forecast. According to the authors, the most convenient and simple way of physical and statistical modeling is the use of feedbacks.

Feedbacks acting in the earth climatic system are the most important regulator of climatic dynamics. Positive feedbacks enhance the initial impact, thereby striving to disrupt the equilibrium state and cause irreversible changes in the ECS. Negative feedbacks contribute to the maintenance of an equilibrium state in the ECS, preserving the stability of the climate.

According to the authors [4], positive feedbacks are of the greatest interest, since they enhance the effect of anthropogenic impact. The authors of this work believe that the study of negative feedbacks is also very important. Understanding the mechanisms of their action will make it possible to assess future climate trends, which is an important aspect of reducing climate risks and a necessary condition for the adaptation of human society to climate change.

Obviously, climatic changes at relatively short time intervals are determined by the planetary circulation regime, namely, by the dynamics of its structural elements. The purpose of this publication is to investigate the dynamics of the structural element of the planetary circulation for the upcoming assessment of climate fluctuations.

Materials and methods

The object of research was the planetary altitudinal frontal zone - a three-dimensional element of the atmospheric circulation on a global scale. Let us point out that its dynamics determines the weather and climatic instability in temperate latitudes. The section of the planetary altitudinal frontal zone was considered at the middle level of the troposphere - on the isobaric surface of AT-500 hPa.

Let us point out that at present this structural object of circulation is increasingly becoming an object of research [7, 10, 11, 14, 20, 21, 23].

To characterize the planetary altitudinal frontal zone, many indicators are used, such as its length, intensity, latitude of the axial isohypse, area, tortuosity, etc. [1, 2, 22]. We point out that some of these characteristics are linearly related to each other, for example, the length and tortuosity, the latitude of the axial isohypse and the area bounded by this isohypse, etc.

In the present study, we selected such characteristics of planetary altitudinal frontal zone that do not exhibit a linear relationship with each other. These characteristics are area and intensity. It should be noted that the correlation coefficient between these values is close to zero.

The area of the planetary altitudinal frontal zone is understood as the circumpolar space bounded from the south by the axial line, or axial isohypse [1, 9]. The values of the center line for each month were selected according to the methodological developments of Yu.B. Khrabrov [22].

The magnitude of the geostrophic wind was taken as the intensity of the planetaryaltitudinal frontal zone. This characteristic was calculated using the well-known formula (1).

$$Vg = \frac{9{,}8}{l}\frac{\partial H}{\partial n} \quad (1)$$

where l is the Coriolis parameter $(l = 2\omega\sin\varphi)$,

ω is the earth angular velocity,

φ is the latitude of the axial isohypse,

H is the height of the geopotential of the isobaric surface, $H = 500$ hPa,

$\partial H/\partial n$ is the derivative of the geopotential height in the direction perpendicular to the tangent to the isohypse.

The data on the selected characteristics of the planetary altitudinal frontal zone are taken from the electronic supplement to the Reference monograph [18]. In [18], the latitude difference is taken as the intensity of the planetary altitudinal frontal zone. We assume that this characteristic is inversely proportional to the wind speed, so the authors of this publication decided to use the geostrophic wind speed as the intensity.

The dynamics of the planetary altitudinal frontal zone was calculated with reference to periods of climatic variability, which were called the Little Ice Age in Europe, the first wave of global warming, the period of stabilization, and the second wave of global warming. Figure 1 shows a graph of changes in the average surface air temperature anomalies in the Northern Hemisphere, plotted according to the site http://www.cru.uea.ac.uk/cru/data/temperature/#datdow). In this graph the time intervals clearly stand out, where average near-surface temperature has the same variability tendency:

1) Little ice age in Europe, shown in the graph by interval since the middle of the XIX century until the end of 1900s.
2) The first wave of global warming was observed since the middle of 1840s until the middle of 1905 until 1940s.
3) The period of stabilization (relative cold snap) that came out in the 1950s-1960s.

4) The second wave of global warming started at the middle of 1970s and continued until the present moment with an essential slowdown.

Let us point out that the authors did not work with the air temperature, but with its anomalies. Practically in all reputable domestic and foreign publications, in the works of highly respected scientists, studies of the temperature regime in the climatic plan are carried out on the basis of air temperature anomalies [3, 6, 19, 24]. We also note that the statistical processing of materials by the stepwise trend method requires the use of anomalies in the value [5, 8, 13]. Base period 1961 - 1990 (site http://www.cru.uea.ac.uk/cru/data/temperature/#datdow).

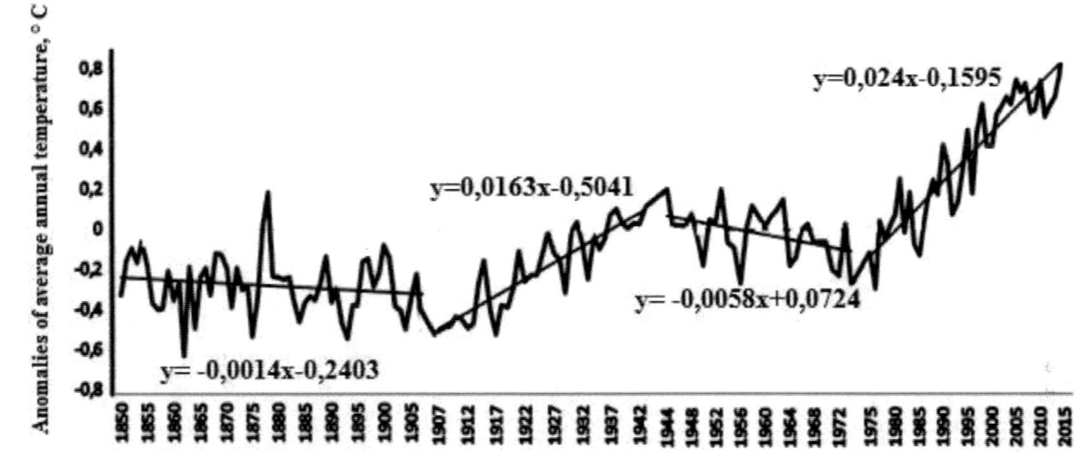

Figure 1. Variability of the average hemispheric air temperature

Trend lines are plotted at each of the four above-mentioned intervals. As can be seen from the Figure, the lines have different slopes, which is clearly from the values of the coefficients of linear trends α, that differ not only in value but in sign as well. The boundaries of the intervals were determined by the stepwise trend method using the Kolmogorov-Smirnov criterion (95% significance level) [5, 8, 13]. To define if the noticed change is climatically significant, we checked those with the synoptic one. We used the confidence interval methods with applying to Student`s statistics and 95% level of significance by the formula (2):

$$\bar{x} \pm t\gamma \frac{s}{\sqrt{n}} \ , \qquad (2)$$

in which $\bar{x}$ - average value; t_γ - Student's statistics, γ - Significance level indicator (95 %);.
s - mean square ; n - number of members of the series.

The above formula calculated the boundaries of the confidence intervals, which are presented in Table 1. In addition, this table contains some statistical characteristics for each of the time series segments.

Table 1. Assessment of the statistical significance of changes in the average annual temperature anomalies in the Northern Hemisphere

Period, year	Statistical characteristics of changes significance			Confidence interval
	α	$\bar{x}$	σ	
1) Little Ice Age (1850-1907)	-0.0014	-0.283	0.147	[-0.508; -0. 315]
2) first wave (1908 – 1943)	0.0163	-0.185	0.194	[-0.240; -0.130]
3) stabilization (1944 – 1974)	-0.0058	-0.020	0.124	[-0.078; 0.038]
4) second wave (1975 – 2016)	0.0240	0.356	0.320	[0.273; 0.439]

According to Table 1, the coefficients of linear trends have an opposite slope in adjacent climatic periods. The absence of overlapping of the boundaries of the confidence intervals allows one to conclude that the change in the surface air temperature in the Northern Hemisphere is statistically significant. S.V. Morozova [17, 25, 26] suggested calling these climatic intervals ***the natural climatic periods*** of the state of the earth climatic system.

In this study, the state of the planetary altitudinal frontal zone was studied in two natural climatic periods - the stabilization period (1949-1974) and the period of the second wave of global warming (1975-2010).

Since changes in the character of atmospheric circulation were noticed in the mid-1990s, the interval of the second wave of global warming was divided into two smaller ones - the first, active phase of the second wave of global warming (1975 - 1995) and the second phase, the phase of warming development (1996 - 2010).

1996 was chosen as the boundary between these intervals, since according to the materials of the Evaluation Report [3] in the mid-90s of the XX century there was a sharp change in the frequency of occurrence of processes corresponding to the three main forms of circulation - *E*, *W* and *C*: an increase in the frequency of occurrence of the *E* and *C* form processes and a decrease in the frequency of *W* form processes. N.K. Kononova [12] showed that since 1995, the greatest total duration of elementary circulation mechanisms has been observed. She [12] found out that since 1995 the greatest total duration of elementary circulation mechanisms has been observed. The sharp change in the prevailing forms of circulation indicates active restructuring processes in the atmosphere. Thus, the boundaries of the identified phases within the second wave of global warming are supported by circulation. Moreover, the application of the Kolmogorov - Smirnov criterion revealed that 1996 was the year of the homogeneity violation.

2. Results

Since the dynamics of the planetary altitudinal frontal zone is considered against the background of the observed climatic variability, and the best indicator of climatic variability is the near-surface air temperature, we will estimate the degree of linear relationship between the selected characteristics of the planetary altitudinal frontal zone and the average near-surface temperature of the Northern Hemisphere.

Table 2 shows the correlation coefficients between the planetary altitudinal frontal zone area and the average hemispheric surface air temperature in the natural climatic periods selected for the study. The significance of the correlation coefficients ($\alpha = 0.05$) was estimated by the formula $/ r / < t_\gamma \sigma_r$ (t_γ is Student's test, σ_r is the standard deviation of the correlation coefficient).

Table 2. The values of the correlation coefficients between the global temperature anomalies and the area of the planetary altitudinal frontal zone (significant correlation coefficients are put in bold).

Month	Period		
	1949 - 2010	1949 - 1974	1975 –2010
January	-0.32	-0.37	**-0.51**
February	-0.20	-0.34	**-0.64**
March	**0.38**	0.19	0.20
April	**-0.63**	-0.35	**-0.62**
May	0.14	-0.38	0.28
June	**-0.63**	**-0.41**	**-0.62**
July	0.03	**-0.49**	-0.17
August	**0.65**	**-0.88**	**0.63**
September	0.12	-0.04	0.04
October	**0.50**	-0.27	**0.50**
November	0.03	-0.33	-0.10
December	**0.50**	0.06	-0.17
Year	-0.18	-0.08	**-0.40**

Correlation coefficient significance	**>0.25**	**>0.38**	**>0.32**

According to Table 2, only half of all correlation coefficients were statistically significant. In addition, the correlation coefficients have different signs. Such heterogeneity indicates instability of connections. The revealed nonstationarity confirms the complexity and nonlinearity of interactions between the selected characteristics, as well as the nonlinearity of interactions within the earth climatic system between its components. We point out that similar conclusions were made when assessing the correlations between changes in the average hemispheric temperature of the surface air layer and the intensity of the planetary altitudinal frontal zone.

To study the nonlinear interactions between the studied characteristics (the area of the planetary altitudinal frontal zone, its intensity and air temperature), we will consider the change in the area and intensity of the planetary altitudinal frontal zone against the background of the observed climatic variability.

Table 3 shows the change in the area of planetary altitudinal frontal zone in two climatic periods as a whole for the year and in the central months of the winter and summer seasons. According to Table 3, during the transition from a colder period to a warmer one (from the stabilization period to the second wave of global warming), an increase in the planetary altitudinal frontal zone area occurred. Moreover, such dynamics is observed both throughout the year as a whole and in the central months of the main seasons of the year [15, 16]. The revealed dynamics corresponds to the displacement of the axial line of the planetary altitudinal frontal zone to more southern latitudes, which results in the expansion of the area of the planetary altitudinal frontal zone. It should be noted that with the seasonal dynamics from the cold to the warm period (from winter to summer), there is a reverse movement - the area of the planetary altitudinal frontal zone shrinks, and the planetary altitudinal frontal zone itself shifts to the north. Thus, it was found out that the climatic dynamics of the planetary altitudinal frontal zone area is opposite to its seasonal dynamics [1, 2, 9].

Table 3. Changing the areas of the planetary altitudinal frontal zone in various natural periods of the earth climatic system state.

Period	Area of the planetary altitudinal frontal zone, mln km^2		
	average annual	January	July
1949-1974 (stabilization)	56.97	62.44	56.15
1975-2010 (second wave of global warming)	57.77 an increase of 1.5%	64.19 an increase of 3 %	59.58 an increase of 6 %

Unfortunately, the graphic representation of the planetary altitudinal frontal zone in two climatic periods is very indistinct and difficult to read due to the large number of superimposed isohypsum. Therefore, the figure is not shown.

Let us consider another characteristic of the planetary altitudinal frontal zone - its intensity. Figure 2 shows the change in the average annual speed of the geostrophic wind in two natural climatic periods of the state of the earth climatic system - stabilization and the second wave of global warming.

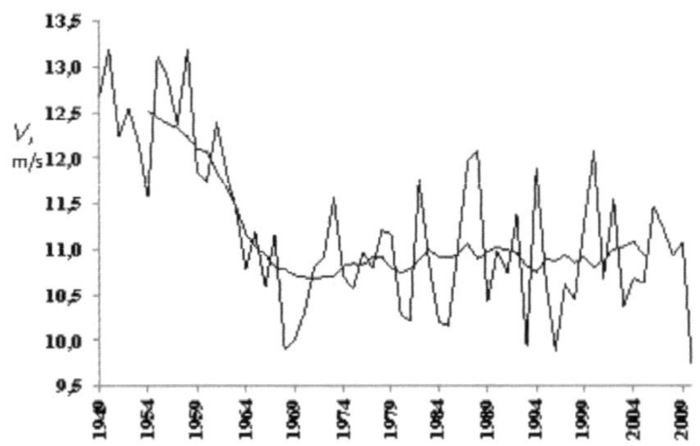

Figure 2. Change in the average annual speed of the geostrophic wind

According to Figure 2, during the transition from the cold to the warm period, the geostrophic wind speed decreases. A similar trend can be traced practically in all months. The table 4 shows the slope coefficients of the linear trends in the geostrophic wind speed for all months of the year (tabl.4).

Table 4. Coefficients of the linear trends

Month	Coefficient values	Month	Coefficient values	Month	Coefficient values	Month	Coefficient values
December	-0,004	March	-0,057	June	-0,071	September	-0,016
January	-0,047	April	-0,031	July	-0,029	October	-0,005
February	-0,010	May	-0,002	August	0,006	November	-0,004

It should be noted that the climatic change in the planetary altitudinal frontal zone intensity is consistent with the seasonal dynamics: atmospheric circulation becomes less intense in warmer periods (the second wave of global warming, summer seasons).

Thus, it is possible to conclude, that during the climatic transition from the cold to the warm period, the intensity of atmospheric circulation decreases, while the planetary altitudinal frontal zone moves to the south and its area expands.

The movement of the planetary altitudinal frontal zone to the south presupposes an increase in the area of the circumpolar space, that is, the area of below zero temperature anomalies. Thus, the rise in hemispheric mean temperatures is inhibited. This conclusion does not contradict the role of the planetary altitudinal frontal zone in climatic processes - the separation of areas of negative and positive temperature anomalies, and smoothing of temperature contrasts between the pole and the equator. The difference is that, in synoptic processes, the contrasts between the pole and the equator are equalized by means of inter-latitudinal air exchange by increasing the degree of meridionality of flows.

Let us check the last assumption on the materials of 1975 - 2010, corresponding to the second wave of global warming, in the development of which a significant decrease in its rate was recorded.

Table 5 shows the change in the area of the planetary altitudinal frontal zone during two phases of the second wave of global warming.

Table 5. Changing the areas of the planetary altitudinal frontal zone.

Period	Area of the planetary altitudinal frontal zone, mln km^2		
	average annual	January	July
1975-1995 (active phase)	58.51	68.5	60.03
1996-2010 (retarding phase)	56.70 decreasing by 3%	58.15 decreasing by15%	58.98 decreasing by2%

IOP Conf. Series: Earth and Environmental Science 1040 (2022) 012021 doi:10.1088/1755-1315/1040/1/012021

According to Table 5 the area of the planetary altitudinal frontal zone decreases during the second wave of global warming.

The analysis of the dynamics of the planetary altitudinal frontal zone area during the second wave of global warming showed that in most months, during the transition from the first phase of warming to the second phase, there is a decrease in the average areas of the planetary altitudinal frontal zone in eight months. At the same time, at the end of the studied interval of the second wave of global warming in some months, an increase in the speed of the geostrophic wind was observed. As an example, we can cite the change in the speed of the geostrophic wind in August (Figure 3), which is the most "quiet" month in relation to the variability of the wind regime.

Figure 3 shows a decrease in the speed of the geostrophic wind from the stabilization period to the second wave of global warming and an increase in the intensity of atmospheric circulation during the second phase of the second wave of global warming. Let us note that a decrease in wind speed during the active phase of the second wave of global warming is indicated in other publications [3, 19, 24].

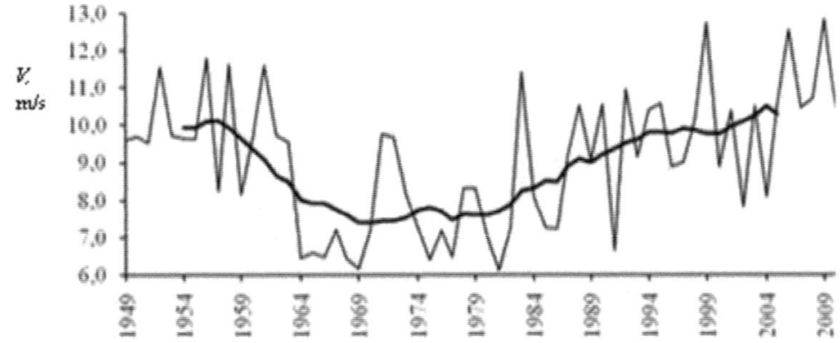

Figure 3. Change in geostrophic wind speed in August

Described dependences of the characteristics of the global circulation object (planetary altitudinal frontal zone) and the air temperature can be represented in the form of negative feedback, reflecting the operation of the natural model (Figure 4). In this figure, the symbols Δt, ΔS, and ΔV denote changes in the average temperature of the hemisphere, the area of the planetary altitudinal frontal zone and the average speed of the zonal wind, respectively.

$$\Delta t>0 \rightarrow \Delta S>0 \rightarrow \Delta V<0 \rightarrow \Delta t<0$$
$$\Delta t<0 \rightarrow \Delta S<0 \rightarrow \Delta V>0 \rightarrow \Delta t>0$$

Figure 4. Negative feedback scheme between the average hemispheric air temperature and the characteristics of the planetary altitudinal frontal zone.

According to the above, the general circulation of the atmosphere can be considered as a factor hindering the development of modern warming and contributing to the preservation of a fragile climatic and ecological balance in the earth climatic system. This conclusion does not contradict the position of the climate theory about the role of the planetary circulation as a mechanism that puts the earth climatic system into a state of equilibrium.

The revealed tendency of changes in the area of the planetary altitudinal frontal zone and the average speed of the zonal wind can determine the nature of the upcoming climatic changes - the onset of the next short period of stabilization against a higher temperature background, followed by an increase in the wind regime in the hemisphere.

The suspension of global warming observed since the mid-10s of the XXI century confirms the existence and effect of the identified negative feedback. Figure 5 shows a short stabilization period, representing climatic fluctuations against the background of warming.

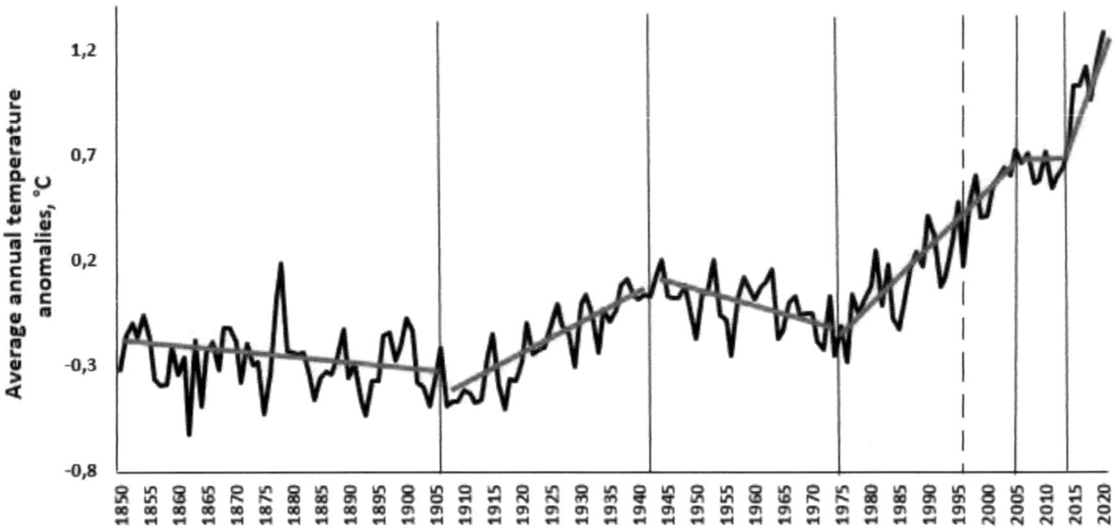

Figure 5. Northern Hemisphere mean temperature changes and natural climatic periods

Thus, we can conclude that using physical and statistical modeling it is possible to predict short-term climatic fluctuations.

3. Conclusions

In the course of the study, the following conclusions were obtained:

1. The object of circulation on a global scale (the planetary altitudinal frontal zone) reacts to climatic changes in the Northern Hemisphere by changing its characteristics - area and intensity.

2. During the transition from a colder period to a warmer one, the area of the planetary altitudinal frontal zone expands and its intensity decreases.

3. The climatic change in the intensity of the planetary altitudinal frontal zone is similar to the seasonal, the climatic change in the area of the planetary altitudinal frontal zone is opposite to the seasonal one.

4. Changes in the characteristics of the planetary altitudinal frontal zone against the background of the observed climatic trends are implemented as a negative feedback.

5. Physical and statistical modeling of processes in the terrestrial climate system makes it possible to predict short-term climatic fluctuations..

4. Acknowledgements

The work is performed within the Program Fundamental Research Institute Geography RAS, project No.0148-2019-0009.

The study was carried out asa part of the IBSS research (registration No.: AAAA-A19-119061-190081-9).

5. References

[1] Baidal M Khanzhina DG 1986 Long-term variability of macrocirculation climate factors. (Moscow)
[2] Bashkirova L I and Ped D A 1996. *Proceedings of the State Medical Center* **328** 19-34
[3] Second Assessment Report of Roshydromet on Climate Changes and Their Consequences on the Territory of the Russian Federation 2014 (Moscow)
[4] Ginzburg AS and Demchenko PF 2019 *Physics of the atmosphere and ocean***55**94 – 113
[5] Gmurman V E 2003 Theory of Probability and Mathematical Statistics (Moscow)
[6] Gruza G V and Rankova E.Ya 2012 Observed and expected climate changes in Russia: air

IOP Conf. Series: Earth and Environmental Science 1040 (2022) 012021 doi:10.1088/1755-1315/1040/1/012021

temperature (Moscow)
[7] Durneva E A and Chkhetiani O A 2021 *Meteorology and Hydrology* **6** 24-33.
[8] Isaev A A 1988 Statistics in meteorology and climatology (Moscow)
[9] Kanter CA 1965 *Climate and weather issues in the Lower Volga region***1** 29-39
[10] Kleshchenko L K and Aristova L N 2007 *Proceedings* of *VNIGMI-MCD* **173** 128-136.
[11] Kleshchenko L K and Rankova E Ya 2007 *Proceedings* of *VNIGMI-MCD* **173** 97-112.
[12] Kononova NK Classification of the circulation mechanisms of the Northern Hemisphere according to B.L. Dzerdzeevsky 2009 (Moscow)
[13] Malinin V N 2007 Statistical methods for the analysis of hydrometeorological information (Saint-Petersburg)
[14] Martynova Yu V and Krupchatnikov V N 2015. *Proceedings of the Russian Academy of Sciences. Physics of the atmosphere and ocean* **51** 346-357.
[15] Morozova S V 2014 *Izvestiya of the Saratov University* **14** 25-30.
[16] Morozova S V 2015 *Proceedings* of *VNIGMI-MCD* **180** 67-75.
[17] Morozova SV The role of planetary circulation objects in global climatic processes 2019 (Saratov)
[18] Neushkin AI, Sidorenkov NS, Sanina AT, Ivanova TB, Berezhnaya TV, Pankratenko NVand Makarova ME 2013 Monitoring of the general atmospheric circulation. North hemisphere. (Obninsk)
[19] Perevedentsev Yu P 2009 Climate theory (Kazan)
[20] Razorenova O A 2016 *Meteorology and Hydrology* **1**, 5-16.
[21] Razorenova O A and Shabanov P A 2020 *Turbulence, Atmosphere and Climate Dynamics*, 95.
[22] Khrabrov Yu B 1957 *Proceedings of the TsIP***63** 3-19
[23] Chichasov G N 2020 *Hydrometeorology and Education* **1**, 9-19.
[24] Intergovernmental Panel on Climate Change, Climate change 2013 (Cambridge)
[25] Morozova SV, Polyanskaya EA, Ivanova GF, Levitskaya NG, Denisov KE andMolchanova NP 2018 *IOP Conference Series: Earth and Environmental Science* **107**
[26] Morozova S V, Polyanskaya E A, Kononova N K, MolchanovaN P and SolodovnikovAP 2019 *IOP Conference Series: Earth and Environmental Science* **381**

IOP Conf. Series: Earth and Environmental Science 1040 (2022) 012022 doi:10.1088/1755-1315/1040/1/012022

Numerical modeling of methane hydrates dissociation in the submarine permafrost

V V Malakhova[1, 2]

[1]The Institute of Computational Mathematics and Mathematical Geophysics SB RAS, Novosibirsk, Russia
[2]Lomonosov Moscow State University, Physical Faculty, Moscow, Russia

malax@sscc.ru

Abstract. In this study, we evaluate the sensitivity of gas hydrate reservoirs associated with submarine permafrost conditions to changes in global climate. We apply numerical simulations to assess the timings of methane hydrate dissociation on the East Siberian Arctic Shelf after flooding the shelf with the seawater. The modeling combines a model of submarine permafrost evolution with a model of methane hydrate dissociation that accounts for the consumption of latent heat during hydrate dissociation. Based on the analysis of the performed experiments, we found that the endothermic reaction is a significant mechanism for slowing hydrate dissociation in frozen sediments. As a result, it additionally increases the lag of the subsea permafrost and hydrates stability zone response to glacial cycles.

1. Introduction

Permafrost is widespread in the Arctic region, both in the continental region and in the vast areas of the shallow shelf of the Arctic seas [1-3]. A large amount of methane can be found in permafrost in a hydrated form [4]. Gas hydrates are crystalline ice-like compounds that form from gas molecules caged in a lattice of water molecules [5]. Methane hydrates are prevalent in the bottom sediments of seas and oceans and exist under stability conditions, which include: the presence of a sufficient amount of methane and water, as well as the required range of temperatures and pressures (thermobaric conditions) [6, 7]. Hydrates are formed at low temperatures and generally high pressures which are typical of relatively shallow depths in oceanic sediments or deeper frozen sediments in the Arctic [8-11]. Temperature increase [12-17] or pressure drop (sea-level drop, glacier collapse) [18-20] is effective enough to dissociate hydrate deposits, as a result, methane is released from the hydrate structure and may enter the atmosphere.

In the sediments of the shallow-water Arctic shelves, in the presence of submarine permafrost, there are thermobaric conditions for the hydrate formation in the methane hydrate stability zone (MHSZ) [4, 21, 22]. It is commonly believed that they formed during the Pleistocene glaciations when sea level was below the current value by up to 130 m. This exposed the shelf to very cold temperatures [1, 23]. Cycles of hydrate formation and dissociation could take place during geological time. For example, such movements of the hydrate-bearing zone are reported in the MacKenzie delta region [24, 25], in Mount Elbert [26]. Insufficient research has been done on the stability and formation of

IOP Conf. Series: Earth and Environmental Science 1040 (2022) 012022 doi:10.1088/1755-1315/1040/1/012022

hydrates associated with relic submarine permafrost. Measurements in boreholes on the Alaska North Slope indicate that the conversion of common free gas accumulations to gas hydrate during climate cooling is the most likely mechanism for the formation of hydrate deposits in this region [9, 10].

To date, in the sediments of the shallow East Siberian Arctic Shelf, the MHSZ begins at depths of 200 - 250 m from the bottom according to the modeling results [1, 23]. With additional freezing of the sediment which occurred during ice ages, the conditions for the methane hydrates formation were met at depths of the order of several tens of meters [1, 23]. In the future, part of the gas hydrates could be retained due to the manifestation of the self-preservation effect at negative temperatures of the ground. The presence of methane hydrates in the upper layers of permafrost under non-equilibrium conditions is considered one of the possible reasons for constant gas manifestations during the operation of wells in gas fields. The thawing subsea permafrost and the destabilization of methane hydrates on the Arctic shelf could lead to emitting large quantities of methane. Significant point emissions have been detected along the East Siberian Arctic Shelf [27, 28].

In [21] it is shown that in the shallow shelf areas (lower 50 m water depth) after flooding (10-5 kyr ago), the hydrate presence conditions in the upper 100 - meter layer of the MHSZ are violated. This suggests that the methane coming from this reservoir is concentrated in the bottom sediments of the shelf. However, in [21] phase transitions of methane hydrates are not considered in the model. This approach neglects the endothermic reaction of hydrate dissociation, which is an important dynamic process that decreases the rate of dissociation [29]. The latent heat of fusion for methane hydrates is 28% larger than for pure water [30]. Based on analysis [29], the endothermic reaction is the dominant mechanism-slowing dissociation in sediments.

We study the permafrost and methane hydrate stability zone evolution in the Laptev Sea region using a numerical model to illustrate how changes in the environment affected hydrate stability. The purpose of the study is to analyze the state of a gas hydrate reservoir formed in shallow permafrost sediment and subsequently found to be in non-equilibrium conditions. We use an approach in which the conditions of stability of methane hydrate are determined, and, in the event of their violation, the degradation of the gas hydrate reservoir is modeled taking into account heat consumption.

2. Model Setup

To study the formation processes of the submarine permafrost and associated methane hydrates, we use a model of the dynamics of permafrost for the last 400 thousand years [23]. The model solves the one–dimensional equation for heat and salt diffusion in the sediment column taking into account phase changes [31]. The distribution of thermal conductivity and heat capacity of the frozen and thawed rocks with depth remains constant during the simulation of the permafrost evolution. Thermal conductivity and heat capacity correspond to the values in the region sedimentary basin [19].

The upper boundary condition is the temporally varying sediment surface temperature. The model of thermophysical processes in bottom sediments is supplemented by the scenario of a climate change on the East Siberian Arctic Shelf for the last 400 thousand years [23]. We used the output of the simulation with an Earth system model of intermediate complexity model CLIMBER-2 forced by time-varying orbital parameters and atmospheric CO_2 content [32]. Sea level is directly simulated by this model. The data for a grid cell, which corresponding to the Laptev Sea shelf is selected. When a shelf is exposed to the atmosphere, this temperature is calculated as a sum of the present-day annual mean surface ground temperature at the nearshore and is a time-varying anomaly by CLIMBER-2. It is assumed that after the flooding of the shelf by the sea, the seawater temperature does not significantly differ from that of today, corresponds to the average annual values for the shelf seas. We assume a seafloor temperature −1.2 °C after the flooding of the shelf [2]. The changes in average monthly surface temperature for the last 400 thousand years were assumed as the upper boundary conditions, figure 1A.

The lower boundary condition is a constant heat flow $60 mWm^2$ which is estimated as typical at the Eurasian shelf out of the rift zones [33]. It is believed that the vertical component of the model domain is large enough so that submarine permafrost does not form here throughout the entire estimated time and it extends to 1500 m.

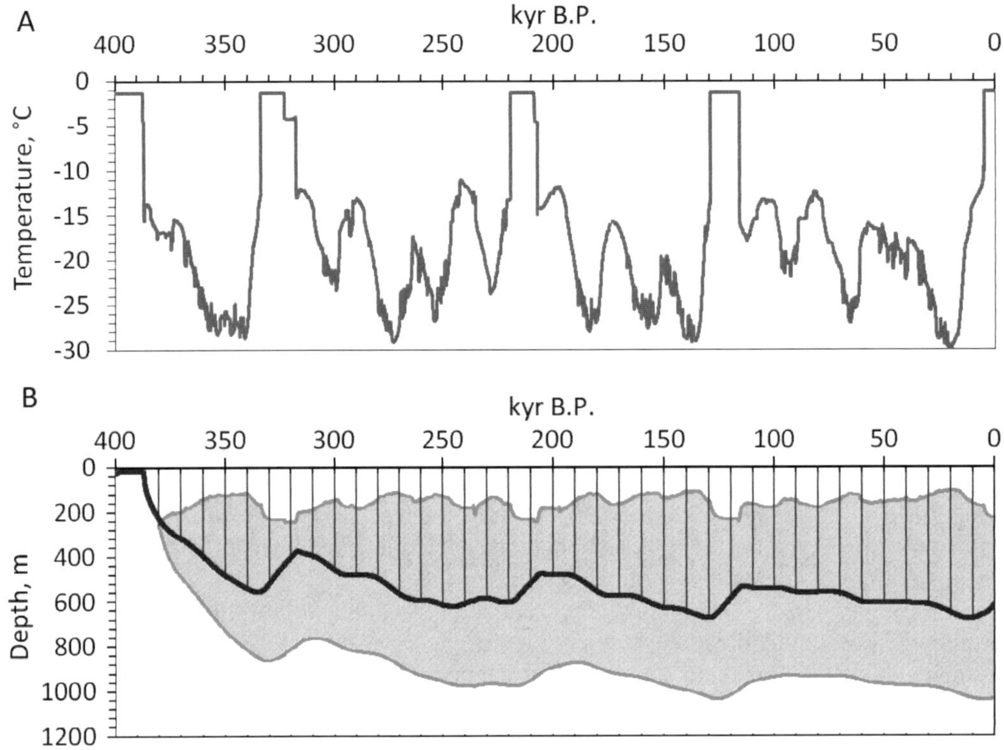

Figure 1. (A) Temperature at the upper boundary of the sea sediments, which are used to force our model for the contemporary shelf depth 10 m. (B) Submarine permafrost evolution (shaded area based) and gas hydrates stability zone (marked in blue) variations in the last 400 ka according to the sediment surface temperature history.

We have combined two numerical models: a model of permafrost dynamics based on the paleoclimatic scenario of changes in temperature and ocean level, and a model of the methane hydrates stability zone. The MHSZ thickness was calculated simultaneously with the sediment temperature as a function of time. The dissociation pressure and temperature of hydrates at depth are obtained from the hydrate model [34].

$$ln(P_H) = \sum_{n=0}^{5} a_n \left(T + T_D\right)^n \quad (1)$$

where P_H is the phase equilibrium pressure (MPa), T is the sediment temperature (K), T_D is the salt-induced temperature depression (K) and a_n are empirical constants for the unfrozen and frozen layers. This shift in the equilibrium temperature can be solved as follows [35]: $T_D = 0.6825T_F$. This temperature depression depends on the freezing pore water temperature (T_F in °C) and is incorporated into the pressure-temperature equilibrium relationship for the methane–hydrate system (1). Methane is assumed to be the only gas component in the system so that all values apply to pure methane hydrate.

The works [20, 21, 29] assume that the subsurface pressure regime is hydrostatic at any time. This condition is not evident in permafrost when sediment freeze during ice ages. Most pores are filled with ice, likely generating pore pressures above the hydrostatic [36]. The configuration of MHSZ strongly depends on the ground temperature and pressure. In this case, the depth of the base of the MHSZ would be deeper and the depth of the top would be less than in the hydrostatic case [21]. Based on these considerations, we assumed the phase curves of methane hydrate under the corresponding lithostatic ground pressure [22]. The ground pressure was calculated as hydrostatic for the thawed layer and as geostatic for the frozen. Variations because of sea-level changes are supposed to propagate instantaneously throughout the sediment column.

Gas hydrate reservoirs of the Arctic shelves are believed originally to be natural gas accumulations converted to hydrate after being placed in the gas hydrate stability zone in response to ancient climate cooling [37]. In numerical experiments, we assume that a gas column accumulates and subsequently is converted to hydrate in sub-permafrost formations in the upper part of the MHSZ.

One of the most important parameters in modeling the dissociation of gas hydrates is the heat of hydrate formation (or enthalpy of dissociation) [6]. It is equal to the amount of heat that is needed to decompose the hydrate into water and gas. The experimentally obtained values for the heat of phase transition of hydrates are in the range of 430–540 kJ/kg [38]. During hydrate formation, heat is released and the enthalpy is positive; during hydrate decomposition, it is negative. This process is considered in the model using the approach [20].

If the conditions for the stability of the hydrate are violated, the calculated temperature of the sediment at the point at which methane hydrate dissociation occurs is corrected by taking into account the latent heat of dissociation of gas hydrate. At the depth where hydrate exists (hydrate saturation $S_H > 0$), the temperature is corrected to T_{DIS} (temperature of the onset of dissociation). At depth where the phase transition occurs, the amount of heat (dQ) per time step equal to:

$$dQ = C \cdot dT$$

dQ is then added to the heat that has accumulated (Q_A) over previous time steps since the start of the phase transition. The equation applies to a unit mass of hydrate, with is the temperature difference between the initially calculated temperature and the temperature after correction for the presence of melting hydrates, C - denoting the isobaric heat capacity of this phase. The hydrate dissociates completely if the accumulated heat exceeds the latent heat of the gas hydrate L_H. The response of the hydrate saturation to the accumulation of heat upon melting is approximated as follows:

$$S_H(t) = S_H(t-1) \cdot \frac{L_H - Q_A}{L_H} \quad (2)$$

where $S_H(t)$ represents the hydrate saturation at time t, the latent heat of melting gas hydrate L_H is 430 kJ/kg [20]. The hydrate saturation decrease as heat accumulates and eventually turns zero once the accumulated heat equals L_H.

3. Results

Numerical experiments were performed for the Laptev Sea shelf with a water depth of 10 m. The following numerical experiments were carried out:

- SH0 - The methane hydrate stability simulation was performed without hydrates in the sediment [21], $S_H = 0\%$.
- SH20 – We assume that the formation of a gas hydrate reservoir in the upper part of the MHSZ occurs during the period of maximum freezing of sediments in the last glacial (20 kyrs ago). Initially, $S_H = 20\%$. The decomposition of the hydrate occurs with the consumption of heat.
- SH40 - Similar to SH20, but initially period $S_H = 40\%$.

3.1 HS0

The model results illustrate the evolution of the permafrost and MHSZ over the past 400 thousand years (kyrs), figure 1. For the shallow shelf, the periods of ocean flooding are relatively short and the permafrost does not disappear in the interglacial period. During oceanic regressions, the depth of the lower permafrost boundary, depending on the glacial cycle, ranges from 540 to 680 m, figure 1B. Methane hydrate can occur within sediment columns across a depth interval of the hydrate stability zone. The subsurface temperature and pressure regime control the stability of the MHSZ. Modeling of MHSZ response to the changing seafloor temperature and the increasing sea level on the Laptev Sea shelf is shown in figure 1B. When the permafrost thickness reaches about 250 m, the formation of the

MHSZ begins. Like the permafrost, the MHSZ does not disappear during the periods of marine transgressions, and its thickness changes insignificantly.

The key obtained result is the depth of the upper boundary of the gas hydrate as it varies with time. Figure 2 shows the stability curve of the equilibrium phase methane hydrate as a function of water depth and its intersection with model temperature profiles obtained for different times. The modeling MHSZ top in the last glacial, 20 kyrs ago, is 103 m below seafloor (figure 2, point 1). The model predicts that the methane hydrates in the upper 60 m of MHSZ dissociated completely within ten millennia after the last glacial (figure 2, point 2). The Holocene sea level rise is modeled with a 10-meter sharp rise that occurred 5 kyrs ago (figure 2, point 2). In the initial period of the transgression (with a corresponding increase in pressure on the bottom), the MHSZ increases over several centuries (figure 2, figure 3A). Subsequently, an increase in temperature in bottom sediments led to a reduction in the zone of stability of gas hydrates and shifted the MHSZ top boundary. The effects of seabed warming are more important and lasting for the shallow shelf than the pressure effects accompanying sea level rise. The current model MHSZ top for water depth 10 m is at 229 m below seafloor (figure 2, point 4).

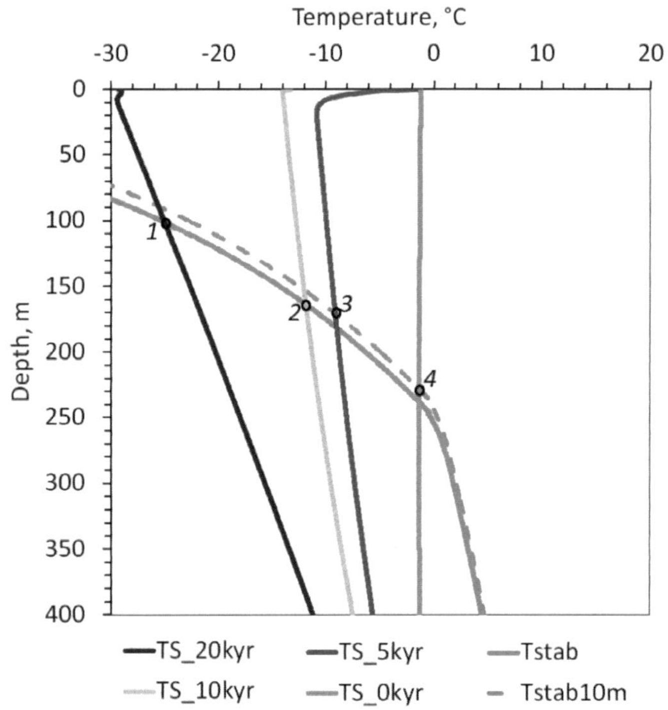

Figure 2. Sediment temperature profiles obtained in numerical experiment HS0 corresponding periods: 20 kyr ago (dark blue line TS_20kyr), 10 kyr ago (TS_10kyr), 5 kyr ago (TS_5kyr), and industrial period (TS_0kyr). Methane hydrate P-T stability (Tstab) in subaerial conditions. P-T stability of methane hydrate (Tstab10m) assuming a water depth equal to 10 m after the flooding of the shelf.
The depth migration of the top of the MHSZ responding to temperature variations is marked by points 1, 2, 3, 4.

3.2 HS20

In SH20, the dissociation of gas hydrate begins when the temperature and pressure conditions are violated and occurs with heat absorption (figure 3). Within the hydrate stability zone, the hydrate saturation in matrix porosity was inferred to be 20%. A change in the hydrate saturation (2) over the calculation period of 20 thousand years is shown in figure 3A. Dissociation takes place at the upper boundary of the deposit, which gradually deepens. Hydrate dissociation slows down under the conditions of permafrost transition to subsea conditions (5ka ago). A methane-hydrate reservoir with a saturation of 20% in a 1 m-thick layer decomposes in about 100 years, figure 3B. Dissociation rates for the HS20 model run are significantly slower than the HS0 approach. Comparing the HS0 and HS20 simulations, the time for complete dissociation is increased by an average of 20% (figure 3A).

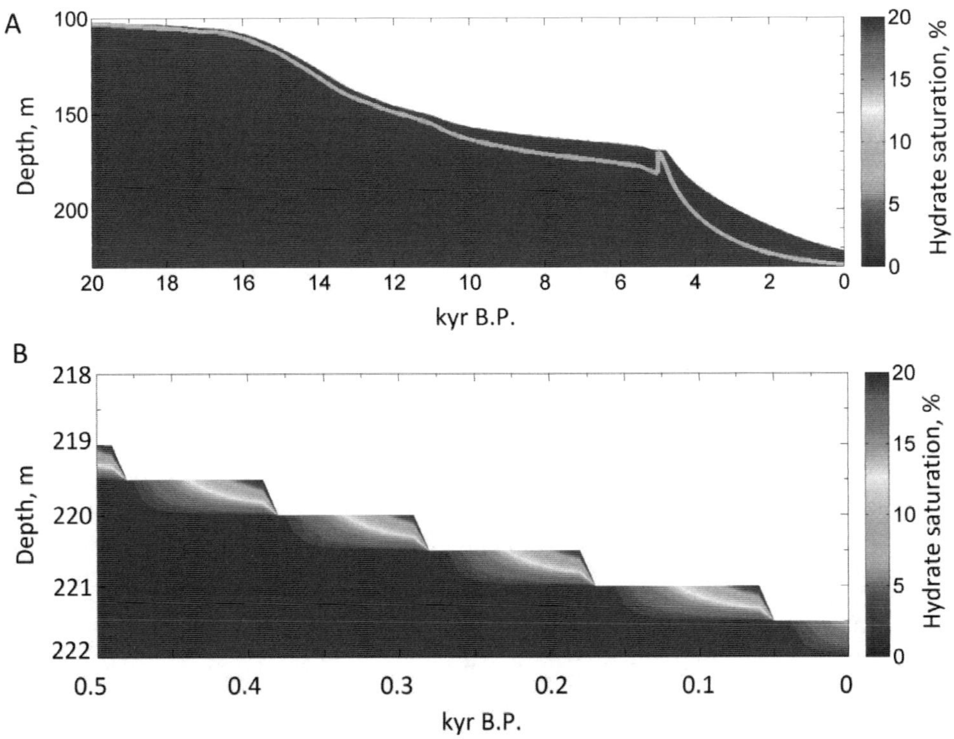

Figure 3. Time variation of the hydrate saturation of bottom sediments during the decomposition of hydrate deposits obtained in HS20. (A) Results for a period of 20 thousand years. The blue line is the depth of the upper boundary of the MHSZ, corresponding to the HS0 experiment. (B) Results for the last 500 years.

There are differences between HS0 and HS20 in terms of both dissociation rate and sediment temperature. The difference arises from the endothermic dissociation reaction. Sediments near the hydrate degradation front are cooled. Estimates of present-day sediment temperature differ substantially between the HS0 approach and HS20 simulation (figure 4). With the transition of permafrost grounds to the submarine state, the sediment temperature rises and becomes higher than minus 2 °C. Under the influence of hydrate decomposition, this process slows down. And near the front of the hydrate decomposition lower temperatures of -3…-4 °C are preserved. The heat of decomposition of the hydrate in HS20 is assumed to be 430 kJ / kg, higher than the heat of melting of ice, which is 320 kJ / kg. The absorption of heat from the formation during the destruction of the hydrate reservoir reduces the heat flux from the lower layers. Such thermal effects can affect the thickness of the permafrost layer, as well as lead to the formation of new permafrost during the destabilization of gas hydrates. Similar areas of submarine permafrost were found on the outer shelf of the Kara Sea in sediment [39].

The evolution of permafrost and MHSZ under the impact of climate change and transgression–regression cycles on the shelf is characterized by a time lag [1, 23]. Malakhova and Eliseev (2020) estimated apparent response time scales via time lags between the temperature at the upper boundary of the sediments, and MHSZ thickness [23]. Apparent response time scales for MHSZ thickness are relatively small for the shallow shelf, from 4 to 5.4 kyr. However, these works used an approach similar to HS0, without implementation of phase transitions of methane hydrates in the model. For the last 5 kyr, the additional lag can be from 250 to 2100 years, depending on time interval.

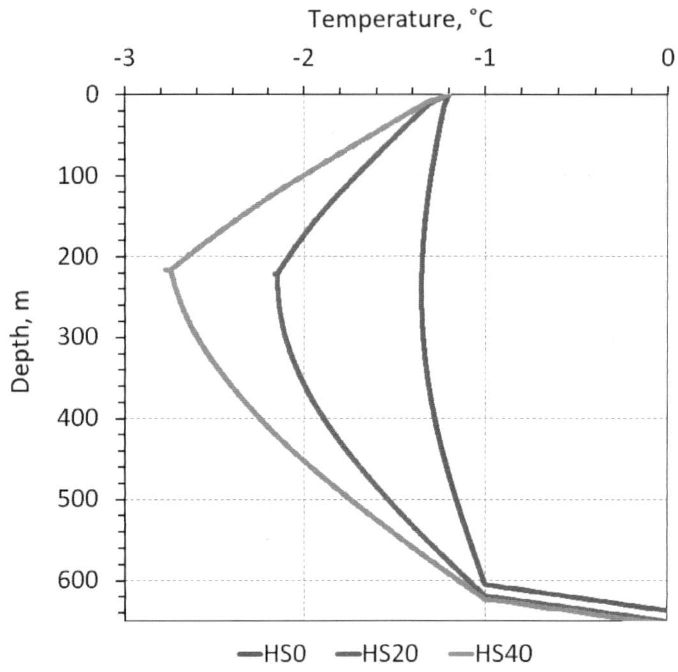

Figure 4. Sediment temperature profiles obtained in numerical experiments HS0 (blue line), HS20 (red line), and HS40 (green line) for the modern climatic conditions

3.3 HS40

The hydrate saturation used in the simulation is entirely conjectural, given the general lack of high-quality seismic lines and the absence of scientific boreholes. The maximum saturation found in part of the Mallik well section [24] is 60%. To illustrate how the modeling is sensitive to hydrate saturation, an alternative level, 40%, was considered. In SH40, within the hydrate stability zone, the hydrate saturation in matrix porosity was inferred to be 40%. MHSZ top boundary migrates downward faster when the saturation is lower (HS20) due to the smaller latent heat released during less saturated hydrate formation. The present-day modeled MHSZ top is at 216 m below seafloor, 6 m higher than in the model with lower saturation (HS20). Due to the assumed 40% hydrate saturation of the pore space during the hydrate decomposition, the sediment temperature decreases, and as it cools, permafrost may form (figure 4). For the last 5 kyr, the additional lag increases by 30-40% and can reach 2900 years.

4. Conclusion

We present an assessment of changes in the gas hydrates stability zone of the Arctic Ocean associated with subsea permafrost conditions. The calculations have been carried out for the interval ranging from 400 kyr ago up to the present. In this study, we evaluate the sensitivity of gas hydrate reservoirs associated with submarine permafrost conditions to changes in global climate during successive glacial-interglacial cycles.

Based on the analysis of the performed experiments, we found that the endothermic reaction is a significant mechanism for slowing hydrate dissociation in frozen sediments. Hydrate decomposition is an endothermic process in which heat is absorbed and the nearby sedimentary layer is cooled. The degradation of hydrates in bottom sediments occurs due to the accumulated heat in the sediments since the heat flux from the surrounding rocks is insignificant. As a result, heat sink occurs, which slows down the propagation of the temperature impulse through the sedimentary column, and reduces the rate of melting of permafrost and methane hydrate deposits in the bottom sediments of the Arctic shelf. The process of degradation of gas hydrate deposits leads to heat absorption, thereby causing the development of frozen rocks.

We found that implementation of phase transitions of methane hydrates in the model delays the subsea permafrost thaw and the MHSZ disappearance during interglacials. As a result, it additionally increases the lag of the subsea permafrost and MHSZ response to glacial cycles. Methane hydrate dissociation occurs even more slowly because the reaction is not only endothermic, it also decreases pore water salinity, and increases the local pore pressure. These effects were not considered in our study. It supports the view that the current changes in the thermal state of the subsea permafrost may be related to the adjustments to the last glacial cycle termination rather than to the contemporary, century-scale climate warming. The process of methane hydrates degradation in the subsea permafrost of the shallow Arctic shelf, which began after the Last Glacial Maximum, is one of the possible reasons for increased methane fluxes in this region [27, 28].

Acknowledgments
An analysis of changes in the hydrate and permafrost characteristics was obtained with the support of the Russian Science Foundation (grant 21-17-00012). A new version of the permafrost model was obtained with the support of the Russian Foundation for Basic Research (grant 20-05-00241).

References
[1] Romanovskii N N, Hubberten H W, Gavrilov A V, Eliseeva A A and Tipenko G S 2005 *Geo-Mar. Lett.* **25** 167–82
[2] Golubeva E, Platov G, Malakhova V, Kraineva M and Iakshina D 2018 *Polarforschung* **87** 195-210
[3] Overduin P, Schneider von Deimling T, Miesner F, Grigoriev M, Ruppel C, Vasiliev A, Lantuit H, Juhls B and Westermann S 2019 *J. Geophys. Res.: Oceans* **124** 3490–507
[4] Ruppel C 2015 *Journal of Chemical & Engineering Data* **60** 429-36
[5] Kvenvolden K A 1993 *Reviews of Geophysics* **31** 173-87
[6] Istomin V A and Yakushev V S 1992 *Gas hydrates in natural conditions* (Moscow: Nedra) p 235 (in Russian)
[7] Sloan E D and Koh C A 2008 *Clathrate Hydrate of Natural Gases*, 3rd Edition (CRC Press: Boca Raton, FL, USA) p 721
[8] Ruppel C D and Kessler J D 2017 *Reviews of Geophysics* **55** 126-68
[9] Boswell R, Rose K, Collett T S, Lee M, Winters W, Lewis K A and Agena W 2011 *Mar. Pet. Geol.* **28** 589–607
[10] Collett T S, Lee M W, Agena W F, Miller J J, Lewis K A, Zyrianova M V, Boswell R and Inks T L 2011 *Mar. Pet. Geol.* **28** 279– 94
[11] Koh C A, Sum A K and Sloan E D 2012 *Journal of Natural Gas Science and Engineering* **8** 132-8
[12] Hunter S, Goldobin D, Haywood A, Ridgwell A and Rees J 2013 *Earth and Planetary Science Letters* **367** 105–15
[13] Malakhova V V, Golubeva E N, Eliseev A V and Platov G A 2018 *IOP Conference Series: Earth and Environmental Science* **211** 012017
[14] Reagan M T, Moridis G J, Elliott S M and Maltrud M 2011 *Journal of Geophysical Research: Oceans* **116** C9
[15] Biastoch A, Treude T and Rupke L H 2011 *Geophysical Research Letters* **38** L08602
[16] Kretschmer K, Biastoch A, Rupke L and Burwicz E 2015 *Glob. Biogeochem Cycles* **29** 610–25
[17] Marín-Moreno H, Minshull T, Westbrook G and Sinha B 2015 *Geochem. Geophys. Geosyst.* **16** 1307–23
[18] Majorowicz J, Osadetz K and Safanda J 2013 *Energies* **6** 5775–806
[19] Gavrilov A, Malakhova V, Pizhankova E and Popova A 2020 *Geosciences* **10** 484
[20] Mestdagh T, Poort J and De Batist M 2017 *Earth-Science Reviews* **169** 104-31
[21] Malakhova V V 2020 *IOP Conference Series: Earth and Environmental Science* **606** 012035
[22] Tinivella U, Giustiniani M and Marin-Moreno H A 2019 *Geosciences* **9** 329
[23] Malakhova V V and Eliseev A V 2020 *Glob. Planet. Change* **192** 103249

[24] Majorowicz J A, Osadetz K and Šafanda J 2012 *J. Geol. Res.* **2012** 879393
[25] Torres M E, Collett T S, Rose K, Sample J C, Agena W F and Rosenbaum E J 2011 *Mar. Pet. Geol.* **28** 332–42
[26] Dai S, Lee C and Santamarina J C 2011 *Mar. Pet. Geol.* **28** 427–38
[27] Shakhova N, Semiletov I, Leifer I, Sergienko V, Salyuk A, Kosmach D, Chernykh D, Stubbs C, Nicolsky D, Tumskoy V and Gustafsson O 2014 *Nature Geoscience* **7** 64–70
[28] Thornton B F, Geibel M C, Crill P M, Humborg C and Mörth C M 2016 *Geophys. Res. Lett.* **43** 5869–77
[29] Stranne C, O'Regan M and Jakobsson M 2016 *Geophys. Res. Lett.* **43** 8703– 12
[30] Rueff R, Dendy Sloan E and Yesavage V 1988 *AICHE J.* **34** 1468–576
[31] Ganopolski A and Calov R 2011 *Clim. Past* **7** 1415–25
[32] Malakhova V V 2018 *Proc. SPIE: 24th International Symposium on Atmospheric and Ocean Optics: Atmospheric Physics* **10833** 108337T
[33] Davies J H 2013 *Geochem. Geophyst. Geosyst.* **14** 4608-22
[34] Moridis G J 2003 *Society of Petroleum Engineers Journal* **32** 359-70
[35] Najibi H, Chapoy A, Haghighi H and Tohidi B 2009 *Fluid Phase Equilibria* **275** 127–31
[36] Betlem P, Senger K and Hodson A 2019 *Mar. Pet. Geol.* **100** 246–62
[37] Behseresht J and Bryant S L 2012 *Earth and Planetary Science Letters* **341** 114-27
[38] Groysman A G 1985 *Thermal Properties of Gas Hydrates* (Novosibirsk, Russia: Nauka) p 94 (In Russian)
[39] Portnov A, Mienert J and Serov P 2014 *J. Geophys. Res.: Biogeosciences* **119** 2082-94

IOP Conf. Series: Earth and Environmental Science 1040 (2022) 012023 doi:10.1088/1755-1315/1040/1/012023

To the theory of convective flows in a rotating stratified medium over a thermally inhomogeneous surface

L Kh Ingel [1,2] **and A A Makosko** [2]

[1] Research and Production Association Typhoon, Obninsk, Russia
[2] Obukhov Institute of Atmospheric Physics of the RAS, Moscow, Russia

e-mail: lev.ingel@gmail.com

Abstract. In a liquid (gaseous) medium above a thermally inhomogeneous horizontal surface in a gravity field, convective flows arise, which can play an important role, in particular, in the dynamics of the atmosphere. An extensive literature is devoted to their theory, but, due to the complexity of the problem and the variety of possible combinations of parameters, not all situations of interest have been sufficiently studied even in the linear approximation. In this work, a theoretical model of such thermal circulations is considered, which is more general than in a number of previous works. An analytical solution is found within the framework of a linear stationary two-dimensional model of convective flows in a semi-infinite stably stratified medium rotating around a vertical axis. The obtained relations allow analyzing the dependences of the components of velocity and helicity on the parameters of the problem. A number of rather general statements about the ratio of different helicity "components" in the considered thermal circulations, in particular, in atmospheric currents with characteristic horizontal scales of the order of hundreds of kilometres, have been proved. Examples of numerical calculations of the vertical distribution of these components are given.

1. Introduction

In a liquid (gaseous) medium above a thermally inhomogeneous horizontal surface in a gravity field, convective flows arise, which can play an important role, in particular, in the dynamics of the atmosphere. An extensive literature is devoted to their theory (see, for example, [1-6] and bibliography in these articles). Much useful information can be obtained already from the analysis of linear problems that make sense for relatively small amplitudes of thermal inhomogeneities (see, for example, [1, 3-6]). In these articles and in a number of other publications, there are developed the linear models of mesoscale flows in a stably stratified rotating medium over an inhomogeneously heated surface – mesoscale atmospheric flows. However, due to the complexity of the problem and the variety of possible combinations of parameters, not all situations of interest have been studied sufficiently. For example, in our recent paper [6], the Ekman-type boundary layer arising at the lower boundary is assumed to be rather thin as compared to other characteristic vertical scales of the problem. In problems of atmospheric dynamics, this condition is not always satisfied with confidence. In this paper, we consider a more general model free from this limitation.

In addition, as far as we know, the helicity of this type of convective flows has been little studied so far. The importance of this characteristic, in particular, for atmospheric currents, has long been discussed in the literature (see, for example, [7-13]). The present pater contains some relevant results.

2. Statement of the problem of thermal circulations over an inhomogeneously heated surface

Here we consider a semi-bounded volume of a stratified medium $z \geq 0$ rotating around an axis z directed vertically upward. By using the Boussinesq approximation, we assume that the density of the medium depends linearly on temperature perturbations (in problems of atmospheric physics, it is convenient to use the deviations of the potential temperature as the corresponding variable [14, 15]) T:

$$\rho = \rho_0 (1 - \alpha T). \tag{1}$$

Here ρ is the density, ρ_0 is the background density value, α is the thermal coefficient if expansion.

The linearized system of equations of dynamics and heat transfer in a rotating coordinate system for a stationary two-dimensional problem has the form [3-6]

$$0 = -\frac{\partial P}{\partial x} + \nu \Delta u + f\mathrm{v}, \qquad 0 = \nu \Delta \mathrm{v} - fu, \qquad 0 = -\frac{\partial P}{\partial z} + \nu \Delta w + \alpha g T, \tag{2}$$

$$\frac{\partial u}{\partial x} + \frac{\partial w}{\partial z} = 0, \qquad \gamma w = \kappa \Delta T. \tag{3}$$

Here u, v, w are velocity vector components of $\mathbf{v}$ along horizontal axes x, y and vertical axis z respectively; $P = p/\rho_0$, p is the pressure disturbance; g is the acceleration of gravity; ν, κ are the exchange coefficients; $\Delta = \partial^2/\partial x^2 + \partial^2/\partial z^2$ is the two-dimensional Laplace operator; f is the Coriolis parameter (the doubled angular velocity of the background rotation). Temperature stratification is assumed to be stable, so that the background potential temperature gradient $\gamma > 0$. Note the coincidence (up to coefficients) of the equations for the vorticity $\partial \mathrm{v}/\partial x$ and the vertical temperature gradient $\partial T/\partial z$, which can be seen easily if we differentiate the second equations (2) and (3) by x and z, respectively, and take into account the continuity equation. Therefore, the vorticity of convective currents arising in the field of Coriolis accelerations is quite simply related to horizontal inhomogeneities of the vertical heat flux.

The stationary two-dimensional horizontal periodic thermal inhomogeneities, as well as non-slip and impermeability conditions are specified at the lower boundary:

$$c\rho_0 \kappa \frac{\partial T}{\partial z} = -Q \cos kx, \quad w = u = \mathrm{v} = 0 \qquad \text{at} \quad z = 0. \tag{4}$$

Here c is heat capacity of medium, and the sense of parameters Q and k is obvious. It is assumed that far from the surface (at $z \to \infty$) all disturbances damp.

3. The system of equations for the amplitudes and its general solution

The problem posed above in a number of respects is close to those considered earlier in the papers [3, 4, 6]. Below, a solution will be found that is free from some of the previously adopted restrictions. We are looking for horizontally periodic solutions in the form

$$u(x,z) = U(z)\sin kx, \qquad \mathrm{v}(x,z) = V(z)\sin kx, \qquad w(x,z) = W(z)\cos kx,$$
$$P(x,z) = \Phi(z)\cos kx, \qquad T(x,z) = \theta(z)\cos kx. \tag{5}$$

The system of equations for the amplitudes has the form

$$-k\Phi = \nu\left(\frac{d^2U}{dz^2} - k^2 U\right) + fV, \qquad 0 = \nu\left(\frac{d^2V}{dz^2} - k^2 V\right) - fU, \tag{6}$$

$$\frac{d\Phi}{dz} = \nu\left(\frac{d^2W}{dz^2} - k^2 W\right) + \alpha g \theta, \qquad kU + \frac{dW}{dz} = 0, \tag{7}$$

$$\gamma W = \kappa\left(\frac{d^2\theta}{dz^2} - k^2\theta\right). \tag{8}$$

By eliminating all unknowns from the last system, except W, we obtain the equation

$$\left(\frac{d^2}{dZ^2}-1\right)^3 W=-\mathrm{Ta}\frac{d^2W}{dZ^2}+\mathrm{R}W, \tag{9}$$

$$\mathrm{R}=\frac{N^2}{\kappa\nu k^4},\qquad \mathrm{Ta}=\frac{f^2}{\nu^2k^4}=\left(\frac{4}{kh_E}\right)^4,\qquad N=(\alpha g\gamma_T)^{1/2},\qquad h_E=\left(\frac{2\nu}{f}\right)^{1/2}.$$

Here, the buoyancy frequency N and the dimensionless variable $Z=kz$ are introduced; the dimensionless parameters $\mathrm{R}>0$, $\mathrm{Ta}>0$ are some analogs of the Rayleigh and Taylor numbers.

We seek a solution to equation (9) in the standard way in the form of a linear combination of exponentials of the type $\exp(\sigma Z)$. The characteristic equation has the form

$$\left(\sigma^2-1\right)^3=-\mathrm{Ta}\sigma^2+\mathrm{R}. \tag{10}$$

Taking into account the damping condition of disturbances at $z\to\infty$, the solution for the vertical velocity is a linear combination of three exponentials

$$w(x,z)=\sum_{j=1}^{3}C_j\exp(k\sigma_j z)\cos kx, \tag{11}$$

where the roots σ_j with negative real parts are selected (here it is assumed that these roots are different); the integration constants C_j are determined from the boundary conditions. From the equation of continuity

$$u(x,z)=-\sum_{j=1}^{3}C_j\sigma_j\exp(k\sigma_j z)\sin kx. \tag{12}$$

The expressions for the perturbations of temperature, pressure, and the vortex component of the horizontal velocity, as it is easy to see, can also contain terms with the fourth exponent $\exp(-kz)$:

$$T(x,z)=\left[C_4\exp(-kz)+\frac{\gamma}{\kappa k^2}\sum_{j=1}^{3}\frac{C_j}{\sigma_j^2-1}\exp(k\sigma_j z)\right]\cos kx, \tag{13}$$

$$\mathrm{v}(x,z)=\left[C_5\exp(-kz)-\frac{f}{\nu k^2}\sum_{j=1}^{3}\frac{C_j\sigma_j}{\sigma_j^2-1}\exp(k\sigma_j z)\right]\sin kx. \tag{14}$$

By differentiating the first equation (2) with respect to z, and the last with respect to x, it is easy to verify the relation

$$C_5=\frac{\alpha g}{f}C_4. \tag{15}$$

4. The simplifications based on a scale analysis

In the general case, the solution is rather cumbersome. But it is useful to bear in mind that the values of parameters R, Ta in the atmosphere are usually quite large, so it makes sense to analyze some relatively simple limiting cases. For example, at $N=10^{-2}$ s^{-1}, $f=10^{-4}$ s^{-1}, $\kappa=\nu=1$ m^2/s (effective coefficients of turbulent exchange), $k=10^{-5}$ m^{-1} (that corresponds to a horizontal half-wave length of about 300 km) $\mathrm{R}\sim10^{16}$, $\mathrm{Ta}\sim10^{12}$. If everything else being equal, $\kappa=\nu=10$ m^2/s, then $\mathrm{R}\sim10^{14}$, $\mathrm{Ta}\sim10^{10}$. The roots of the characteristic equation σ_j in such situations are large in absolute value compared to unity. In [3, 6], there were considered the values of the parameters, for which

$$1 << \mathrm{R}^{2/3} << \mathrm{Ta} << \mathrm{R}. \tag{16}$$

All of these inequalities hold well for the above estimate at $\kappa = \nu = 1$ m²/s. But the intensity of turbulent exchange in the atmosphere can be very different, and often a more adequate estimate is $\kappa = \nu = 10$ m²/s. In this case, the second of inequalities (16) becomes softer, and some of the assumptions made in [3, 6] are not fulfilled strictly. In this paper, we consider the case of the softer inequality $\mathrm{R}^{2/3} < \mathrm{Ta}$ (but $\mathrm{R}^2 << \mathrm{Ta}^3$). In this case, the assumptions about the smallness of some dimensionless parameters cease to be fulfilled, so that the model becomes more complicated. The approximate expressions for three roots of the characteristic equation appearing in (11) — (14) in this case coincide with [3, 6]:

$$\sigma_1 \approx -b, \qquad \sigma_{2,3} \approx -(1 \pm i)a, \qquad a \equiv \left(\frac{\mathrm{Ta}}{4}\right)^{1/4} \qquad b \equiv \left(\frac{\mathrm{R}}{\mathrm{Ta}}\right)^{1/2} = \left(\frac{\nu}{\kappa}\right)^{1/2}\frac{N}{f}, \tag{17}$$

at that

$$a > b; \qquad |\sigma_{2,3}| = \mathrm{Ta}^{1/4} > |\sigma_1| >> 1 \tag{18}$$

(in [3, 6], the first inequality was stronger).

5. The approximate solution

Taking into account the lower boundary conditions, for the coefficients at exponentials, the system of equations is obtained:

$$\sum_{j=1}^{3} C_j = 0, \qquad \sum_{j=1}^{3} \sigma_j C_j = 0, \tag{19}$$

$$C_4 \approx \frac{f^2}{\alpha g \nu k^2}\sum_{j=1}^{3}\frac{C_j}{\sigma_j}, \qquad C_4 - \frac{\gamma}{\kappa k^2}\sum_{j=1}^{3}\frac{C_j}{\sigma_j} \approx \frac{Q}{c\rho_0 \kappa k}. \tag{20}$$

By eliminating C_4 from (20), we obtain

$$\left(1 - \frac{\kappa}{\nu}\frac{f^2}{N^2}\right)\sum_{j=1}^{3}\frac{C_j}{\sigma_j} \approx -\frac{kQ}{c\rho_0\gamma}.$$

For the considered characteristic values of the parameters, the second term in parentheses is much less than unity; therefore, together with two equations (19), we consider

$$\sum_{j=1}^{3}\frac{C_j}{\sigma_j} \approx -\upsilon, \qquad \upsilon \equiv \frac{kQ}{c\rho_0\gamma}. \tag{21}$$

The approximate solution of the system (19), (21) has the form

$$C_1 \approx \frac{b\upsilon}{B}, \qquad C_{2,3} \approx -\frac{b\upsilon}{2B}[1 \pm i(1-\delta)], \qquad C_4 \approx -\frac{f^2}{N^2}\frac{Q}{c\rho_0 \nu k}. \tag{22}$$

Here the dimensionless parameters are introduced

$$\delta = \frac{b}{a} = \left(\frac{4\mathrm{R}^2}{\mathrm{Ta}^3}\right)^{1/4}, \qquad B = 1 - \delta + \frac{\delta^2}{2}.$$

In [3, 6] the first of them was assumed to be a small parameter, that made it possible to significantly simplify the calculations. As it is easy to see, the smallness of this parameter means that $B \approx 1$, and the Ekman scale of the height

$$h_E = \frac{1}{k}\left(\frac{4}{\mathrm{Ta}}\right)^{1/4} = \left(\frac{2\nu}{f}\right)^{1/2}$$

is much less than another significant vertical scale

$$h_b = (k\sigma_1)^{-1} = \frac{1}{k}\left(\frac{\mathrm{Ta}}{\mathrm{R}}\right)^{1/2} = \left(\frac{\kappa}{\nu}\right)^{1/2}\frac{f}{kN}.$$

The present model is free from this simplification – the scales indicated can be comparable. For the sought variables, we get the expressions:

$$u \approx \frac{b^2\upsilon}{B}\left\{\exp(-bkz) - \exp(-akz)\left[\cos(akz) + \frac{2-\delta}{\delta}\sin(akz)\right]\right\}\sin kx,$$

$$\mathrm{v} \approx -\frac{f\upsilon}{\nu k^2}\left\{\exp(-kz) - \frac{1}{B}\exp(-bkz) + \frac{\delta}{2B}\exp(-akz)[(2-\delta)\cos(akz) - \delta\cos(akz)]\right\}\sin kx,$$

$$w \approx \frac{b\upsilon}{B}\{\exp(-bkz) - \exp(-akz)[\cos(akz) + (1-\delta)\sin(akz)]\}\cos kx, \tag{23}$$

$$T \approx \frac{fQ}{B(\nu\kappa)^{1/2}c\rho_0 kN}\times$$

$$\times\left\{\exp(-bkz) + \frac{\delta^2}{2}\exp(-akz)[\sin(akz) - (1-\delta)\cos(akz)] - B\left(\frac{\kappa}{\nu}\right)^{1/2}\frac{f}{N}\exp(-kz)\right\}\cos kx.$$

In the limiting case of small values of the dimensionless parameter δ, these expressions turn into the solution [6]. (Note that a misprint in the sign has crept into the expression for temperature [6] that did not affect further results).

Bearing in mind the calculation of the flow helicity, we also present the expressions for the vertical derivatives of the horizontal velocity components:

$$\frac{\partial u}{\partial z} = -\frac{a^2bk\upsilon}{B}\left\{\delta^2\exp(-bkz) + 2\exp(-akz)[(1-\delta)\cos(akz) - \sin(akz)]\right\}\sin kx,$$

$$\frac{\partial \upsilon}{\partial z} = \frac{fb\upsilon}{\nu kB}\left\{\frac{B}{b}\exp(-kz) - \exp(-bkz) + \exp(-akz)[\cos(akz) + (1-\delta)\sin(akz)]\right\}\sin kx. \tag{24}$$

Note the relative smallness of the second of these derivatives at the lower boundary, since at $z = 0$ all the terms in curly brackets, except for the small first term, are mutually compensated.

Let us estimate the amplitudes of the horizontal velocity perturbations for the typical conditions in the atmosphere. If the heat flux deviation is $Q = 10$ W/м^2, $k = 10^{-5}$ m^{-1}, $\gamma = 3\cdot10^{-3}$ K^{-1}, $\kappa = \nu = 10$ m^2/s, then $\upsilon \approx 3\cdot10^{-5}$ m/s, $b = 100$, $\delta \approx 0.45$, $B \approx 0.65$; the amplitudes u и v reach some meters in second. Strictly speaking, this is beyond the scope of the linear approximation used here, so that the use of the results obtained for smaller disturbance amplitudes is more justified.

The spatial scales included in the obtained solutions, generally speaking, are different significantly. One of the three exponentials included in the solution $\exp(-kz)$ slowly decays with height, the other two exponentials decay much faster – on vertical scales of the order of h_E and h_b. The slowly decaying exponent is not included in the expressions for u and w and makes a relatively small amplitude contribution to the temperature perturbation up to heights of the order of h_E and h_b. But it largely determines the vortex component of the horizontal velocity υ. It is interesting to note that anticyclonic circulation occurs in areas of positive heat inflow (values υ are negative). A detailed physical analysis of the solution is contained, for example, in the article [3], which is limited to a particular case $\nu = \kappa$ and small values of the parameter δ.

Figure 1 shows an example of a solution for speed components at $\kappa = \nu = 1$ m^2/s (the values of other parameters are given above). In this case $\sigma_1 = -100$, $\sigma_{2,3} \approx -(1 \pm i)\cdot 700$, $a \approx 700$, $b = 100$, $h_b = 1000$ m, $h_E \approx 140$ m, $\delta \approx 0.14$, $B \approx 0.9$; k^{-1} >> h_b >> h_E. Velocity components

u, w practically do not go beyond the thickness of the layer h_b (a pronounced maximum is contained in a much thinner Ekman layer – "Ekman inflow" into the heat release region).

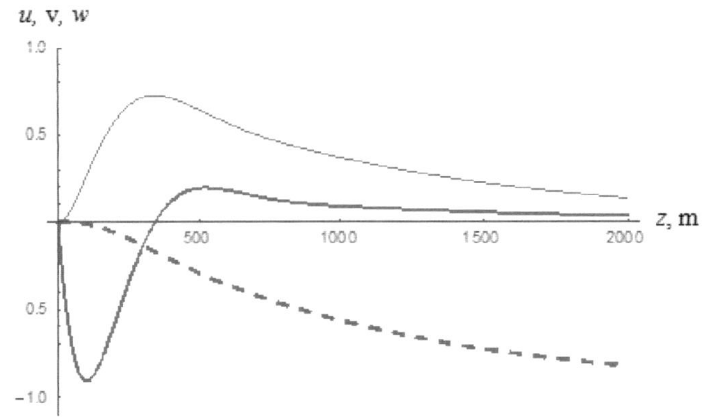

Figure 1. Vertical dependencies of speed components at $\kappa = \nu = 1$ m²/s: u (thick line, normalized to $4b^2\upsilon / B$), v (dashed line, normalized to $f\upsilon / \nu k^2$) on the vertical $kx = \pi / 2$; w on the vertical $x = 0$ (thin line, normalized to $b\upsilon / B$).

Figure 2 shows the example of profiles for the case $\kappa = \nu = 10$ m²/s. In this case $\sigma_1 = -100$, $\sigma_{2,3} \approx -(1 \pm i) \cdot 220$, $a \approx 220$, $b = 100$, $h_b = 1000$ m, $h_E \approx 450$ m, $\delta \approx 0.45$, $B \approx 0.65$; $k^{-1} >> h_b > h_E$. The thickness of the Ekman layer is significantly greater than in the first example; although h_E remains less than h_b, it has the same order of magnitude. The inflow of mass into the region of heat release occurs in a much thicker layer.

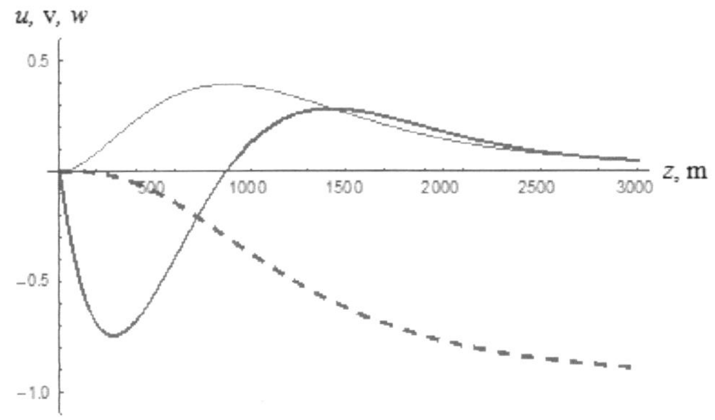

Figure 2. The same as in figure 1, but at $\kappa = \nu = 10$ m²/s and normalization u to $b^2\upsilon / B$.

The thickness of the perturbation region at the lower boundary, on the one hand, depends on rotation and stratification (scale h_b), and on the other hand, on viscosity (scale h_E). In the first of the considered examples, the mentioned scales are quite different, so that boundary layers of different nature can be distinguished. In the second example, these scales are of the same order of magnitude, therefore, effects of different natures "mix".

6. The analysis of thermal circulations helicity

Recall the definition of helicity:

$$H = \mathbf{v} \cdot \mathbf{rot}\,\mathbf{v}. \tag{25}$$

For two-dimensional flows in the rotating coordinate system considered here

$$H = -u\frac{\partial \mathrm{v}}{\partial z} + \underline{\mathrm{v}\frac{\partial u}{\partial z} - \mathrm{v}\frac{\partial w}{\partial x}} + w\frac{\partial \mathrm{v}}{\partial x}, \tag{26}$$

where "azimuthal" components are underlined (see below).

We analyze the solution derived above in this context. Since we are talking about solving a linear problem, and helicity is quadratic in velocity, in this case it makes sense to consider not the absolute values of helicity (in the linear approximation they are obviously very small), but its spatial distribution. In addition, it is of interest to analyze the relative contribution of different terms in (26). For example, in [11-13], where the axisymmetric problem was considered, the contributions of the "radial" component (its analogue is the first term in the right-hand side of (26)), "azimuthal" (underlined second and third terms), and "vertical" helicity (last term) were compared.

Let us first dwell on the dependence of (26) on the horizontal coordinate x . As one can see easily, all the terms in the right-hand side (26) are proportional $(\sin kx)^2$, or (the last) $(\cos kx)^2$. When averaged over the wavelength of the considered horizontal harmonic, this gives the same factor ½ for all terms. The wavelength-averaged values of the last two terms obviously coincide, and the sum of these terms is uniform horizontally, since

$$-\mathrm{v}\frac{\partial w}{\partial x} + w\frac{\partial \mathrm{v}}{\partial x} = V(z)W(z)\sin^2 kx + V(z)W(z)\cos^2 kx = V(z)W(z) \tag{27}$$

The third term in (26) (and, consequently, the fourth) in amplitude should be much less than the second, since near the maxima of the vertical profiles the ratio $|(\partial w/\partial x)/(\partial u/\partial z)| \sim |(w/u)\cdot(kh_E)|$; under the sign of the last modulus, each of the two expressions in parentheses is much less than unit. Thus, from the scale analysis it follows that the main contribution to the helicity integrated over space should be made by the first two terms on the right-hand side of (26).

Some rather general statements can be proved about the mean (vertically integrated) terms in the right-hand side of (26). Let us integrate the second of these terms over the entire vertical:

$$\int_0^\infty \mathrm{v}\frac{\partial u}{\partial z}dz = \sin^2 kx \int_0^\infty V\frac{\partial U}{\partial z}dz = \sin^2 kx\left(UV\big|_0^\infty - \int_0^\infty V\frac{\partial U}{\partial z}dz\right) = -\int_0^\infty u\frac{\partial \mathrm{v}}{\partial z}dz \tag{28}$$

(integration "by parts" was performed and it was taken into account that u vanishes at the lower boundary and decays with height). Thus, the integral contribution of the first two terms into the helicity in (26) is the same. Therefore, the integral contribution of the "radial" and "azimuthal" components is approximately the same.

From the second equation (2) it follows:

$$u = \frac{\nu}{f}\Delta \mathrm{v} \approx \frac{\nu}{f}\frac{\partial^2 \mathrm{v}}{\partial z^2} \tag{29}$$

(in the approximate equality, the smallness of the horizontal derivatives is taken into account in comparison with the vertical derivatives). Hence it follows that the first term in (26) can be approximately represented in the form

$$-u\frac{\partial \mathrm{v}}{\partial z} \approx -\frac{\nu}{f}\frac{\partial^2 \mathrm{v}}{\partial z^2}\frac{\partial \mathrm{v}}{\partial z} = -\frac{\nu}{2f}\frac{\partial}{\partial z}\left(\frac{\partial \mathrm{v}}{\partial z}\right)^2.$$

Consequently,

$$-\int_0^\infty u\frac{\partial \mathrm{v}}{\partial z}dz \approx -\frac{\nu}{2f}\left(\frac{\partial \mathrm{v}}{\partial z}\right)^2\Bigg|_0^\infty = \frac{\nu}{2f}\left(\frac{\partial \mathrm{v}}{\partial z}\right)^2\Bigg|_{z=0} = \frac{f\upsilon^2}{2\nu k^2}\sin^2 kx \tag{30}$$

(the last equality takes into account the second formula (24)).

Figures 3, 4 show the vertical dependencies of the first two terms in the right-hand side (26) and their sums for the first and second numerical examples, respectively. The helicity is generated effectively in the lower layers, where influence of the bottom boundary is significant. In the first example, first of all, this is rather sharply expressed Ekman boundary layer. The essential role of the Ekman layer in this context has long been noted in the literature for flows of various natures (for example, [7, 8]). But, in addition to the Ekman scale h_E, for flows of the nature under consideration, the influence of the lower boundary and the region of helicity generation are also determined by the vertical scale h_b, which does not depend on the viscosity. This is especially clearly seen in the second numerical example.

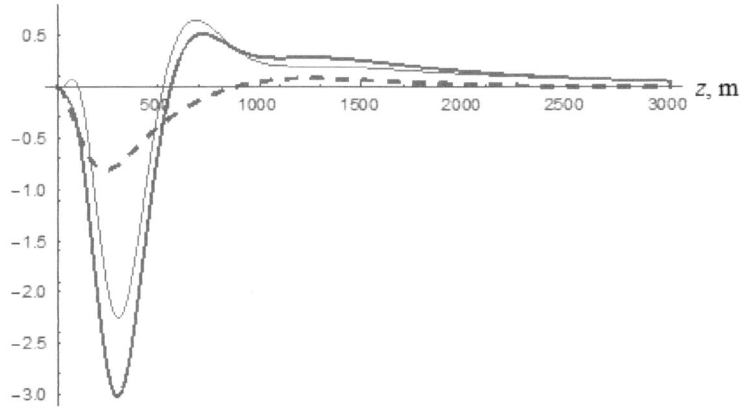

Figure 3. Vertical dependencies of the horizontally averaged helicity components: $\mathrm{v}(\partial u / \partial z)$ (thin line), $-u(\partial \mathrm{v} / \partial z)$ (dashed line) and their algebraic sum (thick line) for the first numerical example. All profiles are normalized to $fb^3\upsilon^2 / 2\nu k$.

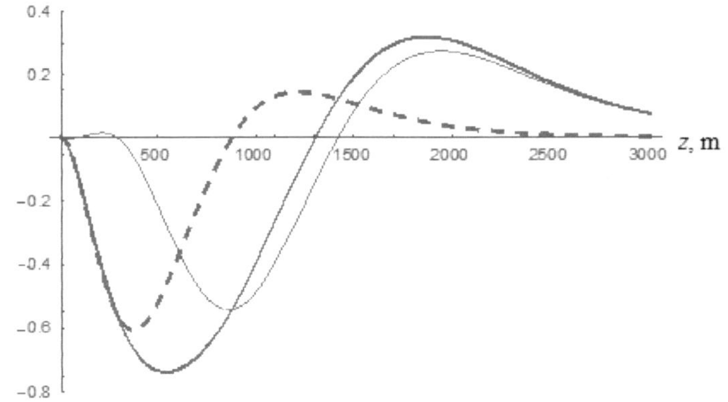

Figure 4. The same as in Fig. 3, but for the second numerical example.

Although the integral contribution of the "radial" and "azimuthal" helicity components turned out to be practically the same, but, as can be seen from the figures, their vertical dependencies differ markedly. The layers with positive and negative helicity in the considered examples compensate each other to a large extent, so that integral (30) over the entire region turned out to be much less in absolute value than the integrals over each of the mentioned relatively thin layers.

7. Conclusion

The considered theoretical model of convective flows over a thermally inhomogeneous surface generalizes some previous studies. In particular, this is free from the assumption that the Ekman boundary layer is relatively thin as compared to other characteristic vertical scales of the problem. This assumption (not always justified) greatly simplified the calculations, since it was associated with the presence of a small parameter in the problem. Rejection of this assumption required a noticeable complication of the calculations.

The obtained relations allow analyzing the dependencies of the components of velocity and helicity on the parameters of the problem. For example, at first glance, one could assume that with the intensification of rotation (an increase in the Coriolis parameter), the generation of helicity should increase. But, as can be seen from the solution, the radial velocity rapidly decreases in absolute value c f – on sufficiently large horizontal scales, rotation prevents radial movements. Thus, helicity decreases in absolute value (with the exception of the Ekman boundary layer, where the radial velocity decreases relatively slowly with f).

A number of rather general statements regarding the ratio of different helicity "components" in the considered thermal circulations, in particular, in atmospheric currents with characteristic horizontal scales of the order of hundreds of kilometers, have been proved. There are given examples of numerical calculations of the vertical distribution of these components. It is shown that the integral contribution of the "radial" and "azimuthal" helicity components in the situations under consideration is practically the same, but their vertical dependences can differ significantly. The horizontal velocity vertical shears, on which the generation of helicity depends, are determined not only by the Ekman scale of the height, but also by the scale h_b that does not depend on the viscosity.

References

[1] Lin Y-L 2007 *Mesoscale Dynamics* (Cambridge: Cambridge University Press)
[2] Gebhart B, Jaluria Y, Mahajan R L and Sammakia S 1988 *Buoyancy-Induced Flows and Transport* (New York: Hemisphere Publishing Corporation)
[3] Perestenko O V. and Ingel L Kh 1990 *Izv. Atmos. Ocean Phys.* **26** 666
[4] Ingel L Kh, Belyaeva M V 2011 *J. Engineering Phys. Thermophys.* **84** 820
[5] Svirkunov P N and Felde E A 1988 *Sov. Meteorol. and Hydrol.* № 1 17
[6] Ingel L Kh and Makosko A A 2019 *Geophys. Res.* **20**, № 3 36
[7] Lilly D K 1986 *J. Atmos. Sci.* 1986 **43** 126
[8] Molinari J and Vollaro D 2008 *Mon. Weath. Rev.* **136** 4355
[9] Kurgansky M V 2017 *Izv. Atmos. Ocean Phys.* **53** 127
[10] Levina G V and Montgomery M T 2014 *J. Phys.: Conf. Ser.* **544** 012013
[11] Han Y, Wu R and Fang J 2006 *Adv. Atmos. Sciences* **23** 504
[12] Sukhanovskii A, Evgrafova A. and Popova E 2017 *Archive Mechan. Engineering*, **LXIV** doi: 10.1515/meceng-2017-0011
[13] Teimurazov A, Sukhanovskii A, Evgrafova A and Stepanov R 2017 *J. of Phys.: Conf. Series*, **899** 022017
[14] Pedlosky J 1987 *Geophysical Fluid Dynamics* (New York: Springer-Verlag)
[15] Gill A E 1982 *Atmosphere-Ocean Dynamics* (New York: Academic Press)

IOP Conf. Series: Earth and Environmental Science 1040 (2022) 012024 doi:10.1088/1755-1315/1040/1/012024

Application of A.M. Obukhov's theory of correlation of vectors for scientific research and engineering calculations of ice drift in the Arctic Ocean.

N E Ivanov[1], D M Demchev[1,2], A V Nesterov[1]

[1]. Arctic and Antartic Research Institute (AARI), St. Petersburg,. Russia.
[2].Nansen International Environmental and Remote Sensing Center (NIERSC), St. Petersburg, Russia.

E-mail: neivanov@aari.ru

Abstract The theory of vectors correlation of Obukhov was used to describe an ice drift in the Arctic Ocean. The main features are defined by an area. The tracks tortuosity in Beaufort Gyre and Transarctic current varies threefold: 1.5 and 4.5. Year to year and seasonal variations indicate the position of the Beaufort Gyre and the speed of the Transarctic drift. The average speed does not exceed 10 cm/s. The drift increases in summer, and in winter it intensifies in Fram Strait. In Fram Strait the velocity month–over–month increases from June to December–March from 2.5 to 6.5 cm/s. Overall correlation drift and wind is 0.95 and 0.85 for ice fields and icebergs. The non-wind drift reproduces the cyclonic circulation between Frantz Josef Land and Novaya Zemlya. The wind drift specifies the coefficients and angles of drift evasion from the wind. The coefficient for total drift is 0.14–0.18, fore wind drift 0.10–0.15, and the angles are 15–30° and 10–15° in the Barents Sea. In the Fram Strait the drift and wind directions are stable, the trend considers the modulus and direction. The drift trend has the direction towards the south. It decreases to 5 % of variance in June and increases to 15% from December to March.

1. Introduction, research methods and materials

The report is devoted to the first major result of A.M. Obukhov — the theory of normal correlation of vectors [1, 2]. It was originally created to describe a system of affine vectors. Its applicability to Euclidean vectors was shown later [3, 4, 5]. This is important for the study of ice drift and its dependence on wind and currents in the Arctic Ocean (AO). The greatest efficiency is achieved when it is used in combination with the vector-algebraic method of random vectors [6] and the correlation method Watanabe [7] and Z.M. Gudkovich [8, 9].

By the definition of N.E. Kochin [10] the velocity model is a Euclidean vector $\vec{V}$ with modulus V, direction φ, and projections V_X, V_Y. Addition according to the parallelogram rule, transformation of coordinates during their rotation, and three kinds of multiplication — scalar, vector, and tensor — are defined. Characteristic $\vec{V}$ as a random vector — probability distribution by gradations of modulus V and by rhumb lines φ are graphically represented by "roses" (in figure. 2, 3 below).

Moments of distribution — mean velocity vector $\vec{m}_{\vec{V}}$ and variance tensor $D_{\vec{V}}$

$$D_{\vec{V}} = \mathrm{M}\{(\vec{V}-\vec{m}_{\vec{V}})\otimes(\vec{V}-\vec{m}_{\vec{V}})\} = \begin{pmatrix} D_{V_X} & K_{V_XV_Y} \\ K_{V_XV_Y} & D_{V_Y} \end{pmatrix} = \begin{pmatrix} \lambda_1 & 0 \\ 0 & \lambda_2 \end{pmatrix} \tag{1}$$

Components (1) — variance and covariance of projections $\vec{V}$. The tensor (1) is brought to diagonal form ($\lambda_{1,2}$) by a turn of coordinates to an angle α, showing the direction of maximum variability $\vec{V}$. Invariants $\lambda_{1,2}$ define the distribution of variance in orthogonal directions. Their combinations are also invariants. Linear invariant $I_1 = \lambda_1 + \lambda_2$ shows the total variance (in modulo and direction). Adding to (1) the ratio

IOP Conf. Series: Earth and Environmental Science 1040 (2022) 012024 doi:10.1088/1755-1315/1040/1/012024

of I_1 and scalar modulus variance $\gamma_V=1-D_V/I_1$ shows the contribution of direction variability to the total variance. The ratio of I_1 to the mean velocity modulus $\nu_{\vec{V}} = \boldsymbol{I}_1^{0.5} / \boldsymbol{m}_{\vec{V}}$ is the coefficient of variation.

Standard deviation is an ellipse $\sigma_{\vec{V}}$ with axes $\lambda_{1,2}$ oriented in the direction of maximum variability α, its elongation shows the invariant $\chi = \lambda_2/\lambda_1$. Graphically, the variability $\vec{V}$ is shown by combining the vector $\vec{\boldsymbol{m}}_{\vec{V}}$ and the ellipse $\boldsymbol{\sigma}_{\vec{V}}$ (see figure 2, 3, 5 below).

The covariance in the system of random vectors — $\vec{V}$ drift and $\vec{G}$ wind is described by tensor

$$COV_{\vec{V}\vec{G}} = \begin{pmatrix} K_{V_X G_X} & K_{V_Y G_Y} \\ K_{V_X G_Y} & D_{V_Y G_Y} \end{pmatrix} = \begin{pmatrix} \lambda_1 & 0 \\ 0 & \lambda_2 \end{pmatrix} + 0.5A \begin{pmatrix} 0 & 1 \\ -1 & 0 \end{pmatrix} \quad (2)$$

In contrast to (1) tensor (2) is asymmetric. The invariant $I_1 = \lambda_1 + \lambda_2$ of symmetric part (2) shows the connectivity of changes of collinear components $\vec{V}$ and $\vec{G}$, and the invariant A (rotation indicator) shows the connectivity of changes of orthogonal components.

The proportional drift dependence on wind was established by famous F. Nansen's Fram expedition [11], H. Sverdrup's Maud expedition [12] and confirmed by subsequent measurements [4, 5, 9, 13—18]. A linear approximation of the dependence is defined by A.M. Obukhov with equation

$$\vec{\boldsymbol{V}} = \boldsymbol{A}_{\vec{G}\vec{V}} \vec{\boldsymbol{G}} + \vec{\boldsymbol{B}}_{\vec{G}\vec{V}}, \quad (3)$$

Regression – tensor $\boldsymbol{A}_{\vec{G}\vec{V}} = \boldsymbol{COV}_{\vec{G}\vec{V}} \boldsymbol{D}_{\vec{V}}^{-1}$, free term is a vector $\vec{\boldsymbol{B}}_{\vec{G}\vec{V}} = \vec{\boldsymbol{m}}_{\vec{G}} - \boldsymbol{A}_{\vec{G}\vec{V}} \vec{\boldsymbol{m}}_{\vec{V}}$. Components $\boldsymbol{COV}_{\vec{G}\vec{V}}$ and $\boldsymbol{D}_{\vec{V}}$ and mean velocities $\vec{\boldsymbol{m}}_{\vec{V}}$, $\vec{\boldsymbol{m}}_{\vec{G}}$ drift and wind are defined from the measurement data.

Correlation is an integral part of regression. In the system of vectors ($\vec{V}, \vec{G}$), it is defined by mutual changes of drift and wind moduli and directions. It is important that the correlation not only measures the "closeness" of dependence, but also allows for interpretation in Euclidean space. In the paper [4] the system of indices of collinear $\boldsymbol{r}_{\uparrow\downarrow}$, orthogonal correlation $r_\perp$ and indicator μ by invariants of tensors $\boldsymbol{D}_{\vec{V}}$ (1) and $\boldsymbol{COV}_{\vec{G}\vec{V}}$ (2) is given by

$$\boldsymbol{r}_{\uparrow\downarrow} = \frac{\boldsymbol{I}_1^{(\vec{V}\vec{G})}}{\sqrt{\boldsymbol{I}_1^{(\vec{V})} \boldsymbol{I}_1^{(\vec{G})}}}$$

$$\boldsymbol{r}_{\perp} = \frac{\boldsymbol{A}^{(\vec{V}\vec{G})}}{\sqrt{\boldsymbol{I}_1^{(\vec{V})} \boldsymbol{I}_1^{(\vec{G})}}}$$

$$\mu(r_{\uparrow\downarrow}^2 + r_\perp^2)^{0.5} \quad (4)$$

If the dependence is defined by changes only to moduli $\vec{V}$ and $\vec{G}$, then $\boldsymbol{r}_\perp = 0$. At joint amplification (weakening) of velocities the collinear correlation $\boldsymbol{r}_{\uparrow\downarrow} > 0$, and at amplification of one vector and weakening of another $\boldsymbol{r}_{\uparrow\downarrow} < 0$. If at a stable modulus the vector $\vec{V}$ is reversed to the right of the vector $\vec{G}$, orthogonal correlation $\boldsymbol{r}_\perp > 0$ and vice versa. The indicator μ shows total correlation and emphasizes the need for a joint analysis of $\boldsymbol{r}_{\uparrow\downarrow}$ and $\boldsymbol{r}_\perp$. For independent vectors μ=0, with deterministic dependence μ=1, and with stochastic dependence μ<1. Estimates (4) were used to clarify drift dependence on wind in the entire AO [15, 17] and in its individual regions [5, 14].

The surface circulation and ice drift in the AO are defined by drift dependence on wind and non-wind currents [9, 13, 16]. Wind coefficients $k=V/G$ and angles of drift deflection from wind $\psi = \varphi_{\vec{V}} - \varphi_{\vec{G}}$

were used to calculate and predict drift dependence on wind. They are usually defined by the total drift. In the paper [5] the dependence of k and ψ on the type of object (field or iceberg) was confirmed; replacing the total drift $\vec{V}$ with the wind component $\vec{U}$ using formulas (3, 6) clarifies the estimates of k and ψ. Non-wind currents are formed by adjusting the water density field to the wind [9]. They define the "non-wind" drift $\vec{C}$ (formula 7) and are quasi-constant over intervals from weeks to the season. When averaged over the season, their contribution to total drift in the AB averages 60% and is increased in the Fram Strait and Greenland Sea. The velocity $\vec{C}$ in general for the AB is increased in summer and decreased in winter. In the Fram Strait, a winter maximum was recorded due to intensification of the Icelandic cyclone.

The correlation method of Watanabe – Gudkovich was used to separate wind and non-wind components in total drift. In this method, each drift projection $\vec{V}$ is defined by a multiple regression with both wind projections $\vec{G}$, and C_X, C_Y are non-wind drift projections $\vec{C}$.

$$V_X=a_XG_X+b_XG_Y+C_X,$$
$$V_Y=a_YG_X+b_YG_Y+C_Y \tag{5}$$

In the absence of measurements, geostrophic wind was used according to the drift theory of N.N. Zubov [19]. Its characteristics are specified in the paper [9]. Despite the differences in velocity vector models (Obukhov — modulus and direction, Gudkovich — projections), calculations of drift dependence on wind by formulas (3) and (5) are numerically the same. Formula (3) defines wind drift by $\vec{B}_{\vec{G}\vec{V}} = 0$ as

$$\vec{U} = A_{\vec{G}\vec{V}}\vec{G}, \tag{6}$$

Then non-wind drift is the difference of mean velocities of total and wind drift

$$\vec{C} = \vec{m}_{\vec{V}} - \vec{m}_{\vec{U}} \tag{7}$$

The method effectiveness is demonstrated (in figure 4 below) by the map of cyclonic circulation between Franz Josef Land and Novaya Zemlya [5]. Formula (6) is important for specifying wind coefficients k of angles of drift deflection from the wind in engineering calculations and forecasts.

Figure 1*a* shows a map of seas and major straits in the AO [20]. Figure 1*b* in the paper [16, 18] shows ice drift scheme in the AO — the Beaufort Gyre in the Amerasian sector, the Transarctic Current from the Chukchi Sea to the Fram Strait, and cyclonic circulations in the Siberian seas. The center position and size of the Beaufort Gyre, ice flow from the AB to the Greenland Sea via the Fram Strait and its transit to the Atlantic via the Danish Strait, ice export from the Siberian seas to the Transarctic Current area, and ice runoff via the Davis Strait. ESV data in figure. 1 *c, d* [17] for December 1982 and January 1983 show marked changes from month to month.

Accordingly figure 1 mean ice drift velocity in the AO does not exceed 10 cm/s. In the entire AB, the transarctic drift is intensified in summer. But in the Fram Strait the drift is intensified in winter due to intensification of the Icelandic cyclone. It is an indicator for the entire AO its variability depends on the pressure gradient between Greenland and Scandinavia.

The drift pattern is stable. Interannual and seasonal differences are evident in the center position and size of the Beaufort Gyre, transarctic drift velocity, and ice runoff intensity through the Fram, Denmark, and Baffin Straits. This is shown by drift maps from satellite (ESV) drift measurements in figure 1*c, d* [17]. In just one month (December 1982, January 1983), center position of the Beaufort Gyre and its intensity changed, and in the Kara Sea there was a drift reversal of about 90°.

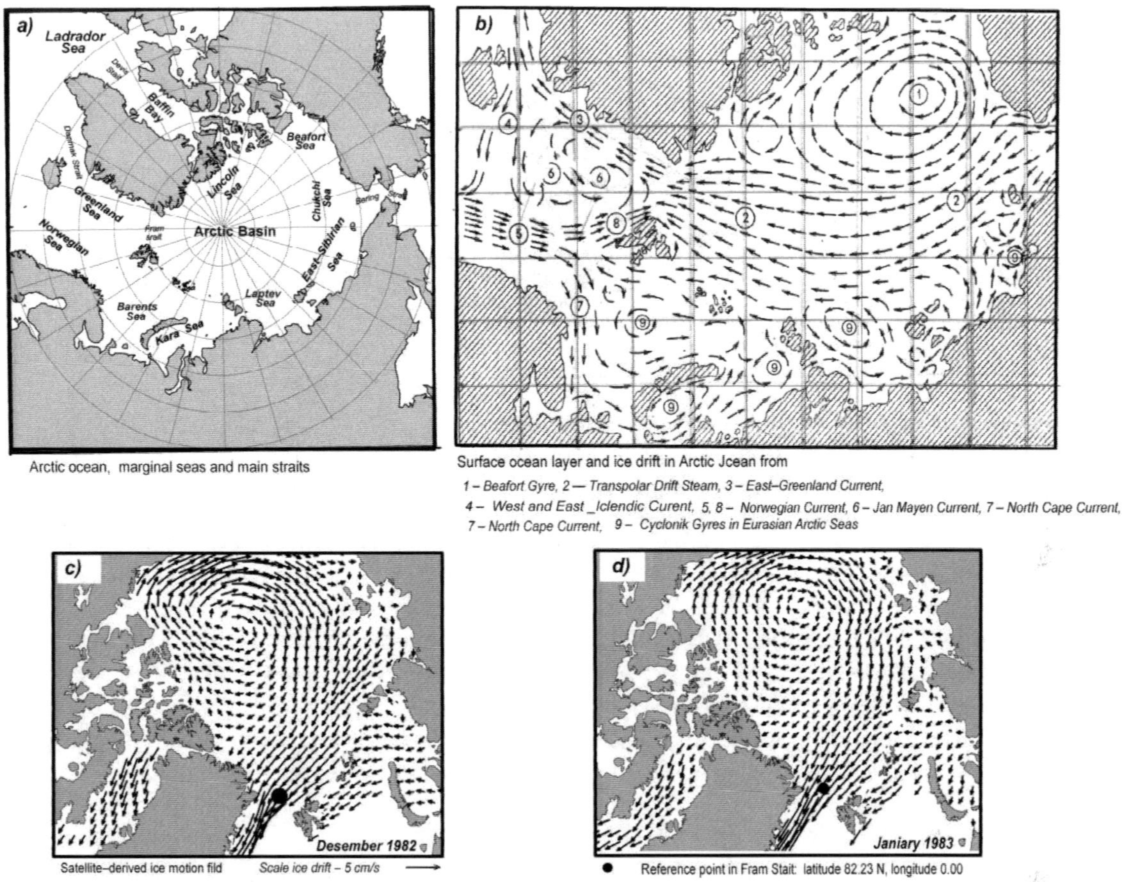

Figure 1. Surface currents and ice drift in the Arctic Ocean: *a* – main Seas and Straits , *b* – general surface circulation pattern in North–European and Arctic Basin [16, 18, 20], *c, d* – satellite-derived sea ice motion in December 1982 and January 1983 [17].

The multiyear variability is described by a trend. The Beaufort Gyre has been shown to intensify in recent decades against the background of warming and ice reduction in the AO, positive trends in the Transarctic drift and ice export from the AB. It was explained by a decrease in the ice extent, ice thickness, and concentration of multiyear ice. The conclusion that wind does not affect the drift trend (for 2000-09 in the Fram Strait drift trend was 24 % and 18 % of variance in winter and summer, and only 2 % and 3 % for wind) needs to be clarified, as these papers use only drift $\vec{V}$ and wind vector $\vec{G}$ speed.

Vector trend $\vec{V}(t)$ is defined as

$$\vec{V}(t) = \vec{m}_{\vec{V}} + \vec{a}(t) + \vec{b} + \vec{\varepsilon}(t) \tag{8}$$

Here, where $\vec{m}_{\vec{V}}$ is the mean velocity, components of trend $\vec{a}$ and free term $\vec{b}$ coefficient vectors are defined by projections of $V_X(t)$, $V_Y(t)$; $\vec{\varepsilon}(t)$ — trend anomalies. Trend contribution to the variance defines the ratio of invariants $\gamma_{\vec{a}} = I_1^{(\vec{\varepsilon})} / I_1^{(\vec{V})}$ of tensors (1) of anomalies and of series itself D_ε. It is important to compare the directions. If the trend $\vec{a}$ and mean velocity $\vec{m}_{\vec{V}}$ are colinear — the trend shows modulus amplification (weakening), and if they are orthogonal — the reversal (to the left or to the right by a difference sign $\varphi_{\vec{a}} - \varphi_{\vec{m}}$).

In Sections 2.1, 2.2 the drift velocity $\vec{V}$ of the North Pole stations, icebergs and ice fields in the Barents Sea were calculated based on GPS measurements of geographic coordinates. Discreteness is 1 day. The fields of mean monthly drift in the entire AO in Section 2.3 were obtained from satellite measurements. There are direct measurements for the wind $\vec{G}$ at the NP. For buoys and ESV data, the wind is defined by reanalysis. Wind drift $\vec{U}$ and non-wind currents $\vec{C}$ were defined according to Obukhov's theory by formulas (3, 6, 7) for total drift $\vec{V}$ and wind $\vec{G}$.

2. Results and discussion

2.1. Drift of the North Pole stations

The Russian North Pole (*NP*) stations made an epoch in the study of hydrometeorological processes in the Arctic ocean. They provide the most accurate data — contact measurements of drift, wind and currents. An overview of the drift at *NP* 1—33 (1937-2004) is given in the paper [16] and *NP* 35 [15]. It is shown that the main features are defined by the region — drift characteristics are different in the Beaufort Gyre and in the Transarctic Current zone. Seasonal features and interannual variability are observed.

The drift *NP* 37, 40 in the Beaufort Gyre, *NP* 36 and Ice Base (*IB*) in the Transarctic Current area is considered here. The drift $\vec{V}$ and wind $\vec{G}$ discreteness is 1 day. Dated tracks are shown in figure. 2*a*, surface circulation diagram is reproduced from Atlas [20]. Drift start dates, drift end dates, duration and region (latitude and longitude extremes) are shown in table 1. Track parameters - length *L*, modulus of resultant displacement R and tortuosity coefficient γ_L=L/R are also given here.

Table 1. Drift of the "North Pole" stations *NP* 36, 37, 40 and the Ice Base (*IB*): period, drift region and track characteristics — length *L*, resultant displacement *R* and tortuosity γ_R

№ NP	Start and finish observation			Boundary region				Parameter trek		
	Start	Finish	Day	Latitude, degree		Longitude, degree		*L*	*R*	γ_L
	—	—		*min*	*max*	*min*	*max*	km for 10 day		—
LB	23.06.2007	14.08.2007	53	81.7	84.9	167.9	172.3	96	66	1.4
NP-36	8.09.2008	23.08.2009	350	82.3	88.6	175.2	334.2	76	54	1.5
NP-37	8.09.2009	31.05.2010	266	80.1	81.9	196.1	221.9	71	16	4.7
NP-40	2.10.2012	6.06.2013	248	81.1	86.3	212.8	231.9	65	16	4.0

Figure 2*a* and table 1 confirm that drift depend primarily on the region. Four *NP* drift in different years and seasons, the duration of their drift (from 50 to 350 days), its length and wind conditions are different. But two groups are distinguished by track shape and elongation of tracks — drift in Beaufort Gyre (*NP* 37, 40) and in Transarctic Current (*NP* 36, *IB*). Its underlined by estimates of tortuosity coefficient γ_L in table 1. For *IB* and *NP* 36 it was 1.5, and for *NP* 37, 40 3 times more – 4.0÷4.7.

Drift characteristics, its wind component and wind itself at the IB and NP 36, 37 in table 2 are represented by estimates of the scalar modulus (mean V_{mean}, maximum V_{max}, and standard deviation σ_V), mean vector (modulus m_V and direction φ_V), and variance ellipse ($\lambda_{1,2}$, total RMS $I_1=\lambda_1+\lambda_2$, elongation $\chi=\lambda_1/\lambda_2$, direction of maximum variability α and direction contribution to variance γ_V and coefficient of variation $\nu=I_1/m_V$). Drift and wind recurrence roses by rumbs and modulus gradation are shown in figure. 2*b*. Figure 2*c* combines average velocity and wind dispersion ellipses for total and wind drift.

Table 2. Variability of total drift velocity (V), its wind component (U) and wind (G) of drifting stations NP 36, 37 and the Ice Base (LB).

№ NP	Drift and wind	Scalar speed			Mean vector		Variance speed vector						
		V_{cp}	V_{max}	σ_V	m_V	φ_V	$(I_1)^{0.5}$	λ_1	λ_2	χ	α	γ_V	ν
		—	—	—	—	degree	—	—	—	—	degree	%	—
LB	V, cm/s	11.1	31.8	7.4	7.6	8	11.0	9.3	5.8	0.62	26	54	1.4
	U, cm/s	10.1	27.7	5.7	5.2	359	10.4	8.9	5.4	0.60	28	70	2.0
	G, m/s	4.1	10.6	2.2	2.1	324	4.2	3.3	2.5	0.75	358	73	2.0
NP-36	V, cm/s	9.0	34.3	6.3	3.1	64	10.5	7.8	7.0	0.90	7	64	3.4
	U, cm/s	8.5	24.1	4.1	2.1	73	9.2	6.8	6.2	0.90	25	80	4.3
	G, m/s	4.8	13.5	2.8	1.2	52	5.2	3.7	3.6	0.98	14	71	4.3
NP-37	V, cm/s	8.1	26.5	5.1	1.8	110	9.5	7.1	6.3	0.88	114	72	5.3
	U, cm/s	7.7	18.4	3.6	1.0	195	8.5	6.4	5.6	0.88	117	82	8.5
	G, m/s	4.8	11.5	2.3	0.6	121	5.3	4.1	3.4	0.83	98	81	8.8

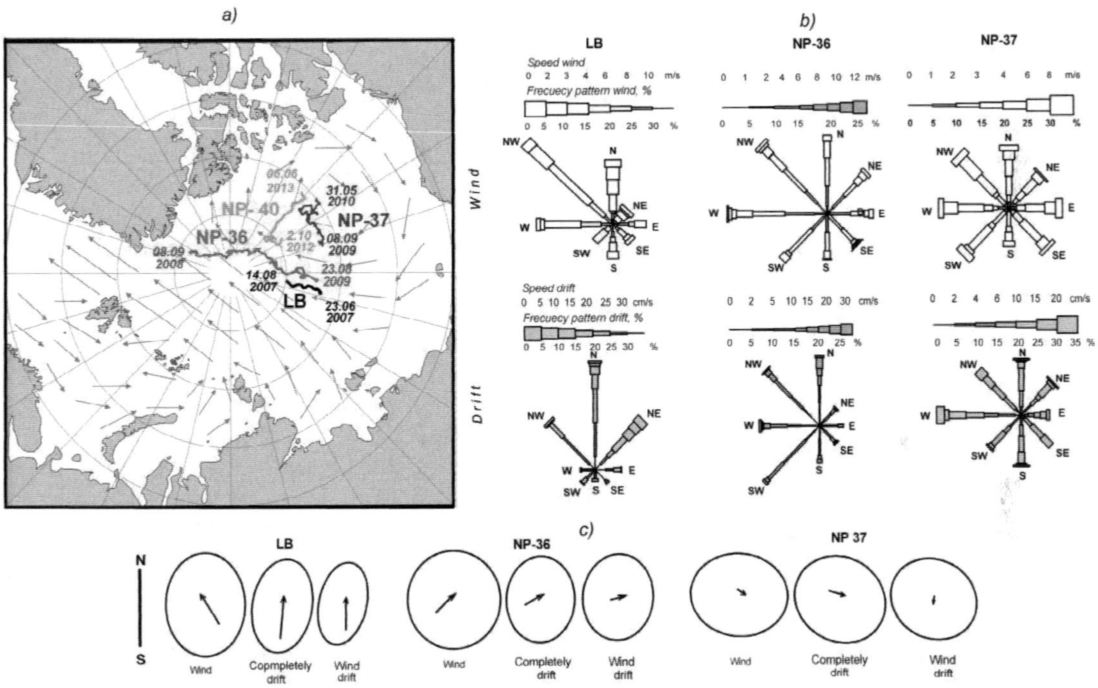

Figure 2. The North Pole drifting station (*IB* and *NP* 36, 37, 40): *a* – sea ice drift line, *b* – frequency pattern wind and ice drift speed (module and direction), c –mean speed $\vec{m}_{\vec{V}}$ and MSE variance ellipse $\sigma_{\vec{V}}$ of wind and drift – completely and wind component.

According to table 2 and figure. 2 for all *NP* (except for *NP* 37 in the Beaufort Gyre) the drift is to the right of the wind according to the theory. The differences between resulting drift velocity vectors $\vec{m}_{\vec{V}}$ and wind drift $\vec{m}_{\vec{U}}$ calculated according to (6), as well as between dispersion ellipses $\sigma_{\vec{V}}$ and $\sigma_{\vec{U}}$, emphasize the significant contribution of the "non-wind" component $\vec{C}$ according to formula (7) to the total drift $\vec{V}$. Drift features in the region of the Transarctic Current and the Beaufort Gyre most clearly show the coefficient of variation ν and elongation of dispersion ellipses χ (table 2, figure. 2*c*). The coefficient ν was 2.0 for IB, about 4.5 for *NP* 36, and increased to 8.5 in the Beaufort Gyre. The coefficient χ is sharply reduced at the IB (0.6), and at the other NPs the distribution of variance in directions is close to circular $\chi\approx0.9$.

Base *LB* drift feature in comparison with the other *NP* station is connected with the increased wind direction stability in June-August 2007. Station *NP* 36 drifted in the same region in summer 2009 (figure. 1). The measurement data showed (in table 3) that in 2007 the frequency of wind directed to the north-west (NW) was 35%, and in W–NW–N sectors it was 70% in total. In 2009, wind distribution by rhumb lines is much more even — from 8% to 20%

Frequency (%) of winds by rhumb lines in June-August based on data for 2007 and 2009

Station	Year	Wind direction							
		S	SW	W	NW	N	NE	E	SE
LB	2007	10	4	17	35	16	1	10	7
NP-36	2009	12	10	18	8	8	8	20	17

2.2. Drift of icebergs and ice fields in the north—eastern part of the Barents sea

Contact measurements of drift and wind at the *NP* stations are the most accurate. But they are insufficient for analysis of climatic variability — there are data only in 1 point at any given time. A good basis for analyzing long-term drift variability in the Arctic basin of the AO is data from American automatic buoys. According to this program, since 1979, GPS sensors have been installed on drifting ice objects in the Amerasian sector of the AB and their coordinates have been used to define drift. The points of buoy installation are random and mobile. The AARI has developed a technology for bringing data to regular grid nodes throughout the AB based on matching drift measurements with the pressure field. It is used to monitor hydrometeorological conditions in the AO.

Here are the drift characteristics of 10 icebergs and 20 ice fields in the northeastern part of the Barents Sea in May-August 2009. They were obtained at the AARI. The expedition was unique in the number of buoys in a limited area (10 icebergs and 20 fields). Its results made it possible to obtain detailed drift estimates and compare the drift of icebergs and ice fields [5].

Parameters of icebergs and fields – *min÷max* range and the specific spread *(max–min)/m* differ markedly (especially for the mass of icebergs and the area of fields). They were 9÷17 and 0.5 for iceberg height (m), 46÷82 and 0.6 for draft (*m*), 150×3700 and 2.7 for mass ($kg\div10^6$); 25÷705 and 3.2 for field area (km^2). The differences of drift between iceberg-field groups are significant (due to significant height and draft of icebergs compared to the fields), but no regular dependence of drift on morphometric characteristics of icebergs and fields was revealed within the groups.

In figure. 3 for synchronous drift (June 2009) of the iceberg and the field probability distributions by 8 rumb lines and gradations of the velocity modulus are represented by wind $\vec{G}$ "roses" and total drift $\vec{V}$. The interoperability of roses of iceberg and field drift with the wind demonstrates drift dependence on wind. Small differences between the field and the iceberg are due to the drift region. Drift roses are more "concentrated" (by rumb lines) compared to the wind. This shows a noticeable contribution of non-wind currents to the drift.

At the bottom of figure. 3, roses are approximated by combining mean velocity and dispersion ellipses, and additionally vector $\vec{m}_{\vec{U}}$ and ellipse $\sigma_{\vec{U}}$ of the drift wind component are shown. They show the right drift turn from the wind and the influence of non-wind currents on the drift. It is manifested by a greater elongation of the drift ellipses $\sigma_{\vec{V}}$ compared to the wind and the difference in velocities of total $\vec{m}_{\vec{V}}$ and wind $\vec{m}_{\vec{U}}$ drift in modulo and even in direction.

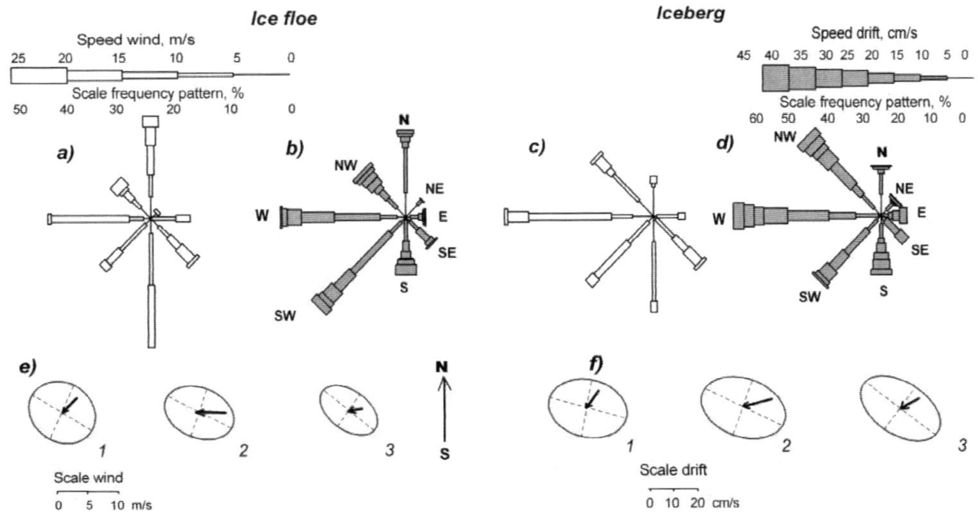

Figure 3. Synchronized the drift ice floe and iceberg from daily measurement in June, 2009: *a, d* – frequency pattern wind speed module and direction (*a, c*) and completely drift (*b, d*); *e, f* – mean speed $\vec{m}_{\vec{V}}$ and variance ellipse $\sigma_{\vec{V}}$ (*e* – ice floe, *f* – iceberg) of wind (*1*), complete drift (*2*) and wind component ice drift (*3*), according to [5].

Table 3 shows for June 2009 the spread (within groups of 10 icebergs and 20 fields) of variability parameters of the scalar velocity V (mean, maximum and RMS), modulus of the mean velocity vector m_V, linear invariant I_1, elongation of dispersion ellipse χ and contribution of direction to the variance γ_V for the total drift and its wind component. It shows that the drift of ice fields differs from the icebergs in its greater velocity and enhanced variability. For the total drift, mean velocity, maximum and mean vector modulus values V_{mean}, V_{max}, m_V were 13-19, 27-40 and 4.5–11 cm/s for fields and 10-15, 22–34 and 3.52–8 cm/s for icebergs, and variability (invariant I_1) was 13.5–20 and 12–16 cm/s. The contribution of direction variability γ_V to the variance according to Table 3 exceeds 50%, the increase of γ_V for fields (65–75 %) compared to icebergs (60–70 %) is due to iceberg draft.

Table 3. Spread (*min–max*) parameter of scalar and vector speed Iceberg and Ice Floe in Barents sea (June, 2009): *a* – complete drift, *b* – wind component.

Complete Drift (*a*) and Wind Ddift (*b*)	Scalar speed drift			Vector speed drift			
	V_{cp} cm/s	V_{max} cm/s	σ_V cm/s	m_V cm/s	$(I_1)^{0.5}$ cm/s	χ —	γ_V %
			Iceberg				
a	9.9÷15.1	21.9÷34.2	4.7÷9.4	3.6÷8.1	11.9÷16.2	0.48÷0.94	60÷70
b	9.3÷13.6	18.2÷27.0	4.0÷6.3	2.9÷5.6	10.2÷14.1	0.42÷0.94	—
			Ice Floe				
a	12.8÷19.2	27.1÷40.2	8.0÷11.8	4.6÷10.8	13.5÷20.1	0.49÷0.92	65÷75
b	11.0÷17.8	24.7÷41.1	6,2÷9.3	3.5÷12.6	12.2÷	0.45÷0.84	—

Table 4 shows mean values *m*, *min–max* ranges and *R=mi–-max* range of colinear $r_{\uparrow\downarrow}$, orthogonal correlation $r_{\perp}$ and indicator μ (4) dependence of mean daily drift velocities of icebergs and fields on the

IOP Conf. Series: Earth and Environmental Science 1040 (2022) 012024 doi:10.1088/1755-1315/1040/1/012024

wind. For icebergs (draft 50–80 m) it is less than for fields (0.85 and 0.95 on average), and the relative spread (*R/m*) for individual ice objects is small (10–15 %). Colinear correlation is many times higher than orthogonal $r_{\uparrow\downarrow} >> r_{\perp}$. The main reason for the correlation is unidirectional changes in wind and drift. In accordance with the theory, positive values $r_{\perp} > 0$ show a rightward drift reversal from the wind, and for icebergs (draft 50–80 m) it is greater than for fields (mean values $r_{\perp}$ 0.20 and 0.08).

Table 4. Correlations speed Drift Iceberg and Ice Floe in Barents sea with Wind, June–August 2009 year.

Iceberg (*a*) and Ice Floe (*b*)	$r_{\uparrow\downarrow}$			$r_{\perp}$			μ		
	m	*min÷max*	*R*	*m*	*min÷max*	*R*	*m*	*min÷max*	*R*
a	0.82	0.77÷0.89	0.12	0.20	0.12÷0.24	0.12	0.84	0.78÷0.91	0.13
b	0.92	0.87÷0.95	0.12	0.08	0.02÷0.22	0.20	0.94	0.88÷0.98	0.10

The "non-wind" drift velocity $\vec{C}$ (6, 7) characterizes the surface circulation. The map $\vec{C}$ in figure. 4 in May–June 2009 was obtained for drift of all icebergs and fields and in July-September for icebergs only (fields melt in july). It shows that detailed (by the number of drifting objects) drift measurements in northeastern Barents Sea clearly reproduce the cyclonic circulation highlighted in this region from multiyear data (figure. 1*b*). It is important that the approximate constancy of the direction $\vec{C}$ in the same region from late May to September confirms the hypothesis of quasi-continuity of the non-wind drift within a season.

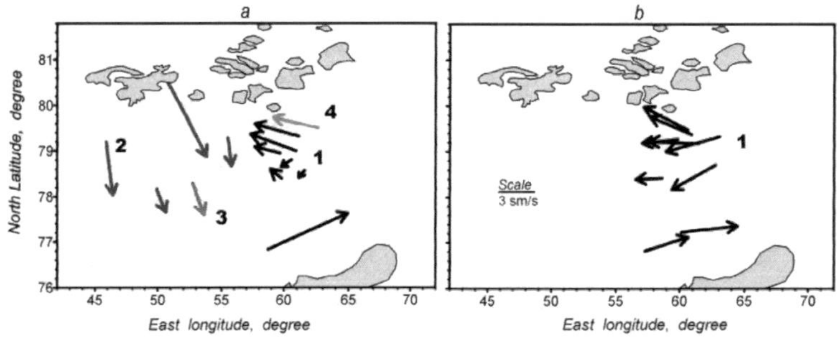

Figure. 4. Non–wind drift iceberg (*a*) and ice floe (*b*) in June (*1*), July (*2*), august (*3*) and September (*4*) in 2009 year between the Franc–Josef Land and Novaya Zemlya, according to [5].

In Arctic hydrocarbon production is actively developing, and icebergs and large ice fields are the most dangerous for economic activities. The Ice Management System makes it possible to identify potentially dangerous objects in advance. To calculate their drift according to wind forecast, accurate estimates of wind coefficients $k=V/G$ and angles of drift deflection from wind are needed. Usually they are defined by the total drift. Replacing the total drift $\vec{V}$ with the wind component $\vec{U}$ using Obukhov's formulas (3, 6) significantly clarifies estimates of *k* and ψ. In the Barents Sea in 2009, k_V coefficients of total drift varied in the range of 0.014—0.18 (median *me*=0.015), for wind drift k_U decreased to 0.010—0.015 (*me*=0.012). For angles of deflection, the difference between ψ_V (14—30°, *me*=18°) and ψ_U (10—15°, *me*=12°) is noticeable in the reduction of median and spread.

2.3. Drift fields in the AO according to satellite (ESV) data

Drift and wind data at the NP stations are the most accurate, but there are only 1 point. The buoy data describe processes only in the Arctic basin. To haracterize modern climatic variability in the entire AO and the adjacent region of the North Atlantic, remote (ESV) drift velocity fields are the best in terms of

IOP Conf. Series: Earth and Environmental Science 1040 (2022) 012024 doi:10.1088/1755-1315/1040/1/012024

spatial coverage. The archive of these fields began in 1979 and is replenished monthly. The drift maps for December 1982 and January 1983 are shown above in figure 1*c,d.*

The most important process for the entire AO is the AB ice ruboff into the Greenland Sea through the Fram Strait. Estimates of the long-term (1979 to 2018 year) variability and trend of ice and wind drift were obtained for a point at coordinates 82.2 °N and 0.0 °d. (figure. 1*c,d*). In accordance with natural seasons highlighted for the AO [16] in Table 5 (in the form of table 2), they are defined for 4 key months: September (ice min), December (ice development), March (ice max), June (ice degradation). In figure 5, average drift velocities $\vec{m}_{\vec{V}}$, wind $\vec{m}_{\vec{G}}$ and dispersion ellipses $\sigma_{\vec{V}}$, $\sigma_{\vec{G}}$ and vectors of corresponding trends $\vec{m}_{\vec{a}}$ are shown in blue. Drift peculiarity in the Fram Strait is stable drift and wind direction (the southern component predominates) at a significant velocity. According to table 5 and figure. 5, average drift velocity m_V in December and March was 6–6.5 cm/s, decreased to 5 cm/s in September, and was minimal at 2.5 cm/s in June. Drift velocity maxima in December and March were 17–18 cm/s, decreased to 14 cm/s in September, and 11 cm/s in June. Decrease in elongation χ for drift relative to wind indirectly demonstrates the contribution of the non-wind current to total drift.

Table 5. Multi–year variance and trend speed month Drift in Wind in Fram strait (82.2° N latitude and 0° longitude) from September, December, March, June 2009 year.

Month	Drift and wind	Speed			Mean vector		Variance speed vector							Trend	
		V_{cp}	V_{max}	σ_V	m_V	φ_V	$I_1^{0.5}$	λ_1	λ_2	χ	α	γ_V	γ_a	a	φ_a
		Speed and variance Drift –cm/s, Wind – m/s; irection – degree; γ_V *and* γ_a – %													
IX	*V*	5.9	14.2	3.0	5.0	171	4.4	3.6	2.6	0.70	168	55	10	1.3	140
	G	2.7	6.6	1.3	1.7	138	2.5	2.0	1.4	0.71	189	75	2	0.3	178
XII	*V*	8.0	17.7	4.3	5.9	198	6.9	6.1	3.2	0.52	184	60	15	2.2	155
	G	4.2	7.8	1.7	3.6	178	2.8	2.3	1.6	0.71	168	65	5	0.5	160
III	*V*	7.9	16.9	4.5	6.4	176	6.4	5.9	2.5	0.42	157	45	17	2.2	170
	G	4.2	8.6	1.8	3.8	177	2.5	2.1	1.4	0.70	162	50	20	0.9	160
VI	*V*	3.3	11.1	2.4	2.3	155	3.4	2.8	1.9	0.45	177	50	5	0.5	180
	G	2.4	5.8	1.1	1.5	112	2.1	1.7	1.2	0.70	174	73	5	0.4	160

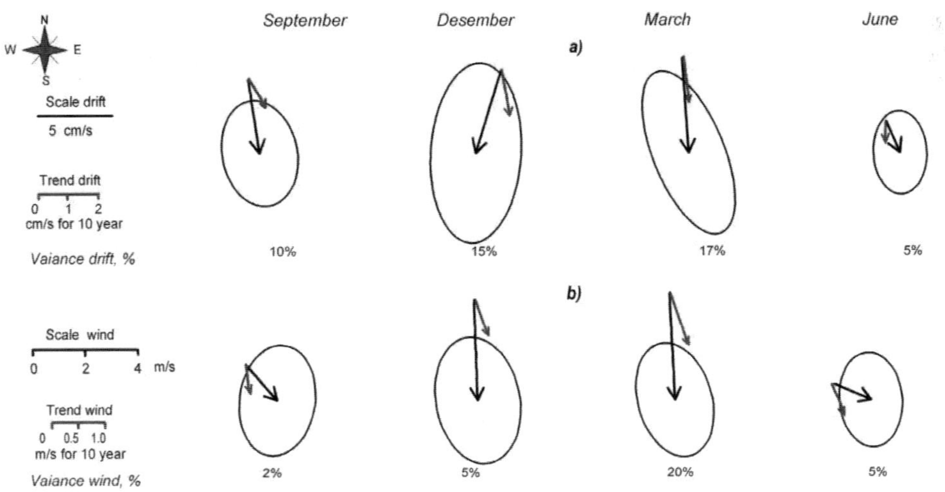

Figure 5. Mean speed $\vec{m}_{\vec{V}}$, variance ellipses $\sigma_{\vec{V}}$ (in black) and vector trend $\vec{a}$ (in blue) of ice drift (*a*) and wind (*b*) and variance trend drift and wind (%).

Table 5 and figure 5 show multi-year trend estimates using formula (8), trend contribution to the variance γ_a and modulus (10–year change) a and direction value φ_a of vector trend coefficient from 1979 to 2018. Drift trend vector $\vec{a}_{\bar{V}}$ and mean drift $\vec{m}_{\bar{V}}$ are oriented predominantly to the south (in March the directions coincide, and in September and December there is a slight deflection $\vec{a}_{\bar{V}}$ from $\vec{m}_{\bar{V}}$ eastward). This confirms the conclusion that in the last decades, the ice drift in the AO is characterized by a positive velocity trend at a relatively stable direction. According to table 5, drift trend variance γ_a is relatively small and depends on the season — 15 % in the cold season (December and March) and 10 % in September. In September and December, wind trend variance is radically smaller (up to 5 %), which indirectly confirms that drift acceleration is caused not by wind, but by a decrease in the total ice extent and a decrease in the concentration of multiyear ice. But in March (ice maximum), the variance of drift and wind trends is increased and comparable (20 %), and in June (at the stage of rapid ice degradation), drift and wind trends disappear weakly (5 %). Therefore, additional studies are needed to clarify the estimates of ice drift trends in the AO.

3. Conclusion

The above results show that the theory of normal correlation of vectors developed by A.M. Obukhov in 1938–40 is relevant for studies of ice dynamics in the AO even at the turn of the XX–XXI centuries in conditions of rapidly changing climatic conditions.

Analysis of ice drift speed in the Arctic Ocean after the year 2000 by the method of A.M. Obukhov confirmed the conclusions of the paper [21] on the intensification of the transpolar drift and the Beaufort Gyre under the conditions of the Arctic current warming. Estimation of wind coefficients and wind drift deviation angles precise the forecast of icebergs and ice fields drift, hazardous to marine activities [5, 14].

Acknowledgments

The reported study was funded by RFBR according to the research project № 18–05–60048.

References

[1] Obukhov A M 1938 *News of the USSR Academy Sciences.* 3 339–370
[2] Obukhov A M 1940 *Scientific notes of Moscow State University.* 45 73–92
[3] Klevantsov Y P Mikulinskaya S M Rozhkov V A. 1996 *Russian Meteorology and Hydrology.* **9** 96–105
[4] Ivanov N E 2004 *Russian Meteorology and Hydrology.* **8** 65–78
[5] Zubakin G K Ivanov N E 2014 *Russian Meteorology and Hydrology***.** **10** 61–72
[6] Belyshev A P Klevantsov Y P Rozhkov V A 1983. *Probabilistic analysis of sea currents* (Leningrad: Gidrometeoizdat) p 264
[7] Watanabe K 1962 *Oceanogr. J.* **11** 23–37
[8] Gudkovich Z M 1965 *Problems of the Arctic and Antarctic.* **21** 56-60
[9] Gudkovich Z M Doronin Y P 2001 *Drift of sea ice* (Saint-Petersburg: Gidrometeoizdat) p 110
[10] Kochin N E 1961 *Vector analysis and beginnings of tensor calculus* (Moscow: Academy of Sciences USSR) p 425
[11] Nansen F 1902 *Scientific Rezalts. London.* **3** 357–386
[12] Sverdrup H U 1928 Maud" 1918–1925. *Scientific Rezalts. Bergen.* p 46
[13] Zubakin G K 1987 *Large-scale variability of the state of the ice cover of the North European Basin. Drift of sea ice* (Leningrad: Gidrometeoizdat) p 160
[14] Zubakin G K Ivanov N E Nesterov A V 2013 *Proceedings of the Twenty third International*

Offshore and Polar Engineering Anchorage. USA, June 30 – July 5. 1072–1079
[15] Ivanov N E Visnevsky A A Sokolov V T 2011 *Problems of the Arctic and Antarctic*. **1** 5-21
[16] Frolov I E Gudkovich Z M at al 2005 *Scientific research in the Arctic* (Saint-Petersburg) p 267
[17] Volkov V A Ivanov N E Demchev D M 2012 *Journal of Operational Oceanography*. **2** 61–71
[18] *Geography of the World Ocean. Arctic and southern oceans*. (Leningrad: Science) p 501
[19] Zubov N N 1945 *Arctic ice* (Moscow) p 360
[20] *Atlas of the Arctic* 1985 (Moscow) p 204
[21] Kwok R Spreen G Pang S 2013 *J. Geophys. Rec.* 118 656–679

IOP Conf. Series: Earth and Environmental Science 1040 (2022) 012025 doi:10.1088/1755-1315/1040/1/012025

Satellite-derived estimations of the clear-air turbulence in the upper troposphere

A F Nerushev[1]**, K N Visheratin**[1] **and R V Ivangorodsky**[1]

[1] Research and Production Association "Typhoon", Obninsk 249038, Russia,
e-mail: nerushev@rpatyphoon.ru

Abstract. The paper presents the results of a study of turbulence zones in the upper troposphere of the Northern Hemisphere based on measurements from European geostationary meteorological satellites in 2007 - 2018. The essence of the method for determining the zones of turbulence is the use of inhomogeneity's of concentration of water vapor as tracers and the use of correlation-extremal algorithms. Turbulence zones with a horizontal mesoscale turbulent diffusion coefficient $K_{ed} \geq 10^4$ m^2/ s can occupy up to 50% of the Northern Hemisphere area visible from satellites. It is shown that in analyzed time interval there is a significant (by 60-120%) increase in areas occupied by relatively weak and moderate turbulence and a slight decrease (by 6-33%) in areas with strong and very strong turbulence. The temporal variability of these zones and their relationship with the characteristics of jet streams and some climatic parameters are analyzed. A statistical model of the temporal variability of the turbulence zone areas has been built. It is shown that the multiple linear regression model with jet stream characteristics as predictors describes up to 68% of variability of zones of strong turbulence. The use of climatic parameters as predictors makes it possible to describe up to 74% of the temporal variability of strong turbulence zones.

1. Introduction

In the atmosphere at altitudes of 6-11 km there is so-called Clear Air Turbulence (CAT). It is mainly associated with jet streams with very strong vertical and horizontal shears of wind speed. The CAT can pose a serious threat to aircraft traveling at specified altitudes. It can cause pitching, manifested in the form of tremors, frequent and small jolts or throws of the aircraft in different directions, including up and down [1]. According to [2, 3], the probability of encountering CAT and orographic turbulence caused by the air flow over the mountain range in the upper troposphere can exceed 80%. Turbulence is costly for airlines: the annual economic damage from it for US carriers varies from 100 to 200 million dollars [3].

Therefore, the problem of diagnosing and predicting atmospheric turbulence is one of the most important problems of aviation meteorology. The difficulty is that CAT zones cannot be visually identified, unlike turbulence zones in clouds. They are also not identified by the aircraft radars. The main sources of information about CAT are reports from pilots, as well as measurements of vertical accelerations, equivalent vertical gust velocities, vortex dissipation velocity using equipment installed on aircraft, which are transmitted within the AMDAR (Aircraft Meteorological DAta Relay) program (https://public.wmo.int/en/programmes/global-observing-system/). It should be noted that this information is mainly of a qualitative nature, identifying turbulence in the form of gradations - weak, moderate, strong, extreme [4]. The calculations of certain predictors (CAT indices) are most often used based on predictive models [1]. It can be stated that at present there are no reliable instrumental methods for determining the characteristics of turbulence in a clear air.

The method we have developed for processing satellite information allows us to identify turbulence zones, including CAT, in the upper troposphere in quasi-real time by the movement and evolution of atmospheric tracers [5]. This article describes shortly the method for determining turbulence zones in the upper troposphere and analyzes the temporal variability of these zones in 2007 - 2018. Their relationship with the characteristics of jet streams and some climatic parameters is analyzed, and simple statistical models are described.

2. Initial data and methods used

The data of SEVIRI radiometer (Spinning Enhanced Visible and Infrared Imager) of the European geostationary meteorological satellites of the second generation Meteosat 8– Meteosat 10 for the period 2007–2018 was used as the initial satellite information. Turbulence zones are determined on the basis of calculating horizontal mesoscale turbulent diffusion coefficient K_{ed} in the upper troposphere. The essence of the method, described in detail in [5, 6], consists in the use of inhomogeneity's of the field of concentration of water vapor as a conservative tracers and the use of correlation-extremal algorithms. These algorithms, based on calculations of the characteristics of spatial and spatio-temporal structural functions, allow us to determine on a single spatial scale the components of the horizontal velocity of tracer, vorticity and the coefficient of horizontal mesoscale turbulent diffusion K_{ed}.

The main elements of the method are:

• The 6.2 μm water vapor channel of SEVIRI radiometer is used. The spatial resolution of the images at the subsatellite point is 3 km, at the boundaries of the calculated area – approximately 2 times worse.

• The values of the coefficient of horizontal mesoscale turbulent diffusion K_{ed} are calculated from 2 images, separated by a time interval of Δt = 15 min at grid points with a step of 10 pixels in the region of 0–60 N, 60W – 60E. The area occupied by turbulence with different ranges of K_{ed} variation is calculated. The calculation results are attributed to the level of the maximum weight function of the radiometer at a wavelength of 6.2 μm (approximately 350 hPa, about 8 km for mid-latitude conditions).

• The results of [7] were used to validate the K_{ed} values from satellite data.

The following characteristics of jet streams were used as predictors for constructing the first version of the statistical model of turbulence variability: the average area of the jet stream (S); maximum wind speed (V_m) on the axis; latitude (φ) and longitude (λ) of the center of the jet stream area; the maximum gradient (shear) of the horizontal wind speed on the cyclonic side (G_c) and on the anticyclonic side (G_a); the effective "life" time of the jet stream (τ) [8].

In the second version of the statistical model the climatic parameters are used as predictors:

- monthly mean quasi-zonal (averaged over the longitudinal region 70W – 70E) tropospheric temperature (T_{tr}) values at different levels according to the NCEP / NCAR reanalysis data [9] and their differences at levels 200 and 500 hPa (ΔT200 and ΔT500) between low (0^0) and high north (80^0) latitudes;
- monthly mean area of Arctic sea ice (S_{ice}) according to NOAA data (ftp://sidads.colorado.edu/DATASETS/NOAA/).

Relationships between turbulence zones and predictors were analyzed using cross-correlation wavelet analysis [10, 11]. Ridge regression and stepwise regression methods were used to assess the collinearity of predictors and to construct multi-regression models. Preliminarily, the time series of turbulence zones as well as used predictors were standardized and a linear trend was excluded from the series.

3. Main results and discussion

3.1 Temporal variability and parameters of turbulence zones

Calculations showed that the K_{ed} values lie mainly in the range ($10^4 - 10^6$) m^2 / s. Figure 1 gives an idea of average turbulized zones for the considered time interval. It can be seen that turbulence zones with

$K_{ed} \geq 10^4$ m^2 / s can occupy up to 50% of the Northern Hemisphere visible from the satellite, and with $K_{ed} \geq 10^6$ m^2 / s – only 0.5%.

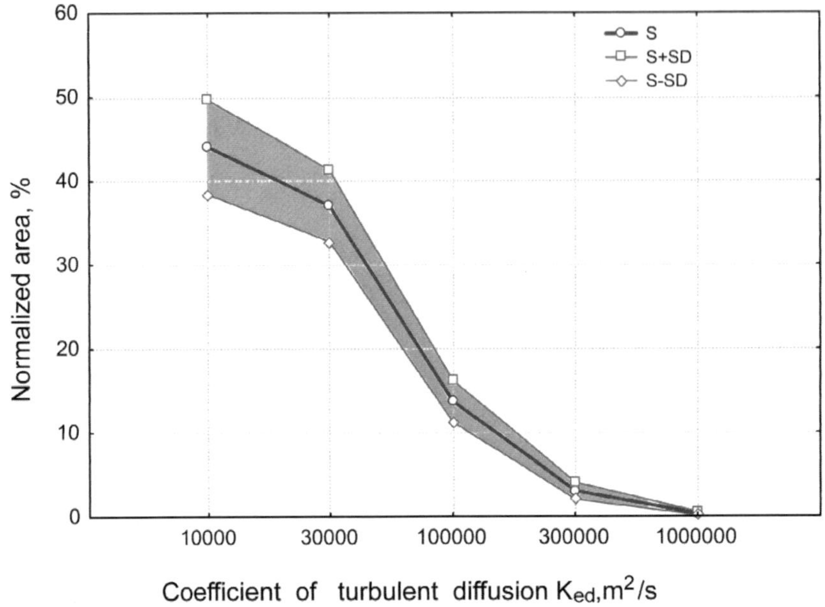

Figure 1. Average value of Northern Hemisphere area occupied by turbulence with different values of K_{ed} in 2007–2018.

Let us consider the variability of the monthly average characteristics of turbulence zones. For further analysis, we introduce gradations of turbulence based on ranges of K_{ed} values (in m^2/s): weak turbulence ($3 \cdot 10^4 > K_{ed} \geq 10^4$), moderate turbulence ($10^5 > K_{ed} \geq 3 \cdot 10^4$), strong turbulence ($3 \cdot 10^5 > K_{ed} \geq 10^5$) and very strong turbulence ($10^6 > K_{ed} \geq 3 \cdot 10^5$). Analysis shows that the largest area (25% of the Northern Hemisphere viewed from a satellite) falls on moderate turbulence, followed by strong (10%), weak (8%) and very strong (2%) turbulence. Almost all areas of turbulent zones are characterized by noticeable interannual and seasonal variability. Since 2007 the weak turbulence area has increased more than 2 times, moderate turbulence - more than 50%. At the same time, the areas of strong turbulence and very strong turbulence decreased by 6% and 33% respectively.

In what follows, we will consider the characteristics of strong turbulence. Turbulence zones in the first approximation can be presented in the horizontal plane by ellipses. According to the calculations, the characteristic parameters of the zones of strong turbulence are as follows: the minimum size is (100 km × 50 km), the maximum size is (2000 km × 600 km), the average size is (600 km × 250 km), and the characteristic "life" time is (0.5 - 3) hours.

3.2. Relationship with jet streams characteristics

There should be a relationship between the parameters of the turbulence zones and the characteristics of jet streams. Such relationships are clearly manifested with the jet stream area (S), the jet stream velocity gradient on the cyclonic side (G_c), and the jet stream center latitude (φ) [12]. Cross-wavelet analysis reveals clearly pronounced co-phase oscillations with a period 12 months for the S and G_c series with the area of turbulence zones. The counterphase oscillations of the φ series and the area of turbulence zones are also clearly manifested. Thus increasing jet stream area and velocity gradient leads to an increase in the turbulence zones area, which is quite natural, and the poleward displacement of the position of the jet stream causes a decrease in the area of turbulence zones [12].

3.3. Relationship with basic climatic parameters

The relationship between strong turbulence area and the temperature characteristics of upper troposphere and Arctic ice area is well traced. Figure 2 shows the results of a cross-wavelet analysis of the area of strong turbulence with the specified climatic parameters, as well as the time course of these characteristics. The arrows show the relationship between the phases of the time series: to the right - in phase, to the left - in counterphase. The degree of correlation of the analyzed series (color scale) is given in relative units. The bold black line marks the boundaries of the region with a confidence interval of more than 90%. Considering the above and taking into account the results of [13], in which the relationship between the characteristics of jet streams and the temperature characteristics of the troposphere and the area of Arctic sea ice is studied, it should be stated that the relationship between the area of strong turbulence and the main climatic parameters acts, from our point of view, indirectly - through the characteristics of the jet stream, primarily through its area and velocity gradient.

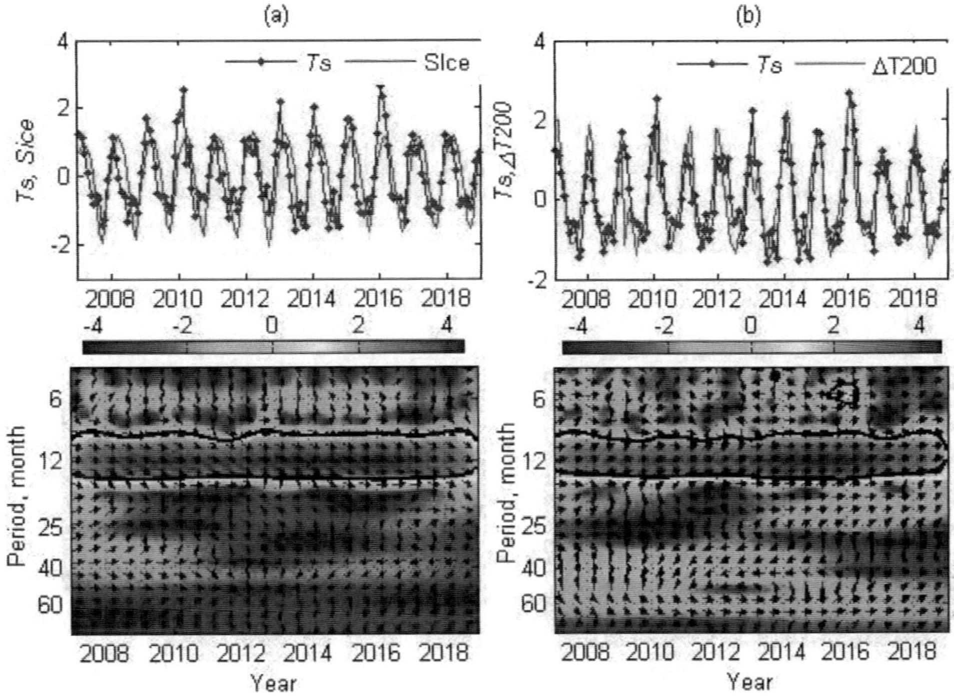

Figure 2. Cross-correlation wavelet transform of strong turbulence Ts and ice area S_{ice} (a), and temperature difference ΔT200 at 200 hPa (b). Upper figures show original time series.

3.4. Statistical model

Based on the multiple linear regression analysis, two versions of statistical model was designed for temporal variability of strong turbulence area. The characteristics of jet streams were used as predictors for constructing the first version of the model. The temperature characteristics of the troposphere and the area of Arctic sea ice were used for the second model. Time series of area of turbulence zones, as well as predictors, were presented as standardized series with removed linear trend. Preliminary calculations by the ridge regression method showed the absence of collinearity of the predictors.

To construct a statistical model the method of stepwise multiple regression was applied. The stepwise multiple regression technique is based on the iterative procedure of selection of possible predictors and estimation of magnitude and significance of the regression coefficients. As a result, the predictors with the largest contribution to the explained variance were included in model and predictors whose contribution was less than 1% were excluded. Figure 3 shows the dependence of the explained variance for predictors with the largest contribution. The model with the jet stream predictors describes variations in strong turbulence with a coefficient of determination R2 = 0.68 and model with the climatic predictors - with R2 = 0.74. The regression coefficients for models are shown in the table.

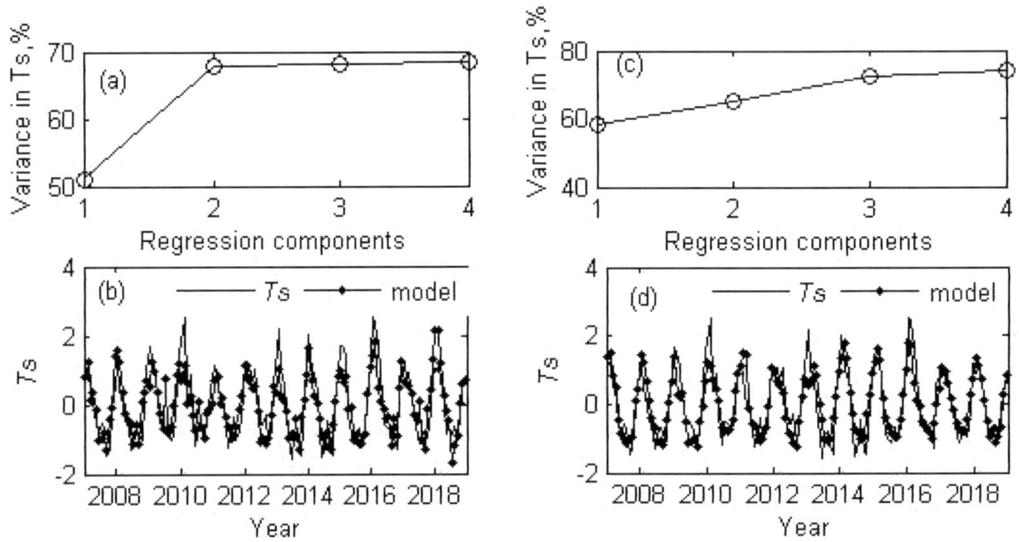

Figure 3. Statistical model. a) and c) - dependence of explained variance in *Ts* from regression components *S*, *φ*, G_c, V_m and ΔT200, ΔT500, S_{ice}, T_{tr}. b) and d) - comparison of experimental data and calculations by model with *S*, *φ*, G_c, V_m and with ΔT200, ΔT500, S_{ice}, T_{tr}.

Table. Parameters of statistical models. In brackets 95% confidence interval

	Jet stream predictors	Regression coefficient
1	Area (*S*)	0,58 (0,24)
2	Center latitude (*φ*)	-0,46 (0,23)
3	Gradient (G_c)	-0,06 (0,07)
4	Maximum wind speed (V_m)	-0,04 (0,10)
	Variance explained	0,68
	Climatic predictors	Regression coefficient
1	ΔT200	0,55 (0,20)
2	ΔT500	0,35 (0,29)
3	S_{ice}	0,35 (0,19)
4	T_{tr}	0,31 (0,27)
	Variance explained	0,74

It can be seen that the quality of both versions of the model is quite high, which indicates the dominant role of the jet stream area and temperature difference in the temporal variability of zones of strong turbulence. It should be noted that the time series analyzed in the article have a prevailing annual harmonic, which could lead to rather high estimates of correlations. To estimate the contribution of long-period variations the approach [14] was used. For this, seasonal variations and oscillations with periods from 3 to 20 months in the series of strong turbulence and predictors were filtered out. The results of these calculations confirm the conclusion about the predominant influence of the area and latitude of jet streams and temperature gradients between low and high latitudes on the area of strong turbulence.

4. Conclusion

The method developed for processing atmospheric sounding data obtained by geostationary meteorological satellites in the absorption band of water vapor with a center at 6.2 μm makes it possible to identify turbulence zones of different intensities in the upper troposphere. The main restriction of the method consist in a limited altitude range (200-500 hPa), determined by the type of the radiometer contribution function in the channel of 6.2 microns, as well as the spatial scale of the determined turbulent characteristics (mesoscale).

Analysis of the relationship between turbulence zones, jet streams characteristics and basic climatic parameters allow us to draw the following conclusions.

1. The average area of zones with a horizontal mesoscale turbulence coefficient $K_{ed} \geq 10^4$ m^2 / s in the upper troposphere in 2007–2018 is about 50% of the northern hemisphere visible from the satellite.
2. The largest area (on average 25%) is occupied by areas with moderate turbulence, in which 10^5 m^2 / s> $K_{ed} \geq 3{\bullet}10^4$ m^2 / s.
3. In the considered time interval there is a significant (by 120-60%) increase in the area occupied by relatively weak and moderate turbulence and a slight decrease (by 6-33%) in areas with strong and very strong turbulence.
4. A regression model with the jet stream characteristics as predictors describes up to 68% of the temporal variability of strong turbulence area.
5. A regression model with the climatic predictors describes up to 74% of the temporal variability of strong turbulence area.

Acknowledgments
The authors are grateful to the colleagues from Planet Research Center for providing satellite data. This work was supported by the RFBR grant 18-05-00831a.

References

[1] Shakina N P and Ivanova A R 2016 *Forecasting meteorological conditions for aviation* (Moscow: Triada Ltd p 312)
[2] Meneguz E, Wells H and Turp D 2016 An automated system to quantify aircraft encounters with con-vectively induced turbulence over Europe and the Northeast Atlantic *J Appl. Meteorol. Climatol.* **55** (5) 1077– 89
[3] Williams P D 2017 Increased light, moderate, and severe clear-air turbulence in response to climate change *Advances in atmospheric sciences*. **34** 576-586.
[4] WMO No. 306: Manual On Codes 2015
[5] Nerushev A F and Kramchaninova E K 2011 Method for Determining Atmospheric Motion Characteristics Using Measurements on Geostationary Meteorological Satellites *Izvestiya, Atmospheric and Oceanic Physics* **47** 9 1104–1113
[6] Nerushev A F and Ivangorodsky R V 2019 Determination of turbulence zones in the upper troposphere based on satellite measurements *Sovremennye problemy distantsionnogo zondirovaniya Zemli iz kosmosa* **16** 1 205–215
[7] Golitsyn G S 2001 An Explanation of the Relative Eddy Diffusion Law in the Atmosphere and on the Ocean Surface *Dokl. Earth Sci.* **381** 8 939–941
[8] Ivangorodsky R V and Nerushev A F 2014 Characteristics jet streams upper troposphere by measurements of the European geostationary meteorological satellites *Sovremennye problemy distantsionnogo zondirovaniya Zemli iz kosmosa* **11** 1 45-53
[9] Kalnay E, Kanamitsu M, Kistler R, Collins W, Deaven D, Gandin L, Saha S, White G, Woollen J, Chelliah M, Janowiak J, Mo K C, Wang J, Leetmaa A, Reynolds R, Jenne R, Kung E and Salstein D 1996 The NCEP/NCAR Reanalysis 40-year Project *Bull. Amer. Meteor. Soc.* **77** 437-471.
[10] Torrence C and Compo G P 1998 A practical guide to wavelet analysis *Bull. Am. Meteorol. Soc.* **79** 61–78
[11] Grinsted A , Moore J C and Jevrejeva S 2004 Application of the cross wavelet transform and wavelet coherence to geophysical time series *Nonlin. Processes Geophys.* **11** 561–566

[12] Nerushev A F, Visheratin K N and Ivangorodsky R V 2020 Characteristics of the upper troposphere wind field according to the satellite measurements and their connection with climatic parameters *IOP Conf. Series: Earth Environ. Sci.* **606** 012041
[13] Nerushev A F, Visheratin K N and Ivangorodsky R V 2019 Dynamics of High-Altitude Jet Streams from Satellite Measurements and Their Relationship with Climatic Parameters and Large-Scale Atmospheric Phenomena *Izvestiya, Atmospheric and Oceanic Physics* **55** 9 1198–1209
[14] Nerushev A F, Visheratin K N and Ivangorodsky R V 2021 Statistical model of time variability of characteristics of northern hemisphere high-altitude jet streams based on satellite measurements *Izvestiya, Atmospheric and Oceanic Physics* **57** 4 401-413

IOP Conf. Series: Earth and Environmental Science 1040 (2022) 012026 doi:10.1088/1755-1315/1040/1/012026

On the ratio of the components of the atmospheric vertical electric current density in fair weather

S V Anisimov, S V Galichenko, A A Prokhorchuk, K V Aphinogenov

Borok Geophysical Observatory of Schmidt Institute of Physics of the Earth RAS, Borok, Yaroslavl region, RF

Corresponding author e-mail: svga@borok.yar.ru

Abstract. This work presents the results of ground-based and tethered balloon observations of altitude profiles of the components of the atmospheric vertical electric current density. The magnitude of the conduction current density was observed in the range 0.2 - 2 pAm^{-2}. The ratio of the negative to positive component of the conduction current density averaged 1.6 outside the layer, where their dependence on the height was noticeable. In the framework of developed numerical model with reasonable values of the parameters it is found that the ratio of the convection current density to the density of the total vertical atmospheric electric current in the atmospheric boundary layer mainly falls in the range 0.2 - 0.6 and tends to increase under strong convection and low electrical conductivity. An exponential parameterization of the dependence of the electromotive force on the ground-level electrical conductivity is proposed.

1. Introduction

The Earth's atmosphere is a weak conductor through fair weather [1, 2] regions of which a downward electric current with a density of the order of 10^{-12} A/m^2 continually flows. This air-earth conduction current is driven by ubiquitous atmospheric electric field, which is due to a variety of worldwide electromagnetic phenomena. Among these are global thunderstorm activity, mesoscale convective systems, electrified clouds, charged precipitation, magnetospheric dynamo, etc. [3 - 9]. In addition to global and large-scale generators, local space charge on aerosols and/or microdroplets can produce significant disturbances in the electric field, thereby affecting the ratio of the components of the atmospheric electric current density [10, 11]. Above the planetary boundary layer, the net vertical electric current density is the sum of the polar conduction current densities due to oppositely directed drift of positive and negative air ions in the electric field and diffusion currents acting against the ion concentration gradients [12]. The balance of the components of the atmospheric vertical electric current within the boundary layer is significantly complicated by the turbulent mixing of space charge arising from the electrode effect at the earth's surface and conductivity inhomogeneities caused by the attachment of ions to aerosol particles of various origins, which are abundant in the lower atmosphere [13 - 16]. Time-dependent and patchy distribution of the rate of exhalation of radioactive radon gas is another reason for the generation of large fluctuations in the ion production rate and eventually in the conductivity, which can enhance charge separation processes in a following way. Small ions, created in a continuous process of air ionization and then advected by a velocity field are involved into a turbulent cascade extending from a scales comparable to the mixed-layer height up to the smallest scales determined by the Schmidt number. This cascade along with inhomogeneous distribution of radioactive nuclei results in a wide range of spatiotemporal scales of fluctuations in polar electrical conductivities. The ratio of polar conductivities considered as a local imbalance of positive and negative components of the conduction current determines a change of charge density in time in a certain volume due to an electric field. Under these circumstances, the convection (otherwise turbulent) current associated with the vertical mixing of space charge by a turbulent flow can comprise a significant fraction of the net electric current density, thereby giving rise an electromotive force (EMF). This issue has attracted attention and first theoretical and modeling estimates indicated that the values of this variable in very unstable boundary layer can reach several tens of percent of the ionospheric potential [17]. Subsequent attempts to assess this EMF acting in fair weather regions of the lower atmosphere over land and sea gave a value of a few tens kilovolts distributed in the first several hundred meters above the earth's surface [18]. Recently performed calculations using contemporary methods for modeling turbulent transport processes in the convective boundary layer gave a smaller range of values not exceeding 20 kV. The results of this simulation made it possible to put forward an assumption about the possibility of the formation of electrical stratification with several series-connected sections, in which the generators can operate in the opposite direction, partially compensating each other [19, 20]. Estimates of the electromotive force based on the data collected during systematic ascents and descents of a tethered balloon with a measuring platform fall in the range of 5 - 55 kV, which is consistent with those obtained earlier, but the localization of the charge separation region did not allow to assume that the generator was associated with convective mixing of space charge [21].

This work aims to estimate variability of the components of the atmospheric vertical electric current density in the atmospheric boundary layer (ABL) above land based on results of balloon-borne observations and a model of turbulent electrodynamics consisted of a large-eddy simulation to reproduce of atmospheric turbulent flow dynamics and the Poisson equation for calculation of electric potential and electric field created by a distribution of air ions and charged aerosol particles.

2. Site and instrumentation

Ground-based and tethered balloon observations of atmospheric electricity was conducted at the mid-latitude Borok Geophysical Observatory site [58°04' N; 38°14' E] located in an environment with a natural electromagnetic, radioactive and aerosol background. The results of observations presented in this work were obtained in the summer-autumn season 2019. A winch-driven helium-filled balloon made of polyester fiber material was used for measuring aloft. The following devices were mounted on the original measuring platform: two electrostatic field mills, two Gerdien chamber type sensors of air polar conductivities, sensor of radon activity Radon Scout PMT, aerosol particle counter AEROTRAK 9306v2, barometric altimeter, GPS receiver, meteorological and telemetric sensors, autonomous controlling and data collection systems, and power supply [21]. The elevation of platform varied from 2 to 500 m depending on wind speed and research program.

Instrumentation for ground-based observations consisted of three electrostatic field mills, one of which was placed on a metal sheet flush with the ground and two others were mounted at a height of 1 m above the earth's surface. Three Gerdien chamber type sensor of polar conductivities were installed on a mast on a heights of 0.35 m, 1.3 m, and 2.1 m. Temperature profiles were measured with a meteorological temperature profiler MTP-5HE.

3. Governing equations and numerical methods

The building blocks of the numerical model for simulation of turbulent electrical processes in the atmosphere are the conservation equations for the momentum, mass, energy, and electric charge, Poisson equation for the electric potential, and Ohm's law in the differential form. Simulations of large-scale turbulent flow field in the ABL are performed with the Parallelized Large-Eddy Simulation Model (PALM) version 6.0 code with adjustable parameters for connecting subgrid turbulent flows of momentum and scalar quantities [22]. The model solves the following system of filtered Navier-

Stokes equations under anelastic approximation with the advection-diffusion equation for the filtered potential temperature

$$\frac{D\boldsymbol{u}}{Dt} + 2\boldsymbol{\Omega}\times\boldsymbol{u} = -\nabla\left(\frac{\delta p}{\rho_0}\right) + g\frac{\delta\theta}{\theta_0}\boldsymbol{k} + \frac{\boldsymbol{F}}{\rho_0}, \tag{1}$$

$$\nabla\cdot\left(\rho_0\boldsymbol{u}\right) = 0, \tag{2}$$

$$\frac{D}{Dt}\left(\frac{\delta\theta}{\theta_0}\right) + \frac{w}{\theta_0}\frac{d\theta_0}{dz} = \frac{Q}{\rho_0 c_p T_0}, \tag{3}$$

$$\frac{\delta\theta}{\theta_0} = -\frac{\delta p}{\rho_0 g}\left(\frac{1}{\rho_0}\frac{d\rho_0}{dz}\right) - \frac{\delta\rho}{\rho_0}, \tag{4}$$

$$\frac{\delta p}{p_0} = \frac{\delta\rho}{\rho_0} + \frac{\delta T}{T_0}, \tag{5}$$

taking into account the Coriolis force and damping internal waves in the overlying weakly stable free atmosphere. The instantaneous resolved flow field $\boldsymbol{u}_r$ is found as a solution on a Cartesian rectangular grid, the cell size of which is determined by the filtration procedure. The anelastic approximation assumes small deviations of density $\delta\rho$, pressure δp and temperature δT from their reference vertical profiles (ρ_0, p_0, T_0) [23]. A dynamic eddy viscosity is used as a structural subgrid-scale model to determine the contribution of unresolved scales to the stress tensor [24].

The contribution of the velocity fluctuations at the subgrid scales to the evolution of scalar fields is modeled by an isotropic divergence-free field of random Fourier modes

$$\boldsymbol{u}'(\boldsymbol{r},t) = \sum_{n=1}^{N}\left[\boldsymbol{a}_n\cos(\boldsymbol{k}_n\boldsymbol{r} + \omega_n t) + \boldsymbol{b}_n\sin(\boldsymbol{k}_n\boldsymbol{r} + \omega_n t)\right]. \tag{6}$$

with a geometric distribution of the lengths of random isotropic wave vectors $\boldsymbol{k}_n$ in the range from the geometric mean width of the LES filter k_c to the inverse Kolmogorov scale k_η, taking into account the spectral flux of the turbulent kinetic energy (TKE) in the inertial range [20, 25, 26]

$$E_{SGS} = \int_{k_c}^{k_\eta} C_K\varepsilon^{2/3}k^{-5/3}dk\,. \tag{7}$$

This allows its conjugation with the rate of dissipation of the TKE in the explicitly resolved LES scales

$$\varepsilon = k_c\left(\frac{2E_{SGS}}{3C_K}\right)^{3/2}. \tag{8}$$

A significant advantage of this approach is the high accuracy of the reproduction of the scaling of the scalar relative dispersion [26].

In the vicinity of the atmosphere-earth boundary an anisotropy and inhomogeneity of statistical parameters of turbulence is pronounced. Moreover, near-wall turbulence statistics is intermittent and strongly non-Gaussian. In order to generate subgrid-scale near-wall turbulent flow field $\boldsymbol{u}'$, a second-order Lagrangian stochastic model is involved which takes into account leptokurtic statistics of random accelerations and variations in the instantaneous dissipation rate of the TKE along Lagrangian trajectories [27]. In a weakly stable free atmosphere, a first-order Lagrangian stochastic model is used, which is coupled to LES in terms of turbulence statistics and the subgrid TKE dissipation rate [20].

The balance equations for the concentration of small ions are

$$\frac{Dn^{\pm}}{Dt} = q - \alpha n^{\pm} n^{\mp} - n^{\pm} \sum_{j} \int \beta_j^{\pm}(D_a) f_j^{\pm}(D_a) dD_a \pm \nabla(\mu^{\pm} n^{\pm} \nabla \phi) + \nabla(D_m^{\pm} \nabla n^{\pm}), \quad (9)$$

where q is the ion pair production rate, α is the recombination rate, $\beta_j^{\pm}$ are the size dependent ion-aerosol flux coefficients accounting for the attachment rate of small ions to aerosol particles with the effective diameter D_a and carrying j elementary charges (j can be equal to zero), $f_j^{\pm}$ are the aerosol particles size distribution functions, $\mu^{\pm}$ the mean mobilities of small ions, ϕ is the electric potential, and $D_m^{\pm}$ is the molecular diffusivity of small ions. Polydisperse atmospheric aerosol is described by a three-mode particle size distribution function, which depends on the sources of particles and their evolution.

The ion pair production rate q is calculated as the amount of ion production rates from cosmic rays (q_c), soil radiation (q_s) and the contribution of alpha- and beta-particles emitted in the sequence of decays of radon and thoron, releasing from the soil into the atmosphere and their short-lived daughter products

$$q = q_c + q_s + \sum_{i} \frac{\lambda_i R_i E_i^{(\alpha)(\beta)}}{E_p^{(\alpha)(\beta)}}, \quad (10)$$

where $E_p^{(\alpha)}$ and $E_p^{(\beta)}$are the energies spent by alpha- and beta-particles respectively on a single ionization act, R_i are the concentration of radionuclides in the decay chain of radon isotopes, and λ_i are their decay constants [28].

The evolution of the atmospheric distribution of radioactive inert gases radon and thoron, as well as their daughter products, is calculated based on the exhalation rate and the numerical model of turbulent dispersion. The balance equations for the concentration of charged aerosol particles (large ions) are [29]

$$\frac{DN_j^{\pm}}{Dt} = n^{+} \beta_{j-1}^{+} N_{j-1}^{\pm} - n^{+} \beta_j^{+} N_j^{\pm} + n^{-} \beta_{j+1}^{-} N_{j+1}^{\pm} - n^{-} \beta_j^{-} N_j^{\pm}, \quad (11)$$

with the ion-aerosol flux coefficients derived in [14]. The number of balance equations for charged aerosol particles is determined by the product of the number of considered charge states of particles by the number of size fractions obtained as a result of discretizing the particle size distribution function.

The numerical solution of the Poisson equation

$$\Delta \phi = -\frac{e}{\varepsilon_0} \left\{ n^{+} - n^{-} + \sum_{j} j N_j^{\pm} \right\} \quad (12)$$

obtained by the conjugate gradient method solving a system of discrete Poisson equations defines the distribution of the electric potential in the computational domain using the time-evolving space charge density distributions interpolated on the grid. At domain's side walls periodic boundary conditions are set. The earth's surface is considered to be an ideally conducting plane; therefore, the equipotentiality boundary condition is set on the earth's surface. The potential at the upper boundary of the computational domain is determined by the density of the total vertical electric current J and the electrical resistance of the atmospheric column R_e above the domain

$$V_t = V_i - JR_e. \quad (13)$$

The connection between the values of the total current density J and the EMF of a generator operating in this column is specified out through Ohm's law for the section of the circuit containing the generator [19]

IOP Conf. Series: Earth and Environmental Science 1040 (2022) 012026 doi:10.1088/1755-1315/1040/1/012026

$$V_{\mathrm{i}} = JR + \mathrm{EMF}\,, \tag{14}$$

where V_{i} is the potential drop between the lower ionosphere (at an altitude of about 60 km) and the earth's surface (ionospheric potential), R is the total electrical resistance of the considered atmospheric column.

Polar components of the conduction current density are found from

$$\boldsymbol{j}_c^{\pm} = e\mu^{\pm} n^{\pm} \nabla \phi\,. \tag{15}$$

The continuity equation relates the time variation of the space charge density with the local change in space of the components of the total current density

$$e\frac{\partial}{\partial t}\left(n^{+} - n^{-} + \sum_{j} jN_j^{\pm}\right) + \nabla\left(\boldsymbol{j}_c^{\pm} + \boldsymbol{j}_t + eD_m^{+}\nabla n^{+} - eD_m^{-}\nabla n^{-}\right) = 0\,, \tag{16}$$

where $\boldsymbol{j}_t$ is the density of turbulent electric current defined as

$$\boldsymbol{j}_t = e\left(n^{+} - n^{-} + \sum_{j} jN_j^{\pm}\right)\left(\boldsymbol{u}_r + \boldsymbol{u}'\right). \tag{17}$$

Effects associated with molecular diffusion becomes noticeable only at significant gradients of ion concentrations that can be achieved very close to the earth's surface. Simple estimates show that at distances from the surface comparable to the Kolmogorov scale the density of diffusion electric current becomes comparable to the value of the conduction current density and requires taking into account. The turbulent and molecular fluxes of ions to the earth's surface in this neighborhood are treated in the same way as in [19].

In this study we have simulated 12 cases with different surface vertical turbulent sensible and latent heat fluxes, aerosol particle concentration, and radon exhalation rate. Specific values of input parameters are presented in table 1.

Table 1. Simulation parameters. z_i is the mixed-layer height, L is the Monin-Obukhov length. The reference value of aerosol particle concentration taken as a relative unit corresponds to 9.17×10^9 particles per m^3

	Sensible heat flux (Wm^{-2})	Latent heat flux (Wm^{-2})	$-z_i/L$	Aerosol load (rel. units)	Radon exhalation rate ($mBqm^{-2}$)
S1	50	100	46.9	1	4.2
S2	100	200	126.8	1	4.2
S3	150	300	181.8	1	4.2
S4	100	200	126.8	0.25	4.2
S5	100	200	126.8	0.5	4.2
S6	100	200	126.8	2	4.2
S7	100	200	126.8	1	2.1
S8	100	200	126.8	1	10.5
S9	100	200	126.8	1	21
S10	100	50	99.6	1	4.2
S11	200	100	206.6	1	4.2
S12	300	150	269.1	1	4.2

IOP Conf. Series: Earth and Environmental Science 1040 (2022) 012026 doi:10.1088/1755-1315/1040/1/012026

4. Results and discussion

4.1. Polar components of the conduction current density

A total of 51 high-altitude profiles of the total conduction current density and its polar components were analyzed that were observed during 8 days of fair weather from August 16 to September 12, 2019. Figure 1 shows an example of temporal changes in vertical profiles of the atmospheric vertical conduction current density with the polar components observed in the afternoon of August 20, 2019 along with the time base of temperature vertical distribution over the period of balloon-borne observations. The magnitude of the conduction current density for the entire time of observations fell into the range 0.2 - 2 pAm^{-2}. In most cases the positive conduction current density measured near the earth's surface was greater or equal than the negative one. However, an episode was registered when the negative component exceeded more than twice the positive one near the earth's surface. The ratio of the negative component to positive one averaged 1.6 outside the layer where their noticeable change with height was observed. A possible reason for such a difference in the polar components of the conduction current density is that during measurements polar conductivities by the Gerdien's chamber, small ions with a mobility less than the threshold are not detected by the sensor, but can still contribute to the conductivity. Since the core of the distribution function of negative small ions by mobility is located in the region of larger values of the argument than of positive ones, this leads to a greater underestimation of the positive conductivity and, hence, the positive component of the conduction current density. In quasi-static conditions, the change in the conduction current density with height indicates the operation of the generator, which either moves positive charges against the electric field or negative charges along it. Traditionally, the nature of this generator, which we look at in the next section, is associated with convective transport of a positive space charge from the electrode layer upwards [17]. Similar altitude profiles in which the magnitude of the conduction current density starting from the surface decreases with height to a certain level, and then remains constant, were also observed over the ocean [18]. However, we also observed another type of altitude profiles with the maximum value of the conduction current density not near the earth's surface, but at a certain height located from about 50 to 120 m. In these cases, the value of the current density at the maximum could exceed 2 times the values at the surface and at higher altitudes, which were approximately equal.

Numerical simulation results show for all considered cases approximate equality of the polar components and the presence of a maximum of the conduction current density located in the range of heights from 20 to 40 m. The maximum value of the conduction current density among simulations made belongs to case with maximal sensible heat flux S12 and its ratio to the total vertical current density is 1.6, while the minimum value of the peak is realized in case with maximal radon exhalation rate S9 and its ratio to the total vertical current density is 1.35. The main difference between the simulation results and the measured profiles is the thickness of the atmospheric layer, in which there is a nonzero gradient of the polar components and the total conduction current density. In this sense, the model better reproduces the cases of strong convection with an insignificant wind shear, which were observed in [18]. Factors increasing electrical conductivity contribute to a decrease in both the conduction current density and its gradient in the upper part of the boundary layer. An increase in the total surface turbulent heat flux as well as in the Bowen ratio results in a more vigorous updraughts causing a decrease in the total vertical current density in line with the mechanism described in [19, 20].

4.2. Turbulent electric current density and EMF

Neglecting the diffusion current density, the absolute value of quasi-static conduction current density in a layer of thickness *h* where the convective turbulent generator acts is equal to

IOP Conf. Series: Earth and Environmental Science 1040 (2022) 012026 doi:10.1088/1755-1315/1040/1/012026

$$j_c = j_t + \frac{V_i - \int_0^h \sigma^{-1} j_t dz}{R}. \tag{18}$$

The second term represents the magnitude of the density of the total atmospheric vertical electric current and is always positive. As can be seen from equation (18), the change in the local conduction current density with respect to the total downward fair-weather current density is determined by the contribution of the convection current density and the EMF generated by this current divided by the total atmospheric columnar electrical resistance R, wherein both the EMF and R are the function of the conductivity.

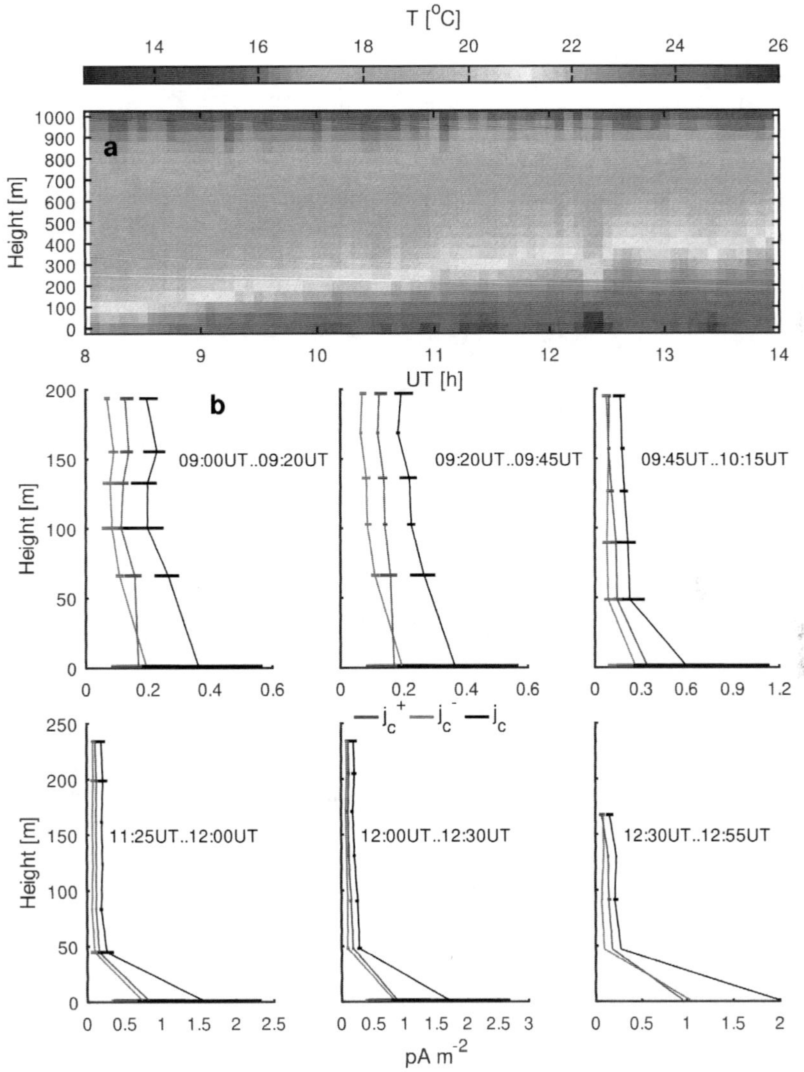

Figure 1. Evolution of average vertical profiles of (a) temperature according to MTP-5HE data, (b) total conduction current density (black) and its polar components (blue and red) on August 20, 2019

Figure 2a shows the vertical dependence of the ratio of the convection current density to the total electric current density in the ABL calculated in the numerical model. With a reasonable choice of parameters this quantity mainly falls in the range 0.2 - 0.6 and tends to increase under strong convection and low electrical conductivity. This effect is understandable, since an increase in vertical turbulent fluxes and/or a decrease in conductivity leads to an increase in the distance overcome by the space charge until it dissipates. Figure 2b shows calculated EMF with respect to ground-level conductivity. Sinulations S2, S4-S9 differing in aerosol particle concentration or radon exhalation rate can be approximated by the exponential law

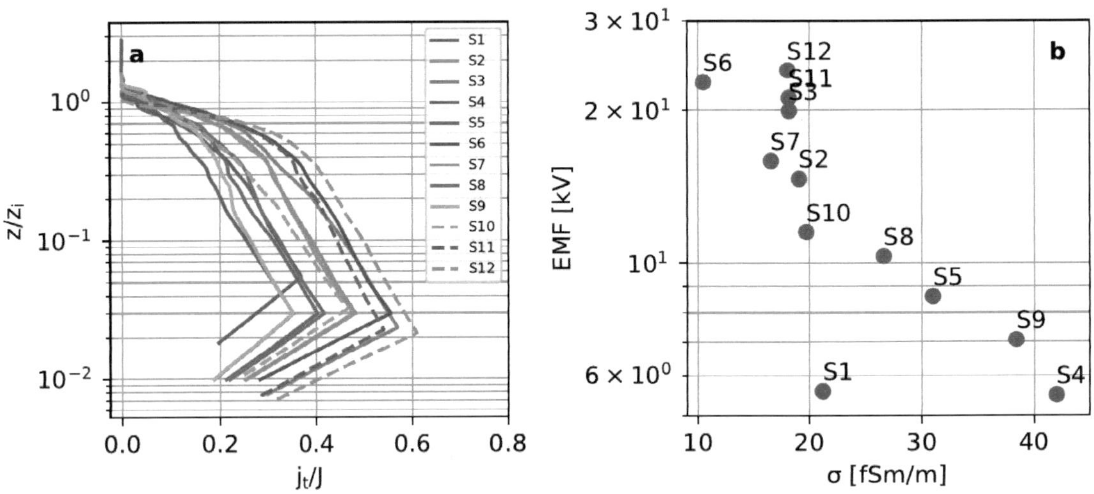

$$\mathrm{EMF} = \mathrm{E}_0 \exp\left(-\frac{\sigma}{\sigma_0}\right), \tag{19}$$

Figure 2. Simulation results: a) average vertical profiles of the ratio of the convection current density to the total electric current density in the atmospheric boundary layer, b) scatterplot of EMF vs. air electrical conductivity near ground surface.

where E_0 = 33 kV and σ_0 = 23.5 fSm^{-1}. Moderate convection cases S1 and S10 give lower EMF values, while strong convection cases S11 and S12 show higher EMF. All the calculated EMF are within the range 5.5 - 24 kV, which corresponds the range from 2.75% to 12% of the ionospheric potential V_i taken equal 200 kV. Thus, taking into account the total area of the globe occupied by convection, the contribution of the fair-weather convective generator to maintenance of the ionospheric potential can be expected at the level of several percent of its value.

5. Conclusions

This work presents the results of ground-based and tethered balloon observations of altitude profiles of the components of the atmospheric vertical electric current density. The magnitude of the conduction current density was observed in the range 0.2 - 2 pAm^{-2}. The ratio of the negative component of the conduction current density to positive one was 1.6 on average outside the layer where their noticeable dependence on the height was observed. In the framework of developed numerical model with reasonable values of the parameters it is found that the ratio of the convection current density to the total atmospheric electric current density in the boundary layer mainly falls in the range 0.2 - 0.6 and tends to increase under strong convection and low electrical conductivity. A numerical model for electrodynamics of the ABL is presented that based on the LES model PALM to reproduce of atmospheric turbulent flow dynamics combined with the kinematic simulation of subgrid-scale scalar transport and the Poisson equation for calculation of electric potential and electric field distribution created by air ions and charged aerosol particles. The model is capable of evaluating the polar components of the conduction current density as well as the convection current density for various regimes of turbulence, aerosol loading, and radon exhalation rates. Results of numerical simulation indicate approximate equality of the polar conduction current density components and the presence of a maximum of the conduction current density located in the range of heights from 20 to 40 m. It is found that factors increasing electrical conductivity contribute to a decrease in both the conduction current density and its vertical gradient in the upper part of the boundary layer. An increase in the total surface turbulent heat flux as well as in the Bowen ratio results in an increase in the EMF and a decrease in the total vertical current density. The EMF of fair-weather convective generator with respect to the ground-level conductivity is approximated by the exponential law. Based on estimates obtained it is suggested that the contribution of the fair-weather convective generator to maintenance of the ionospheric potential can be expected at the level of several percent of its value.

References

[1] Israelsson S 1978 *Pageoph.* **116** 149-58
[2] Harrison R G and Nicoll K A 2018 *J. Atmos. Sol.-Terr. Phys.* **179** 239-50
[3] Mühleisen R 1977 *The global circuit and its parameters Electrical processes in atmospheres* eds H Dolezalek and R Reiter (Darmstadt: Dr Dietrich Steinkopff Verlag) pp 467-76
[4] Gringel W, Rosen J M and Hofmann D J 1986 *Electrical structure from 0 to 30 kilometers The Earth's electrical environment environment* eds E P Krider and R G Roble (Washington: Natl. Acad. Press) pp 166-82
[5] Odzimek A, Baranski P, Kubicki M, Jasinkiewicz D 2018 *Atmos. Res.* **209** 188-203
[6] Siingh D, Gopalakrishnan V, Singh R P, Kamra A K, Singh S, Pant V, Singh R and Singh A K 2007 *Atmos. Res.* **84** 91-110
[7] Soula S and Georgis J F 2013 *Atmos. Res.* **132-133** 264-77
[8] Williams E 2009 *Atmos. Res.* **91** 140-52
[9] Williams E and Mareev E 2014 *Atmos. Res.* **135-136** 208-27
[10] Anisimov S V, Mareev E A, Shikhova N M, Sorokin A E, Dmitriev E M 2005 *Atmos. Res.* **76** 16-28
[11] Anisimov S V, Galichenko S V, Aphinogenov K V and Prokhorchuk A A 2018 *Izv. Physics of the Solid Earth* **54** 764-74
[12] Shreve E L 1969 *J. Atmos. Sci.* **27** 1186-94
[13] Hoppel W A, Anderson R V and Willet J C 1986 *Atmospheric electricity in the planetary boundary layer The Earth's electrical environment* eds E P Krider and R G Roble (Washington: Natl. Acad. Press) pp 149-65
[14] López-Yglesias X and Flagan R C 2013 *Aerosol. Sci. Tech.* **47** 688-704
[15] Jaenicke R 1990 *Russian Chem. Rev.* **59** 959-72
[16] Adams P J and Seinfeld J H 2002 *J. Geophys. Res.* **107** 4370
[17] Willett J C 1979 *J. Geophys. Res.* **84** 703-18
[18] Markson R, Sedláček J, Fairall C W 1981 *J. Geophys. Res.* **86** 12115-21
[19] Anisimov S V, Galichenko S V, Aphinogenov K V, Prokhorchuk A A, Kozmina A S 2019 *Atmos. Res.* **228** 86–94
[20] Anisimov S V, Galichenko S V, Prokhorchuk A A, Aphinogenov K V 2020 *Atmos. Res.* **244** 105035

[21] Anisimov S V, Galichenko S V, Aphinogenov K V, Klimanova E V, Prokhorchuk A A, Kozmina A S, Guriev A V 2021 *Atmos. Res.* **250** 105355
[22] Maronga B, Banzhaf S, Burmeister C, Esch T, Forkel R, Frölich D, Fuka V, Gehrke K F, Geletič J, Giersch S, Gronemeier T, Groβ G, Heldens W, Hellsten A, Hoffmann F, Inagaki A, Kadasch E, Kanani-Sühring F, Ketelsen K, Ali Khan B, Knigge C, Knoop H, Krč P, Kurppa M, Maamari H, Matzarakis A, Mauder M, Pallasch M, Pavlik D, Pfafferott J, Resler J, Rissman S, Russo E, Salim M, Schrempf M, Schwenkel J, Seckmeyer G, Schubert S, Sühring M, von Tils R, Vollmer L, Ward S, Witha B, Wurps H, Zeidler J, Raasch S 2020 *Geosci. Model. Dev.* **13** 1335-72
[23] Bannon P R 1996 *J. Atmos. Sci.* **53** 3618-28
[24] Germano M. Piomelli U, Moin P, Cabot W H 1991 *Phys. Fluids A* **3** 1760-65
[25] Flohr P, Vassilicos J C 2000 *c***407** 315-49
[26] Malik N A, Hussain F 2021 *Phys. Fluids* 33 035135
[27] Reynolds A M 2006 *Phys. Fluids* **18** 075107
[28] Anisimov S V, Galichenko S V, Aphinogenov K V, Makrushin A P and Shikhova N M 2017 *Izv. Physics of the Solid Earth* **53** 147-61
[29] Anisimov S V, Galichenko S V, Aphinogenov K V and Prokhorchuk A A 2018 *Bound.-Layer Meteorol.* **167** 327-48

IOP Conf. Series: Earth and Environmental Science 1040 (2022) 012027 doi:10.1088/1755-1315/1040/1/012027

Three decades of remote sensing of NO_2 vertical distribution and column content at the A. M. Obukhov Institute of Atmospheric Physics

A N Gruzdev and A S Elokhov

A. M. Obukhov Institute of Atmospheric Physics, Pyzhevsky per. 3, 119017 Moscow, Russia

E-mail: a.n.gruzdev@mail.ru

Abstract. The paper reviews the results of 30-year spectrometric measurements of vertical distribution and column content of NO_2 at the Zvenigorod Scientific Station (ZSS) of the A. M. Obukhov Institute of Atmospheric Physics and presents analysis of NO_2 long-term trends and interannual variations. The station is located in the western Moscow region. It is part of the Network for the Detection of Atmospheric Composition Change (NDACC). Measurements are done in the morning and evening twilight by zenith-scattered visible solar radiation. Vertical NO_2 profiles are retrieved and column NO_2 contents are calculated by profile integration. Long term trends and inter-annual variations of NO_2 are analysed by multivariate regression method taking into account autocorrelation of data in a wide range of time scales. Seasonally dependent estimates of NO_2 trends and interannual variations due to the quasi-biennial oscillation, El Nino–Southern oscillation, North Atlantic Oscillation, and the 11-year solar cycle are obtained. Most pronounced (in per cent units) trends and circulation and solar cycle effects are peculiar to the winter–spring period of the year. Mutual influence is assessed between the estimates of the trend and the solar cycle effect in stratospheric NO_2.

1. Introduction

Nitrogen oxides play a key role in the photochemical balance of atmospheric ozone in the stratosphere and troposphere [1, 2]. Nitrogen oxide, NO, and nitrogen dioxide, NO_2, are major contributors to nitrogen oxides. NO and NO_2 are generally in quick photochemical equilibrium between each other, and their ratio depends mainly on ozone content and light intensity. However their sum is more conservative and can be a tracer of atmospheric motions. In fact, stratospheric TO_2 serve as a good indicator of large-scale dynamic processes in the atmosphere [3].

NO_2 has diffusive absorption band in visible spectral range and can be measured remotely. Background NO_2 concentrations provide quite weak absorption of solar radiation, and stratospheric NO_2 measurements are usually done during twilights when differential absorption by NO_2 is maximal due to slant solar ray geometry [4].

Regular remote spectrometric NO_2 measurements at the Zvenigorod Scientific Station (ZSS) started in 1990. The station is a part of the Network for Detecting Changes in the composition of the atmosphere (NDACC). Three decades of NO_2 measurements at the station give possibility to analyze, along with inter-annual variations, also long term changes in NO_2.

IOP Conf. Series: Earth and Environmental Science 1040 (2022) 012027 doi:10.1088/1755-1315/1040/1/012027

2. Measurement method shortly

Zvenigorod station is located in a rural area 50 km west to Moscow. The measurements are carried out by zenith-scattered solar radiation in the visible wavelength range during the morning and evening twilights. The method used allows retrieving the vertical distribution of NO_2 and determining the column NO_2 content in the atmosphere by integrating the profile. The method of observations and retrieval of NO_2 contents from measured spectra are described in detail in a few works [5–7].

Figure 1 shows examples of vertical NO_2 profiles obtained under conditions of clean and polluted surface layer of the atmosphere. The values above the surface level are the NO_2 contents in 5-km thickness layers while the value at zero altitude is the NO_2 content in the near-surface layer, the thickness of which is a priori unknown. Note that the horizontal axis is broken. Figure 1 shows that the NO_2 content in the atmospheric surface layer (ASL) during pollution episodes can be an order of magnitude larger than the total column content in the stratosphere. The NO_2 content of the ASL under unpolluted conditions is close to zero.

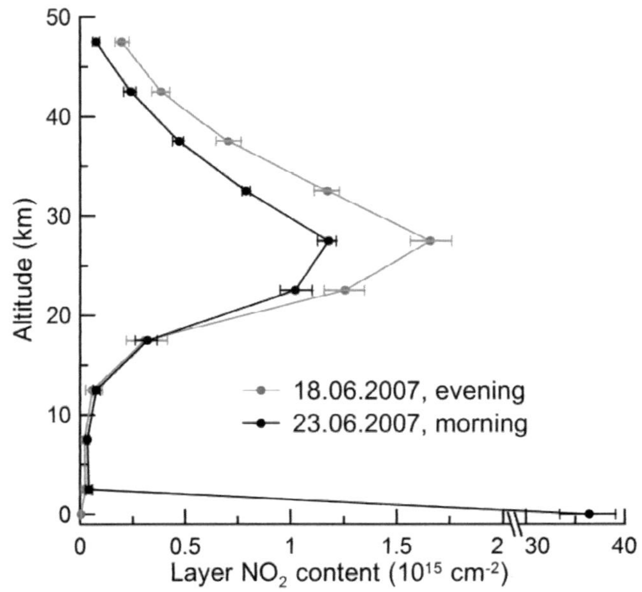

Figure 1. Examples of vertical NO_2 profiles and their standard deviations measured under polluted (black) and unpolluted (gray) near-surface layer. The horizontal axis is broken.

3. Measurement data

Figure 2 shows altitude-time distributions of monthly mean concentrations of NO_2 above the ASL in morning and evening and the evening–morning difference. The difference reflects only a part of the NO_2 diurnal cycle [5]. Strong annual and diurnal variations are evident in figure 2 which are related mainly to photochemical processes. Interannual variations are also seen. The NO_2 decrease just after 1992 are due to heterogeneous chemistry on stratospheric aerosol particles generated after the Mt. Pinatubo eruption.

Daily values of the column NO_2 content above the ASL and the NO_2 content in the ASL are shown in figure 3. The column NO_2 content is mainly contributed by stratospheric NO_2 since the NO_2 abundance in the troposphere is small (Figure 1). Stratospheric NO_2 in evening is typically larger than in morning (Figures 2c and 3a). This is not always a case for near-surface NO_2 since the NO_2 abundance in the ASL is often perturbed by pollution episodes (Figure 3b).

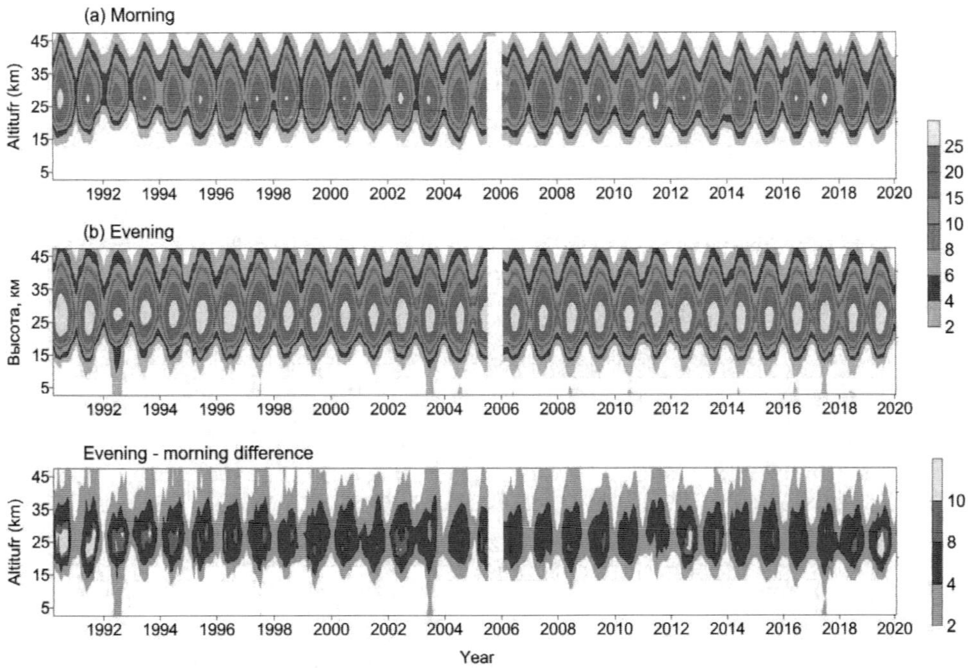

Figure 2. Altitude-time cross-sections of (a) the morning (a) and (b) evening monthly mean NO_2 concentrations and (c) the evening–morning difference of the NO_2 concentration. Units are 10^8 molecules/cm^2.

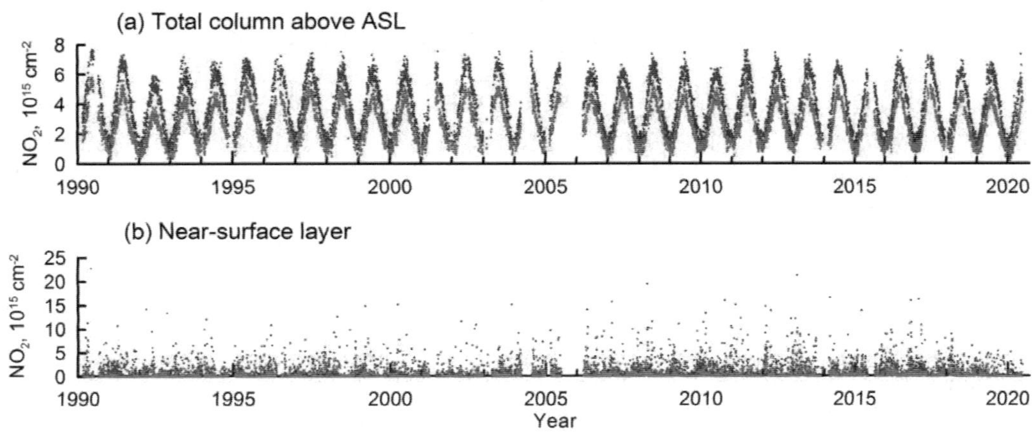

Figure 3. (a) Column NO_2 content above the atmospheric surface layer and (b) NO_2 content in the atmospheric surface layer in morning (red) and evening (blue).

IOP Conf. Series: Earth and Environmental Science 1040 (2022) 012027 doi:10.1088/1755-1315/1040/1/012027

4. Analysis method

Analysis is done with a multivariate regression method. The regression model includes as predictors the following proxies of long term and interannual variability: the free member, the linear term (linear trend), the quasi-biennial oscillation (QBO), the El Nino–Southern Oscillation (ENSO), the 11-year solar activity cycle, the North Atlantic Oscillation (NAO), and stratospheric aerosol.

The NAO proxy is presented by the NAO index (http://www.cru.uea.ac.uk/cru/data/nao/). Aerosol proxy is presented as atmospheric aerosol optical depth (AOD) for the northern hemisphere Studies (http://data.giss.nasa.gov/modelforce/strataer/).

The oscillating predictors, that are the indices of the QBO, solar activity, and Nino 3.4, are presented by two indices: by actual one and by an orthogonal index obtained from the actual index by shifting forwards in time. A pair of orthogonal predictors of the same proxy allows us to describe and estimate a lagged response of a dependent variable [8].

The QBO proxy in the model is presented by the zonal velocity of the stratospheric equatorial wind at 40 hPa level (http://www.geo.fu-berlin.de/en/met/ag/strat/produkte/qbo/index.html) and by the same velocity but shifted forwards by six months. The solar cycle proxy is presented by the solar radio flux at wavelength 10.7 cm, $F10.7$ (https://www.spaceweather.gc.ca/solarflux/sx-5-mavg-en.php), and by the save index but shifted forwards by 37 months. The ENSO proxy is presented by the index of sea surface temperature anomaly in the Nino 3.4 region in the equatorial East Pacific (https://www.esrl.noaa.gov/psd/data/climateindices/list/) and by the same index but shifted forwards by ten months. The above shifts provide orthogonality of the indices in the pairs for the period of NO_2 observations at the ZSS.

All regression coefficients in the regression model are expanded into Fourier pairs corresponding to the annual and semi-annual harmonics in order to account for seasonal dependence of trends and other effects on NO_2.

To solve the system of the regression equations, the method is used taking into account the serial correlation (autocorrelation) of the residual on long time scales [9]. The serial correlation is due to memory in a system and affects confidence intervals of regression estimates [10].

5. Results of analysis

First results of analysis of interannual variations and trends in NO_2 vertical distribution were reported in [8]. The new features of the analysis in this, conference, paper are as follows. First, the series of NO_2 measurements in this work is slightly longer. Second, the Nino 3.4 index is now presented nit by the averaged temperature itself, but by normalized anomaly of sea surface temperature in the equatorial East Pacific. And third, and most important, is that, in this paper, two sets of the NO_2 trend estimates are obtained and compared to each other, which are the trends estimated with taking into account the solar cycle influence on NO_2 and the trends estimated when ignoring the solar influence. The reason is that the maximum of the last solar cycle was very low, and the $F10.7$ index as well as the sunspot number have negative linear trend for the period of the NO_2 observations. Therefore, the solar activity index and the linear term in the regression model are not statistically independent, and regression estimates of the linear trend and the solar cycle effect in NO_2 can mutually influence each other. Below we show primarily the results obtained with accounting for the solar cycle effect. Its influence on linear trend estimates is outlined in the end of the paper.

Figure 4 shows linear trends in morning and evening NO_2 concentrations as functions of month and altitude above the ASL. Trends in near-surface NO_2 are not depicted in figure 1 because of their peculiar character that could make it difficult to perceive the cross-section in figure 1. Trends in the near-surface NO_2 content at t5he ZSS are analyzed in [11] together with trends in near-surface aerosol concentration.

The trends in stratospheric NO_2 above about 25 km altitude are negative, and the trend below this level in the summer period is positive (Figure 4). The tropospheric trends in winter and spring are negative.

The evening trend in tropospheric NO_2 has positive values in late spring and summer (Figure 4b). The probable cause is vertical (turbulent, convective) mixing, which leads to the spread of nitrogen oxides from the polluted ASL into the overlying layers of the troposphere. Such an effect is not detected in morning (taking into account the statistical significance) (Figure 4a) due to the fact that the boundary layer of the atmosphere is statically more stable in the morning than in the evening. We note that the trend in near-surface NO_2 at the ZSS is generally positive (see [11] and Figure 5a).

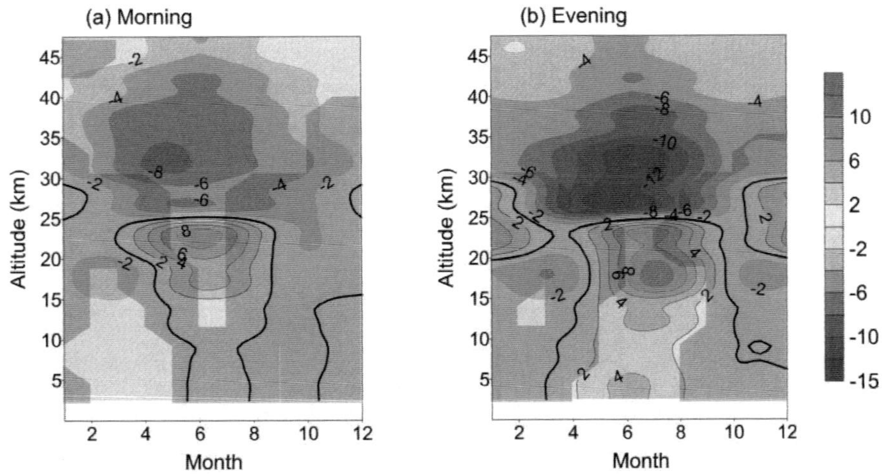

Figure 4. Altitude-month cross-sections of the linear trend in (a) morning and (b) evening NO_2 contents above the atmospheric surface layer. Units are 10^7 molecules/cm^3 per decade. Areas of a statistically significant trends are not darkened.

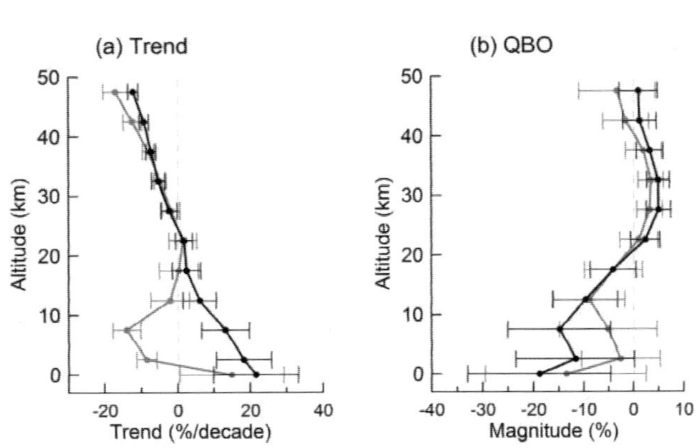

Figure 5. Vertical profiles of the annual estimates of (a) the linear trend and (b) the QBO effect in morning (gray) and evening (black) NO_2 concentrations.

Figure 5a shows altitude profiles of the annual trend estimates derived from morning and evening measurements. Note that figure 5 also includes the ASL. The morning and evening annual trends are close to each other (in per cent values) above 20 km and in the ASL. However the tropospheric trends are fundamentally different, in accordance with the features of the altitude-seasonal distributions in figure 4.

Figure 5b shows altitude profiles of the annual estimates the QBO effect on NO_2. The phase lag of the effect relative to the QBO in the 40 hPa stratospheric zonal wind velocity (not shown) is 1–3 months. The magnitude of the QBO effect is the range of NO_2 oscillations when the equatorial QBO changes between its westerly and easterly phases. The magnitude of NO_2 oscillations due to the QBO is 2–5% in the middle stratosphere. There the maxima (minima) of the NO_2 oscillations follow, with the lag, the maxima of the westerly (easterly) wind velocity at the 40 hPa level in the equatorial stratosphere.

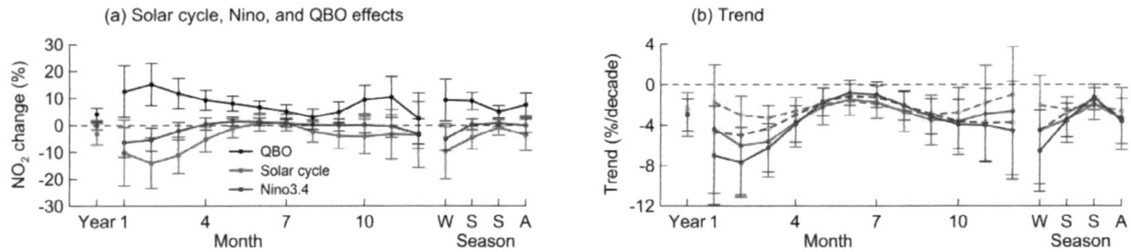

Figure 6. (a) Annual (dots on the left side), monthly (curves in the middle), and seasonal (curves on the right side) estimates of the QBO (black), solar cycle (red), and Nino 3.4 (blue) effects on the morning column NO_2 content above the ASL and their 95% confidence intervals. The QBO and solar cycle effects are magnitudes of associated NO_2 changes. The Nino 3.4 effect is a NO_2 response to change in the Nino 3.4 index by one standard deviation. The lags of the effects are specified in the text. (b) Annual (dots on the left side), monthly (curves in the middle), and seasonal (curves on the right side) estimates and associated 95% confidence intervals of morning (red) and evening (blue) linear trends of the column NO_2 content above the ASL. Closed dots and solid curves correspond to the estimates derived with taking into account the solar cycle effect, while open dots and dashed curves correspond to estimates derived without taking into account the solar cycle effect. The confidence intervals are only shown for trends derived with taking into account the solar cycle effect.

The magnitude of the QBO effect in the tropopause layer (~10–12 km) and in the troposphere is negative, and only evening estimates of the tropospheric effect are statistically significant. The magnitude of the QBO-related oscillations in the evening NO_2 concentrations is of 10–20% (however oscillations in absolute units are weak because of small NO_2 concentration, see Figure1), and the maxima (minima) of the NO_2 oscillations there follow, with the lag, the maxima of the easterly (westerly) equatorial wind velocity 40 hPa.

Figure 6 presents annual, monthly, and seasonal estimates of linear trends and quasi-cyclic variations in the column NO_2 content above the ASL. We note that, due to small NO_2 concentration in the troposphere, Figure 6 reflects mainly changes in stratospheric NO_2. According to figure 6, the morning and evening trends as well as the QBO, solar cycle, and ENSO effects in column NO_2 maximize (in per cent units) in the late winter–early spring period.

The annual estimate of the QBO effect on the morning column NO_2 contents is statistically significant throughout the year (Figure 6a). The effect lags the 40 hPa equatorial QBO by a few months in summer and leads it in autumn through spring. The magnitude of the column NO_2 oscillations, in relative units, is about 10% in early spring. However, in absolute units, the oscillations are larger in summer than in winter because of the strong seasonal variation with summer maximum (Figure 3a).

Variations in morning column NO_2 due to the ENSO-related variations in sea surface temperature in the equatorial east Pacific, which lag by ten months after the Nino 3.4 index, are most pronounced and statistically significant in winter (Figure 6a). The negative sign of the effect denotes that column NO_2 decreases approximately in a year after El Nino episodes and increases after La Nina episodes.

The solar cycle effect in NO_2 is statistically significant in the winter–spring period when it is approximately antiphase to the 11-year solar cycle (Figure 6a). The solar-related changes in column NO_2 in this period is of order of 10% according both to morning and evening NO_2 data. The column NO_2 contents during solar cycle maxima are, on average, less than during solar minima.

The NAO effect on NO_2 is most pronounced in spring (not shown). The amplitude of the associated inter-annual variations of the spring column NO_2 contents is about 2% per one standard deviation of the NAO index, and column NO_2 decreases at positive phase of the NAO, that is when the surface air pressure gradient and, as a consequence, zonal wind in the North Atlantic is strengthened.

The linear trends in column NO_2 are negative throughout the year, and the strongest (in relative units) trends are observed in the winter–spring period (Figure 6b). If the possible influence of the 11-year solar cycle on NO_2 is taken into account, the strongest, February, trend is –6% ÷ –8% per decade (see solid curves in Figure 6b). If the solar cycle influence is ignored, the winter–spring trends diminish (see dashed curves) but are confidently determined. The summer trends as well as annual trends do not undergo noticeable changes.

The comparison analysis of the NO_2 trends with and without accounting for the solar cycle effect shows that although, due to the presence of a linear component in solar activity for a few past solar cycles, the solar cycle effect can probably induce a long term NO_2 trend, this influence does not have a defining value. Along with this, one could allow influence of the linear trend on estimates of the solar cycle effect. Comparison of figures 6b and the "solar" curve in figure 6a shows that the decadal NO_2 changes in winter–spring associated with the solar cycle are significantly larger than the decadal trends. Therefore the solar cycle effect in this period cannot be essentially induced by the linear trend. However we cannot exclude such an influence in summer when a solar cycle effect has not been detected in column NO_2. We can assume that the solar cycle might actually affect stratospheric NO_2 in summer.

6. Discussion and conclusions

The effects of the QBO, ENSO, NAO, and the s11-year solar cycle on the stratospheric NO_2 are generally most pronounced in late winter and/or early spring. For the same season, the most significant (in per cent units) negative trend in stratospheric column NO_2 has been obtained, due to a decrease in the NO_2 concentration in the middle and upper stratosphere. This common seasonal feature indicates the important role of circulation mechanisms in the interannual and long-term changes in NO_2 during this period. In particular, the changes in the NO_2 content can occur as a result of the large-scale evolution of the stratospheric polar vortex. The beginning of spring is characterized by the final stratospheric warming, accompanied by the weakening and destruction of the polar vortex, and a rapid photochemical increase in the stratospheric NO_2 content. Therefore, the NO_2 content during this period may be particularly sensitive to factors affecting the stratospheric vortex.

Comparison of stratospheric NO_2 trends estimated with and without accounting for solar cycle influence on NO_2 as well as comparison of magnitudes of the trend and the solar cycle effect on NO_2 show some interplay between estimates of the trend and the solar cycle effect. However the interplay does not affect critically the trend and the solar cycle effect. One important conclusion from the comparison analysis is that both the linear trends and the solar cycle effect in column (stratospheric) NO_2 are actual.

Acknowledgments

Equatorial stratospheric wind velocities are provided by Freie Universität Berlin. Solar 10.7 cm flux data are provided by the Natural Resources Canada. Data of the Nono3.4 index are provided by the NOAA Physical Sciences Laboratory. NAO data are available at the site of the Climate Research Unit, University of East Anglia. AOD data are prepared by the NASA Goddard Institute for Space.

This work was supported by the Russian Foundation for Basic Research, project No, 20-05-00274.

References

[1] Brasseur G P and Solomon S 2005 *Aeronomy of the Middle Atmosphere* (Dordrecht: Springer)

[2] Gradeel T E and Crutzen J P 1993 *Atmospheric Change. An Earth System Perspective* (New York: W. H. Freeman and Company)

[3] Gruzdev A N, Kropotkina E P, Solomonov S V, and Elokhov A S 2017 Winter–spring anomalies in stratospheric O_3 and NO_2 contents over the Moscow region in 2010 and 2011 *Izvestiya, Atmos. Oceanic Phys.* **53** 195–203

[4] Noxon J F, Whipple E C, Hyde R S 1979 Stratospheric NO_2. 1. Observational method and behavir at mid-latitude *J. Geophys. Res.* **84** 5047–5065

[5] Elokhov A S and Gruzdev A N 2000 Nitrogen dioxide column content and vertical profile measurements at the Zvenogorod Research Station *Izvestiya, Atmos. Oceanic Phys.* **36** 763–777

[6] Gruzdev A N and Elokhov A S 2010 Validation of Ozone Monitoring Instrument NO_2 measurements using ground based NO_2 measurements at Zvenigorod, Russia *Int. J. Remote Sens.* **31** 497–511

[7] Gruzdev A N and Elokhov A S 2011 Variability of stratospheric and tropospheric nitrogen dioxide observed by visible spectrophotometer at Zvenigorod, Russia *Int. J. Remote Sens.* **32** 3115–3127

[8] Gruzdev A N and Elokhov A S 2021 Changes in the column content and vertical distribution of NO_2 according to the results of 30-year measurements at the Zvenigorod Scientific Station of the A. M. Obukhov Institute of Atmospheric Physics, Russian Academy of Sciences *Izvestiya, Atmos. Oceanic Phys.* **57** 91–103

[9] Gruzdev A N 2019 Accounting for autocorrelation in the linear regression problem by an example of analysis of the atmospheric column NO_2 content *Izvestiya, Atmos. Oceanic Phys,* **55** 65–72

[10] Draper N R and Smith H 1998 *Applied Regression Analysis* (New York: John Wiley & Sons)

[11] Gruzdev A N, Isakov A A, Elokhov A S, and Anikin P P 2021 Submicron aerosol and nitrogen dioxide in the atmospheric near-surface layer at the Zvenigorod Scientific Station of the A.M. Obukhov Institute of Atmospheric Physics RAS: Thirty years of measurements *IOP Conf. Ser. Earth Environ. Sci.* This issue

IOP Conf. Series: Earth and Environmental Science 1040 (2022) 012028 doi:10.1088/1755-1315/1040/1/012028

Atomic-molecular effects in geophysical hydrodynamics

Y D Chashechkin
IPMech RAS, 101/1 prospect Vernadskogo, Moscow, 119526, Russia

E-mail: yulidch@gmail.com

Abstract. To calculate the dynamics and structure of flows, a system of fundamental equations of fluid mechanics with equations of state for the Gibbs potential and density of an inhomogeneous medium is applied. The complete solution of the system describes ligaments, waves, vortices, jets, wakes, and other types of flows. Calculations of flow patterns around obstacles are consistent with the experiment. Observations of the processes of merging a freely falling drop with a target fluid revealed that the finest components are formed during the direct generation of ligaments by atomic-molecular processes. The involvement of a scaled and parametrically invariant system of fundamental equations permits the study of unsteady energetic flows and more accurately describes their dynamics and structure in the whole range of scales from microscopic to global.

1. Introduction

Developed techniques of optical observations helped to establish that all flows of fluids or gases are characterized by a "fine structure" in which relatively thick components are separated by thin high-gradient interfaces. The fine structure of liquids and gases is recorded over the entire observable range of scales. A striking example of the structures of outer space are photographs of the "light echo" of the V838 Monoceros star flash. Fibers, deformed rings, thin streaky structures have been observed for several years in interstellar space on spatial scales of several light-years [1].

Wavy, vortex, annular, and jet elements are expressed in the atmospheres of stars and planets, including the Earth. The very fact of the existence of structures is paradoxical since dissipative factors should lead to the smoothing of large gradients. The prevalence of the fact of the very existence of fine structures in fluid flows of various scales, from galactic to microscopic, indicates the universality of the continuously operating mechanism of their formation.

Theories of fluid flows are mostly constructed to describe the phenomena of only one selected class that are jets, wakes, waves, vortices, etc. in a limited range of parameters (exist subcritical and supercritical conditions), as well as in their own temporal and spatial scales. The only scale-invariant system of equations is a set of equations for the transfer of density, matter, momentum, and energy, given back in the middle of the last century in the first edition of the treatise [2] and later reproduced in modern monographs [3, 4]. The system has not yet been completely investigated as the closed system of connected equations taking into account the condition of compatibility.

The theory of fluid flows based on the "continuous medium" concept, although, as is known, all bodies are composed of atoms, molecules, clusters, and other elementary structural formations. Considering the controversy of two approaches that are continuous description in macroscopic mechanics and discrete interactions in atomic-molecular physics, it is natural to search for an "agent" – a flow component that connects processes at opposite boundaries of the scale range. These flow components would be limited by two scales that are a small thickness specified by atomic-molecular

IOP Conf. Series: Earth and Environmental Science 1040 (2022) 012028 doi:10.1088/1755-1315/1040/1/012028

processes and a large length determined fluid volume as well as by a characteristic velocity and the time of evolution (existence) of the system.

The framework of macroscopic hydrodynamics is the system of fundamental equations, the symmetries of which correspond to the basic principles of physics expressed in conservation laws. Balance transport equations [2-4] in relation to the theory of flows determine the basic physical quantities that characterize a fluid as a continuous medium and its flows caused variations of the fluid parameters. The relationships between physical quantities are determined by equations of state, which were first found empirically [5] however later were derived from the principles of thermodynamics [6, 7]. The rational theory takes into account the requirements of engineering mathematics that is the axiomatic science for the principles of choosing the content of symbols, rules of operations, and setting accuracy control criteria. The accuracy concept here means the degree of compliance of the results obtained to the basic axioms of the theory.

2. The system of governing equations and definition of a fluid flow

The set of fundamental quantities for which the conservation laws are formulated includes the following variables – mass (or the specific mass – density), as well as specific momentum and energy. The free mobility of atoms and molecules in the undisturbed state of fluid forbids introducing the tensor of inertia (the mutual position of particles is continuously changing) and predetermines the special role of energy. The energy – a scalar quantity which is a universal parameter applied at all range of scales of physical phenomena extending from sub-nuclear scales to cosmic distances – needs to be used in describing the fluid state and dynamics of its flows.

Energy, including kinetic, potential, and internal components, is the basis of the modern definition of the equation of state of a continuous medium and, at the same time, one of the measures of motion, plays a special and insufficiently studied role in fluid mechanics. The non-mechanical part of the fluid energy is characterized by thermodynamic potentials that are proper internal energy, enthalpy (thermal function), free energy, and free enthalpy (Gibbs potential) [6]. All potentials related by Maxwell's relations are equivalent. However, in the theory of flows, the Gibbs potential the derivatives of which explicitly determine the equilibrium density of the medium and entropy, and indirectly - all other thermodynamic quantities, is most widely used [7]. All thermodynamic quantities that temperature, pressure, the concentration of impurities can be expressed through derivations of the Gibbs potential.

The concept of energy gives room to take into account all types of the mechanism of its transport in fluids as fast resulting of the direct action of atomic-molecular processes, as common transport with a flow velocity or with waves with their intrinsic group velocity and the slowest dissipative processes. The mobility and the combining (clustering) atoms and molecules of a continuous medium mean that the whole or only some of the energy transfer mechanisms can act in fluid flows. The fastest direct atomic-molecular interactions manifest themselves in process of releasing available potential internal energy in the chemical reactions or eliminating the free surface in coalescing fluids. Independent transport of energy takes places with the velocity of a macroscopic flow and with the group velocity of waves (acoustic, capillary, gravitational surface and internal, inertial, and hybrid), as well as slow dissipative transport, affecting both the dynamics and the spatial structure of the fields of physical quantities [9]. Taking into account its special role, energy is included in the definition of the state of a liquid, gas, or plasma in the theory of flows.

To describe flows in the approximation of a continuous medium in fluid mechanics, the balance equations for the transfer of matter (the continuity equation), momentum (Navier-Stokes equation), and energy are postulated [2–4]. The intrinsic properties of an inhomogeneous medium include flow induced by diffusion on topography [10]. Forced flows are created by the action of gradients of pressure (force), temperature (convective), concentration (gravitational), the joint action of some or all types of disturbances (multicomponent or double diffusive convection), the motion of solids and gas volumes.

The dynamics of the fields of physical quantities that are density, momentum, temperature, and concentration of dissolved substances in a liquid with equations of state for density ρ as derivatives of the Gibbs potential $G = G(\mathbf{x}, t)$ in a field of uniform gravity with acceleration $\mathbf{g}$ is described by solutions of a system of fundamental equations, which are presented here. The simplest form of the system without taking into account the global rotation of the liquid and the action of sources of density ρ, temperature T, and concentration of impurity S was selected [2–4]

$$\begin{cases} \dfrac{\partial \rho}{\partial t} + \dfrac{\partial p_i}{\partial x_i} = 0, \ \rho = \rho\left(P(\mathbf{x}, t), S_i(\mathbf{x}, t), T(\mathbf{x}, t)\right), G = G(\mathbf{x}, t) \\ \dfrac{\partial (p_i)}{\partial t} + \left(\dfrac{p_j}{\rho} \nabla_j\right) p_i = -\nabla_i P + \rho g_i + \nu \Delta(p_i) \\ \dfrac{\partial \rho T}{\partial t} + \nabla_j \cdot (p_j T) = \Delta(\kappa_T \rho T) \\ \dfrac{\partial \rho S_i}{\partial t} + \nabla_j \cdot (p_j S_i) = \Delta(\kappa_S \rho S_i) \end{cases} \quad (1)$$

Here symbols ∇ and Δ are the Hamilton and Laplace operators, ν, κ_T κ_S are the coefficients of kinematic viscosity, temperature and concentration diffusivities. The fluid velocity is a local kinematic parameter $v_i = p_i / \rho$, which is defined as the ratio of two invariant quantities, one of which is momentum $\mathbf{p}$ and the second is density ρ.

The form of equations of state for thermodynamic potentials, density, and other physical quantities is selected taking into account the composition of the medium and the energetics of the flows under study. In the equations of system (1), momentum dissipation is taken into account, the appropriate choice of source functions allows one to take into account the external fluxes of energy and matter. In this case, most models use the assumption that the density of the medium is constant. However, in real conditions, the density and other physical characteristics of the state of a continuous liquid (gas, plasma) are non-uniformly distributed due to the action of gravity and other factors causing the movement and stratification of liquids. System (1) itself serves as the definition of physical quantities characterizing both the state and the flow of fluids.

The physical properties of a continuous medium (liquid, gas, plasma) determine the equations of state - thermodynamic potentials and their derivatives - density, pressure, entropy, temperature), as well as kinetic coefficients in the equations of molecular transfer. The special role of density in mechanics as a measure of inertia, gravitation, and thermodynamic processes predetermines the choice of the form of the basic equation of state in the form of the density $\rho = \rho(P, T, S_i)$ dependence on pressure P, temperature T, the concentration of solutes, and suspended particles S_i.

A *fluid flow* is defined as an intrinsic or forced self-consistent transfer of momentum $\mathbf{p}$, matter with the density ρ, and the total energy E_t, which causes changes in the pattern of physical quantities $\mathbf{p}, E_t, \rho, P, T, S_i$ fields in space, ensures the interactions of the flow components and the entire flow with external bodies (gaseous, liquid or solid).

3. Classification of dynamic components of flows

Taking into account the variability of the density permits to find solutions of system (1) in a three-dimensional setting and to classify flow components on the basis of complete solutions of the linearized system. The solutions are constructed by methods of the theory of singular perturbations [8], taking into account the compatibility condition that determines the rank of the complete system, the order of its linearized version, and the degree of algebraic (dispersion) equation. Periodic fluid flows with real frequency ω and complex wave vectors $\mathbf{k}$ are considered. Substitution of the solution in the form of plane waves transforms system (1) into algebraic equations and construction of dispersion relations. In a medium with stable unperturbed vertical profiles of temperature $T_0(z)$, salinity $S_0(z)$, and density

$\rho_0(z)$, which are characterized by the intrinsic scales of buoyancy $\Lambda_T = \left|d\ln T_0(z)/dz\right|^{-1}$, $\Lambda_S = \left|d\ln S_0(z)/dz\right|^{-1}$, $\Lambda_\rho = \left|d\ln \rho_0(z)/dz\right|^{-1}$.

The dispersion equation for system (1), which takes into account the action of all dissipative factors, has the form [11]

$$D_\nu(k,\omega)\cdot F(k,\omega)=0$$

$$F(k,\omega) = -D_\nu(k,\omega)D_{\kappa_T}(k,\omega)D_{\kappa_S}(k,\omega)\left(k^2 + i\frac{k_z(\Lambda_T+\Lambda_S)}{\Lambda_T\Lambda_S}\right)+$$

$$+D_{\kappa_T}(k,\omega)\left(\frac{\omega k_z}{\Lambda_S}D_\nu(k,\omega)-N_S^2k_\perp^2\right)+D_{\kappa_S}(k,\omega)\left(\frac{\omega k_z}{\Lambda_T}D_\nu(k,\omega)-N_T^2k_\perp^2\right) \quad (2)$$

$$D_\nu(k,\omega) = -i\omega+\nu k^2 \quad D_{\kappa_T}(k,\omega) = -i\omega+\kappa_T k^2,\ D_{\kappa_S}(k,\omega) = -i\omega+\kappa_S k^2,$$

$$k^2 = k_x^2+k_y^2+k_z^2,\ k_\perp^2 = k_x^2+k_y^2$$

The complete solution of the system (2) for periodic flows of viscous fluids always contain two types of perturbations that are thin ligaments, and more long waves [8]. They are described by families of singular and regularly perturbed solutions, respectfully. In general, other (non-periodic) components can also be present in the flow patterns.

Ligaments (ligands, interfaces, shells, envelopes, fascia, fibers, threads, filaments, ...) are thin and extended components of flows. In linear models, they are described by the group of singularly perturbed solutions of the system of fundamental equations [11]. The transverse scales of ligaments δ are determined by the dissipative properties of the medium (coefficients of kinematic viscosity ν, temperature κ_T or salinity diffusivities κ_S, or their combinations) and the characteristic values of temporal variability - the time of the process formation Δt, the frequency of the wave ω or the velocity of momentum and energy transfer U: $\delta_T^\nu = \sqrt{\nu\Delta t}$, $\delta_\omega^\nu = \sqrt{\nu/\omega}$, $\delta_U^\nu = \nu/U$. The length of the ligaments is determined by the lifetime of the process under study. There are many types of ligaments in fluids, their number of which is determined by the form of the equation of state and is equal to four if only the variability of density and viscosity is taken into account. When heat conduction effects are taken into account there are six ligaments. And if the equation for salinity diffusion are additionally included in the analysis there are eight different ligaments [11]. In the experiment, the ligaments correspond to thin shells, interfaces, and fibers that visualize the structure of the flows.

Wave is a process in which the parameters of local temporal variability of physical fields (frequency ω) and of instantaneous spatial structure (wavenumber $\mathbf{k}$ or wavelength λ) are related by a functional (dispersion) relationship $\omega=\omega(\mathbf{k},\mathbf{kA},...)$, which may include the amplitudes of disturbances $\mathbf{A}$. Waves are described by a group of regularly perturbed solutions of the fundamental system.

A **vortex** is an unsteady flow component with a relatively high vorticity value $\boldsymbol{\omega}=\mathrm{rot}\,\mathbf{u}$. In a vortex, free solids in the bulk or on the surface of the fluid are transported by the flow and simultaneously twist around their own axis. The compact continuous "liquid particles" are broken down by the ligaments into distinguishable fibers in fluid flows.

The structure of the solution to equation (2) and system (1) depends on the shape of the source. In the case of matching the symmetries of the wave source and the pattern of a conical beam of periodic internal waves, which is emitted by a horizontal disk with a radius R oscillating in the vertical direction, neglecting the effects of diffusion, equation (2) is simplified and takes the form

$$\kappa_S\nu\left(k^2+k_j^2\right)^3 - i\omega\left(\nu+\kappa_S\right)\left(k^2+k_j^2\right)^2-\omega^2\left(k^2+k_j^2\right)+N^2k^2=0 \quad (3)$$

Complete periodic solution of equation (3) for $\omega=\mathrm{const}$ is constructed by methods of singular perturbation theory [8] under the assumption that the coefficients of viscosity and diffusion are small, which for the given problem takes the form $\delta_N^{\nu}, \delta_N^{\kappa_S} << \lambda \sim R$. Regular roots describe the structure of a beam of conical internal waves

$$k_1 \approx -k\cot\theta + \frac{i(\nu+\kappa_S)k^3}{2\omega\sin^3\theta\cos\theta} \tag{4}$$

where $\theta = \arcsin(\omega / N)$ is the angle of inclination of the wave cone to the horizon (as well as the direction of the group velocity of the waves). The regular root k_1 in this problem have an analogue in an ideal fluid.

Singular solutions k_2 and k_3 have a different nature. One of them depends only on the coefficient of kinematic viscosity and characterizes the components of the flow with a transverse length scale $\delta_{\omega}^{\nu} = \sqrt{2\nu / \omega}$ [11]

$$k_2 \approx \frac{1+i}{\delta_{\omega}^{\nu}} \tag{5}$$

The corollary to the assumed assumption of azimuthal decreases in the order of the dispersion equation (3.6) and the loss of one solution describing the isopycnic (Stokes) singular component. In this geometry, the intrinsically inherent component is transformed to the form of the Stokes periodic flow[1], since only on the horizontally placed moving source motion in the periodic boundary layer is isopycnic.

In the general case, when the radiating plane is inclined at an arbitrary angle to the horizon, the determining dispersion equation has a higher (eighth) order. Its roots describe a larger number of singular flow elements, which include internal, Stokes velocity and diffusion boundary layers. Their parameters depend on the characteristics of the medium (ν, κ_S, N), the wave frequency ω and the geometry of the problem (the angle of inclination of the emitting surface to horizon φ. The next root of equation (3.6) defines a singular solution with thickness $\delta_{\omega}^{\kappa_S} = \sqrt{2\kappa_S / \omega}$

$$k_3 \approx \frac{1+i}{\delta_{\omega}^{\kappa_S}} \tag{6}$$

Thus, the viscous and diffusion boundary layers in this problem completely split.

The imaginary parts of the roots (4) - (6) of equation (5.8) are chosen from the condition for the damping of perturbations at infinity $\operatorname{Im} k_j > 0$ [11].

The intrinsic scales of the structural components determine the requirements for the experimental techniques. The choice of the size of the observation field, the number of recorded parameters are defined by spatial and temporal characteristics of large flow components that are by sizes and periods of waves, vortices, wakes, jets and so on. The spatial and temporal resolutions of instruments are defined by parameters of the smallest components that are ligaments. These conditions were taken into account when developing the stands of the USF "HPC IPMech RAS" for studying the dynamics and structure of stratified flows, waves and vortices [12].

Calculated and observed images of periodic internal waves generated by an oscillating horizontal disc and sphere are presented in figure 1. Fine ligaments form envelopes of the wave beams. They are almost invisible in the fields of basic variables like velocity or density but clear manifest themselves in fields of their derivatives. In the experiment, the schlieren instrument helps to visualize these flow components.

Experimentally observed waves of finite amplitude, vortices, and some other structural components of real flows are products of nonlinear interactions of solutions of the linearized model. Moreover, **all** components of the flows directly interact with each other: waves with waves, waves with ligaments, as well as ligaments with ligaments [13]. As a result of interactions, new groups of **all** components of complete solutions arise with their own temporal and spatial characteristics.

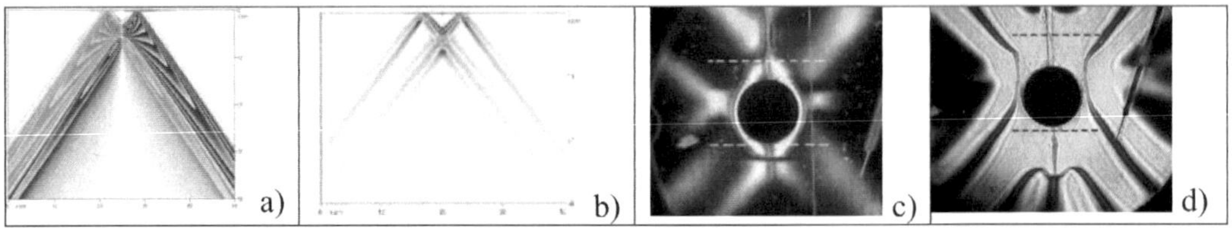

Figure 1. Calculated pattern of flow in the central cross section of the conical periodic wave beam produced by vertically oscillating disc (R = 4 cm? $\omega = 1\ \text{s}^{-1}$, T_b = 5.2 s): a, b) – the horizontal component of velocity and its second derivative at t = 0; c, d) –Schlieren images of periodic flows produced by oscillating sphere D = 4.5 cm, T_b =11.2 s

Nonstationary ligaments, by virtue of the nature of their formation, connect processes at the atomic-molecular level with large structural components, such as waves, vortices, jets, the formation time of which is determined by their eigenvalues (for a wave it is the period).

4. Calculation and schlieren visualization of stratified flow patterns around obstacles

When studying certain types of flows, system (1) is simplified while preserving its basic properties. In particular, when analyzing the flows of an isothermal fluid with a high heat capacity flow around bodies, the effects of compressibility and thermal diffusivity are usually neglected and the following system of equations is used [14].

$$\begin{cases} \operatorname{div} \mathbf{v} = 0, \\ \dfrac{\partial \mathbf{v}}{\partial t} + (\mathbf{v}\nabla)\mathbf{v} = -\dfrac{1}{\rho_{00}}\nabla P + \nu\Delta\mathbf{v} - s\cdot\mathbf{g}, \\ \dfrac{\partial s}{\partial t} + \mathbf{v}\cdot\nabla s = \kappa_S \Delta s + \dfrac{v_z}{\Lambda}, \\ \rho = \rho_{00}\left(\exp(-z/\Lambda) + s\right). \end{cases} \quad (7)$$

Here $S(\mathbf{x},t) = S_0(z) + s(\mathbf{x},t)$ is the total salinity (stratifying component), including the coefficient of salt contraction, $s(\mathbf{x},t)$ is its perturbation, $\mathbf{v} = (v_x, 0, v_z)$ is the induced velocity, P is the pressure minus the hydrostatic part, The equation of state of the medium takes into account only the dependence of density on salinity (stationary undisturbed vertical profile is $\rho_0(S_0(z))$, ρ_{00} is the density on the horizon of the body center) with the boundary conditions of no-slip for velocity and no-flux for salinity [9]. In an unperturbed stratified medium, thin layered diffusion-induced flows start to form at the edges of the fixed body. They are transformed into extended ligaments with the beginning of self-motion of a body of neutral buoyancy.

The compensation flow induced by diffusion on the plate is further taken as the initial conditions

$$\mathbf{v}\big|_{t\le 0} = \mathbf{v}_1(x,z),\ \ s\big|_{t\le 0} = s_1(x,z),\ \ P\big|_{t\le 0} = P_1(x,z),\ \ v_x\big|_{\Sigma} = v_z\big|_{\Sigma} = 0,$$
$$\left[\frac{\partial s}{\partial \mathbf{n}}\right]_{\Sigma} = \frac{1}{\Lambda}\frac{\partial z}{\partial \mathbf{n}},\ \ \mathbf{v}\big|_{x,\zeta\to\infty} = (U,0,0), \quad (8)$$

where U is the flow velocity, (ξ,ζ) are directed along and the outward normal to the obstacle surface.

The system of equations (7) and boundary conditions (8) characterizes a set of parameters with the dimension of length, including the scale of buoyancy Λ, length L or height h of the body. The set of temporal scales includes the period of buoyancy T_b and intrinsic transit time for distance equal the size of the body $T_U^L = L/U$ marking the change in the sign of the pressure gradient in the laboratory coordinate system.

The set contains intrinsic microscales determined by the properties of the medium – dissipative coefficients (kinematic viscosity ν or diffusion κ_S) and the buoyancy frequency N that are $\delta_N^\nu = \sqrt{\nu/N}$ и $\delta_N^{\kappa_S} = \sqrt{\kappa_S/N}$. The structure of the formula for δ_N^ν is similar to the thickness in the Stokes flow $\delta_\omega^\nu = \sqrt{\nu/\omega}$ over an oscillating plane with frequency ω [2]. Here, in the expressions for microscales instead of frequency ω the buoyancy frequency N, that is the frequency of natural oscillations of a particle of a stratified fluid displaced from the equilibrium horizon, was included. The group of large scales of dynamic nature includes the length of the attached internal waves $\lambda = UT_b$ and $\Lambda_\nu = \sqrt[3]{g\nu}/N = \sqrt[3]{\Lambda\left(\delta_N^\nu\right)^2}$ which is the combined viscous-wave scale. The transverse dimensions of the ligaments characterize the Prandtl $\delta_U^\nu = \nu/U$ and Peclet scales $\delta_U^{\kappa_S} = \kappa_S/U$, which are determined by the dissipative coefficients and the obstacle velocity.

Modern computer technology has resources to construct numerical solutions of systems (1) and (2) in a wide range of parameters, in particular for the Reynolds number $1 < \mathrm{Re} < 100000$. Calculations of perturbations and a schlieren photograph of the flow pattern near a free wedge, self-propelled under the action of diffusion-induced flows [10], are shown in figure 2 (local the excess is blue). The wedge moves from right to left at a velocity of about 0.8 cm/hour.

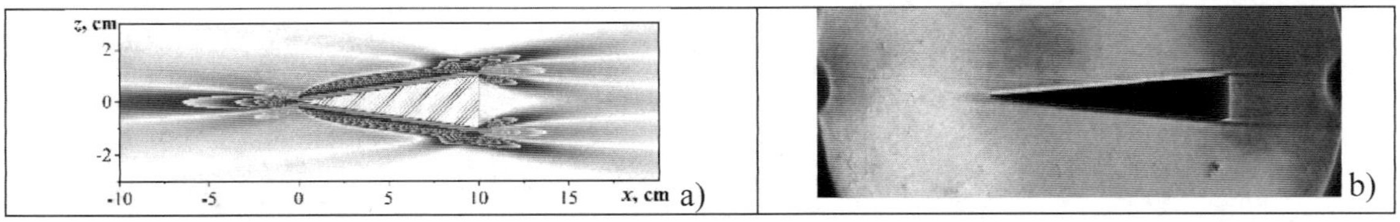

Figure 2. Calculation and schlieren photograph of the patterns of disturbances near a free wedge on the horizon of neutral buoyancy: *a*) - the field of disturbances of the vertical component of the density gradient (L = 10, h = 2 см, T_b = 6.3 s); *b*) - photo of a self-propelled wedge in the tank filled with solution of common salt (T_b = 7 s, "vertical slit - Foucault knife")

The results of the numerical solution of system (2) with boundary conditions (3) with uniform motion of the vertical band are shown in figure 3 [14]. In a creeping flow with a slow motion of the body, both leading (inclined rays) and internal waves attached behind the body are expressed with ligaments limiting the density trace and contacting with the back side of the plate (figure 3 *a*).

In the regime of intense wave disturbances, both waves and an extensive family of ligaments are presented, forming a thin-layered wake (figure 3 *b*). Moreover, both in the calculations and in the experiments, the fine structure of the anticipatory disturbance is visualized, indicating the existence of ligaments both behind and in front of the body. In the experiments, the phase surfaces of the attached internal waves are deformed by the wake flow somewhat more strongly than in the calculations; the split pattern of interfaces is more pronounced.

The set contains intrinsic microscales determined by the properties of the medium - dissipative coefficients (kinematic viscosity or diffusion) and buoyancy frequency. The structure of the formula is similar to the estimation of the thickness in the Stokes flow over a plane oscillating with frequency [1]. Here, instead of the buoyancy frequency, that is the frequency of natural oscillations of a particle of a stratified fluid displaced from the equilibrium horizon, was included in the expressions for microscales. The group of large scales of dynamic nature includes the length of the attached internal waves and the viscous-wave scale. The transverse dimensions of the ligaments characterize the Prandtl and Peclet scales, which are determined by the dissipative coefficients and the obstacle velocity and.

With a further increase in the band velocity, the general structure of the flow is somewhat transformed (the number of visible phase surfaces of internal waves, which deviate in the direction of body motion, change, the geometry of interfaces changes) and the degree of expression of individual components is disturbed. The strongest structural changes are observed in the wake behind the body.

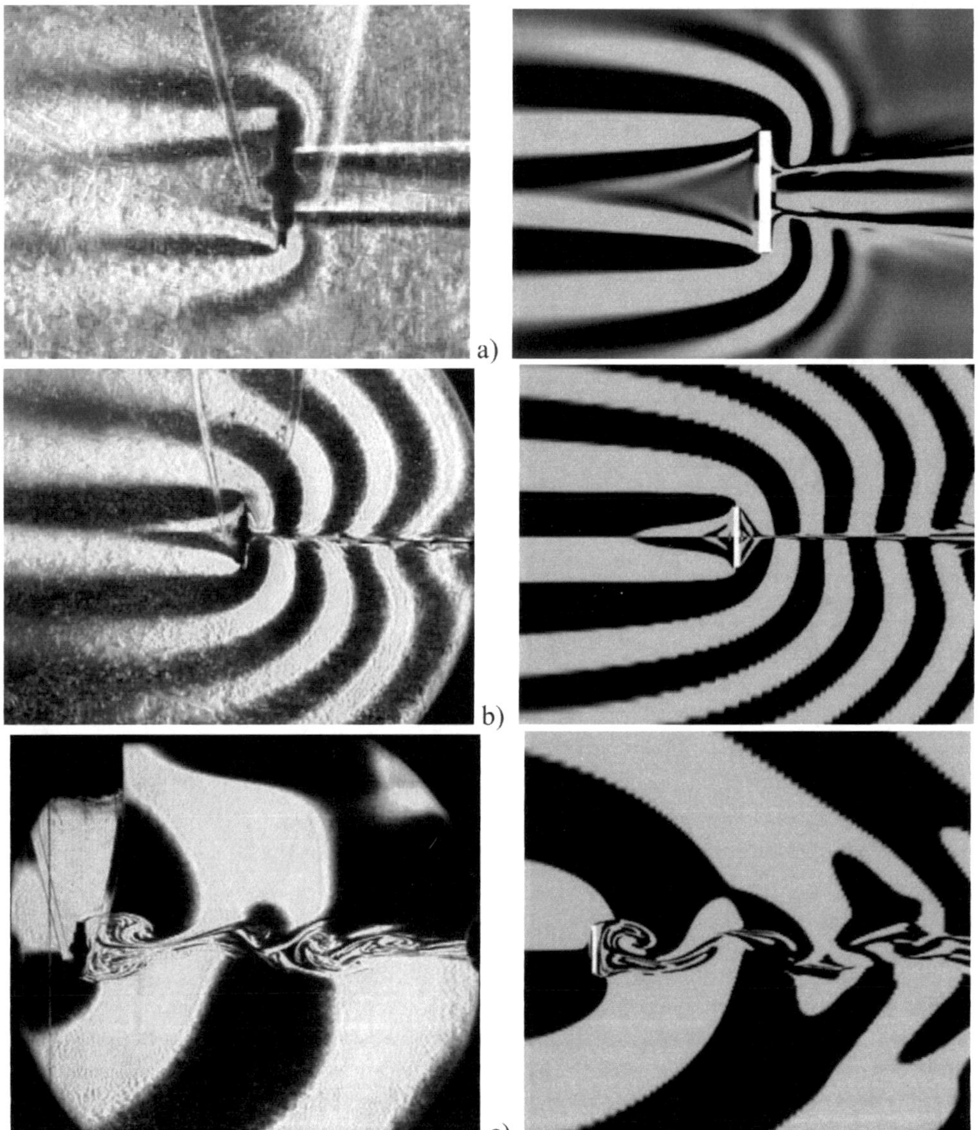

Figure 3. Schlieren images of the flow pattern ("vertical slit - Foucault knife" method) and calculations of the horizontal component of the density gradient fields around a moving band ($h = 2.5$ cm, $T_b = 12.5$ s): *a – c*) – $U = 0.03, 0.18, 0.75$ cm/s; $\mathrm{Re} = Uh/\nu =$ 7.5, 45, 187.5

The length of the attached internal waves increases proportionally the band velocity $\lambda = UT_b$. In the central part of the density field pattern in figure 3 *b*, in the regime of intense generation of internal waves (Froude number $\mathrm{Fr} = U/Nh = 0.14 < \mathrm{Fr}_{cr}$, $\mathrm{Fr}_{cr} = 1$), thin layered ligaments are expressed, separating the attached waves in antiphase in the upper and lower half-spaces. They can be traced along the entire length of the track, starting from the bottom of the plate, in the form of a wavy structure. The phase

IOP Conf. Series: Earth and Environmental Science 1040 (2022) 012028 doi:10.1088/1755-1315/1040/1/012028

surfaces of the attached waves become inclined, their shape noticeably differs from that calculated within the framework of the linear approach.

The flow structure changes radically with a further increase in velocity, as shown in figure 3 *c*. Since in this case the conditions are subcritical $\mathrm{Fr} = 0.6 < \mathrm{Fr}_{cr}$, both the amplitude and the length of the attached waves continue to increase with velocity grows. The structure of the wake radically changes - an oscillating bottom vortex is adjacent to the trailing edge of the plate, generating elongated vortex structures in a thin density wake.

In the patterns of the density gradient fields shown in figure 3, all the structural components of the stratified flow are expressed that are ligaments, drawing the fine structure of the flow, waves, compact vortices. Calculations show that two types of upstream waves can form in front of the body. Long waves ahead the body coupled with attached waves. Short waves are placed at the edges of the band.

Only long waves are represented in the schlieren image, which is due to the shading of some of the structures in the Foucault method. In the experiment in figure 3 c, the bottom vortex, which forms a meandering density wake, moves up and down the rear side of the band. The shape of a tortuous fine-structured wake with a vortex tip adjacent to the bottom of the plate reflects the pattern of displacement of the region of separation of ligaments from the bottom vortex in this regime. Due to the superposition of different-scale components, the patterns of all flows are unsteady; however, some parts of the flow structure evolve especially rapidly.

All illustrations shown in figure 4 were obtained by solving system (7) with physically justified boundary conditions (8) without invoking additional hypotheses, equations, or constants.

Both systems of equations, complete (1) and reduced (2) ones, are parametrically and scale-invariant. However, since the complete solutions of these systems contain a large number of independent multi-scale functions, their overlays, which are represented in the configuration space as a "flow pattern", have different forms in remote ranges of parameters. Limitations of the sensitivity and resolution of instruments, mesh sizes, and time intervals form natural tools for denoting the "boundaries of mode regions" of a system of equations. However, the system of fundamental equations does not contain internal boundaries and does not need the introduction of additional scales, which was noticed by G.G. Stokes.

The differences in kinetic coefficients provide a significant difference in the transverse scales of ligaments of different nature associated with the effects of viscosity, thermal and salinity diffusivities, and, consequently, the difference in the patterns of physical fields on the micro- and macroscales. In a number of phenomena, ligaments provide the formation of stable boundaries between individual elements of a complex flow structure. The ligaments are most clearly expressed in drop impact flows, in which a fast conversion of the available potential energy into other forms during the coalescence of liquids and the elimination of the free surface take place. The elimination can be accompanied by a slow transformation of mechanical energy of a fluid motion into surface potential energy during the formation of a new free surface of the liquid.

Hydrodynamics of a drop impact

In a fluid with a free surface, the density and other characteristics of the medium in the bulk of the liquid and in the near-surface layer with a thickness of the order of the molecular cluster size $\delta_\sigma \sim 10^{-6}$ cm differ markedly. The anisotropy of atomic-molecular interactions is manifested in the existence of available potential surface energy, chemical and other types of internal energy, which can be transformed into thermal, the mechanical energy of fluid flows, and into the work of creating a new free surface. The differential of the Gibbs potential dG_σ in a near-surface layer with an area S_σ и and thickness of the order of the molecular cluster size δ_σ, cm has the form [9]

$$dG_\sigma = -sdT + VdP - S_\sigma d\sigma, \tag{9}$$

where s is entropy, $V = 1/\rho$ is specific volume, S_σ is area of the free surface and σ is surface tension coefficient.

Even greater changes in the atomic-molecular structure of matter are observed directly at the liquid-gas interface, where water decomposes into ionic clusters [9]. Here, in a layer with a thickness of several molecular sizes $\delta_\sigma \sim 10^{-7}$ cm, the thermodynamic potential includes terms that depend on the chemical potential μ_i and the concentration differential of the components

$$dG_\sigma = -sdT + VdP - S_\sigma d\sigma + \mu_i dS_i , \qquad (10)$$

The inhomogeneity of the distribution of thermodynamic potentials and the action of various mechanisms of energy transfer (fast direct atomic-molecular, with a flow of matter, with the group wave velocity and slow diffusion) are clearly manifested in the evolution of the flow pattern of a freely falling drop plunging into a liquid.

The photographs of the flow patterns illustrate the complex texture of the surface of the cavity bottom and crown in the process of merging transparent liquids with water. The coalescence of the water drop with water is given in figure 4 *a* or a yellowish concentrated aqueous solution of ferrous sulfate drop with water shown in figure 4 *b*.

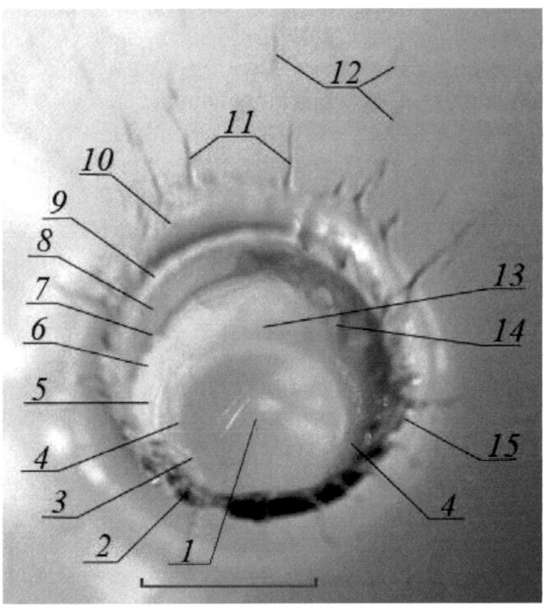

Figure 4a. Drop of water coalesces with target water ($D = 0.43$ cm , $U = 3.7$ m/s, $\rho = 1$ g/cm^3, $\sigma = 73$ g/c^2, $\mu = 0.01$ g/(c·cm), $E_\sigma = 4$ μJ $E_k = 266$ μJ

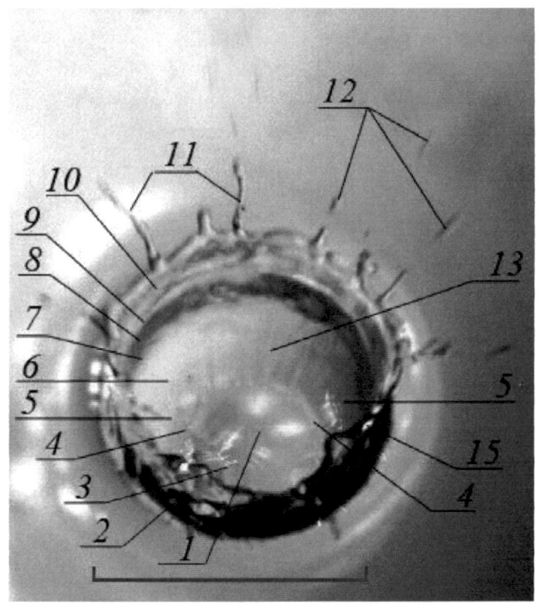

Figure 4b. Drop of saturated solution of ferrous sulfate coalesces with target water ($D = 0.43$ cm $U = 3.5$ м/с, $\rho = 1.18$ г/см^3, $\sigma = 75$ g/c^2, $\mu = 0.02$ g/(c·cm) , $E_\sigma = 4.2$ μJ $E_k = 313$ μJ

Designations: *1* is coalescing drop, *2* is the apex of a crown with teeth, *3 is* a wake of a spray droplet impact on the surface of a drop *1*, *4* is the boundary of the region of fluid confluence, *5* is annular capillary waves at the bottom of the cavity, circling the confluence region, *6* is the bottom of the cavity, 7 is boundary the cavity and the crown, 8, 9 are wall and upper edge of the crown, *10* is veil, *11* are thorns, *12* are small droplets (sprays), 14 are fine jets (ligaments, trickles) at the bottom of the cavity, *14* is 3D texture of the crown wall, *15* are teeth on the apex of the crown.
The marker length is 1 cm.

The contact line *4* of the coalescing drop 1 with the target fluid is not smooth and consists of smooth depressions and pointed ridges [15]. The surface of the cavity wall is distorted by narrow radial jets *13* (*ligaments*) and diverging annular capillary waves *5* $\lambda = 0.03$ cm long, which are emitted by a moving line *4*.

The minimum width of ligaments $\Delta w \approx 0.012$ cm observed near their source, which is tips of ridges on the contact line and it monotonically, increases with the distance from which (figure 4). Small ring structures *3* on the surface of droplet *1* in figures 4, 4 are wakes of fallen droplets escaping from the tips of thorns *11* impacts [15]. Separate sections of veil *10*, adjacent to the inner wall of crown *2*, are inclined towards the center (in figure 4 at 4 and 7 o'clock), the main part directed outward (figure 4). Accordingly, the thorns *11* are mainly directed outward, as well as the droplets *12* (sprays) flying out from their tips [8].

Examination of the videos shows that thin fast fine jets *13*, forming line structures at the bottom of the cavity *6* (figure 4), emerge from the crests of the contact line *4*, pass the boundary of flow components *7*, form a line texture *13* of the crown wall *8*, penetrate the veil *10* and protrude in the form thorns *11*. The flow pattern is complicated by inhomogeneities *14* that are three-dimensional capillary waves traveling from the upper edge of the crown *2* to the bottom of the cavity *6*.

The formation of fast fine jets (trickles) is supported by the conversion of the available potential surface energy into other forms when the free surfaces of the coalescing liquids are eliminated. Transformed energy creates large positive perturbations in the thin layer where two fluids are merged rather fast, accelerate this layer and induced its decay on separated trickles. Forming fine fast jets are rather stable. They propagate along the bottom of the cavity, wall of the crown, and form thorns on the rim of the external veil, from the tip of which small droplets are successfully pushed out.

The complexity and fast restructuring of the droplet impact flow patterns indicates the influence of all mechanisms of energy transfer - with a fluid flow, with the group velocity of various waves, with diffusion momentumless transfer, and fast direct atomic-molecular interactions (for example, in the conversion of the available potential surface energy during the coalescence of fluids), on the dynamics and structure of flows.

5. Conclusion

The dynamics and structure of liquid and gas flows are characterized by a system of fundamental equations that is analogs of conservation laws. The physical quantities included in the system are observable, that is the error can be estimated during measurements. The involvement of a scaled and parametrically invariant system of fundamental equations has room for the study of unsteady energetic flows and more accurately describes their dynamics and structure in the whole range of scales from microscopic to global.

The sets of functions obtained taking into account the compatibility condition of the system of equations, which is characterized by the rank, order of the linearized version, or the degree of the characteristic (dispersion) equation, constitute a complete solution, containing regular and singular perturbed functions describing wave, vortices, and family of ligaments.

In an adequate hydrodynamic experiment, it is necessary to register the field of flows containing basic large-scale components that are waves, vortices, wakes, jets and resolve the microstructural components of the processes described by ligaments, control the completeness of the method, and evaluate the data errors.

The experiments were carried out at the stands of the USF “HPC IPMech RAS”, with the support of the Russian Science Foundation (project 19-19-00598).

References

[1] *Light Echo of V838 Monoceros Star flash* http://www.astronet.ru/db/msg/1254974
[2] Landau L D and Lifshitz E M 1987 *Fluid Mechanics vol 6 Course of Theoretical Physics* (Oxford: Pergamon Press) p 593
[3] Müller P 2006 *The Equations of Oceanic Motions* (Cambridge: CUP) p 291

[4] Vallis G K 2017 *Atmospheric and Oceanic Fluid Dynamics* 2nd ed (Cambridge: CUP) p 946
[5] Mendeleeff D I 1875 *On elasticity of gases* (Saint Petersburg) p 262 (in Russian)
[6] Gibbs J W 1902 *Elementary Principles in Statistical Mechanics* (New York: Charles Scribner's sons) p 207
[7] Feistel R 2018 Thermodynamic properties of seawater, ice and humid air: TEOS-10, before and beyond *Ocean Sciences* **14** 471
[8] Nayfeh A H 2011 *Introduction to Perturbation Techniques* (Wiley-VCH) p 533
[9] Chashechkin Y D 2019 Evolution of the fine structure of the matter distribution of a free-falling droplet in mixing liquids *Atmospheric and Oceanic Physics* **55**(3) 285
[10] Dimitrieva N F and Chashechkin Y D 2018 Fine structure of stratified flow around a fixed and slow moving wedge *Oceanology* **58**(3) 340
[11] Chashechkin Y D 2018 Singularly perturbed components of flows – linear precursors of shock waves *Math. Model. Nat. Phenom.* **13**(2) 1
[12] Unique Science Facility “Hydrophysical complex for modeling hydrodynamic processes in the environment and their impact on underwater technical objects, as well as the transport of impurities in the ocean and atmosphere (USF “HPC IPMech RAS”)” http://www.ipmnet.ru/uniqequip/gfk/#equip
[13] Chashechkin Y D 2021 Conventional partial and new complete solutions of the fundamental equations of fluid mechanics in the problem of periodic internal waves with accompanying ligaments generation *Mathematics* **9**(6) 586
[14] Chashechkin Y D and Zagumennyi I V 2020 Visualization of stratified flows around a vertical plate: laboratory experiment and numerical simulation *Int. J. Comp. Meth. Exp. Meas.* **8**(2) 148
[15] Chashechkin Y D and Ilinykh A Y 2021 Drop decay into individual fibers at the boundary of the contact area with the target fluid *Doklady Physics* **66**(2) 20

IOP Conf. Series: Earth and Environmental Science 1040 (2022) 012029 doi:10.1088/1755-1315/1040/1/012029

The ability of climate models to reproduce the weakening of the annual air temperature cycle over the central part of the Russian Plain

G A Alexandrov[1,2,*]

[1] A.M. Obukhov Institute of Atmospheric Physics, Russian Academy of Sciences, Pyzhevsky 3, Moscow 119017, Russian Federation
[2] This work was supported by the Ministry of Education and Science of the Russian Federation (agreement no. 075-15-2020-776)

[*] Corresponding author: g.alexandrov@ifaran.ru

Abstract. Natural variability plays a large role in local climate that makes it difficult to specify the causes of the weakening of the annual cycle of surface air temperature in the central part of the Russian Plain. This study is to check whether this could be caused by the global warming by using the outputs of climate models. For this purpose, the changes in monthly surface air temperature simulated for the period from 1950 to 2010 were used to calculate the coefficients of correlation between the annual mean of the air temperature and the amplitude of its annual cycle. The negative correlation demonstrated by some models allows us to attribute, at least with a low confidence, the weakening of the annual temperature cycle in the central part of the Russian Plain to the global warming. The model disagreement on the sign of the correlation calls for the study of model ability to reproduce the changes in zonal and meridional atmospheric circulation over the eastern part of the European continent.

1. Background

The global warming manifests itself not only in the growth of annual means of the surface air temperature (MAT) but also in changes of its annual cycle. Therefore, considering the amplitude-phase characteristics of the surface temperature annual cycle [1] is essential for climate change detection and verification of climate models. The amplitude of annual temperature cycle (AAT) is one of these characteristics.

The tendency to AAT decrease is observed in the Northern Hemisphere since 1950 [2]. It reflects the seasonal asymmetry of global warming: winter-time air temperature grows faster than summer-time air temperature. The negative correlation of the AAT and MAT changes is widespread in the Northern Hemisphere and has a theoretical explanation [3].

However, extrapolation of this tendency to the future is valid only under the assumption that the conditions governing the negative sign of the correlation of AAT and MAT changes [3] will be maintained. Since there is no solid theoretical ground for such assumption, one cannot exclude that the sign of the correlation may change to positive under some climate change scenarios.

Moreover, it is not completely clear so far whether the current combination of the conditions causing AAT decrease can be considered as typical. Thus, for example, the large part of the AAT downward tendency could be explained by the natural variability of the atmospheric circulation over the North Atlantic and Europe [4]. Due to this reason, the ability of climate models to reproduce the regional AAT trends is essential for predicting the global warming impact on the annual cycle of the surface air temperature.

This study is to evaluate the ability of the modern climate models [5] listed in Table 1 to reproduce the negative correlation of the AAT and MAT changes in the central part of the Russian Plain.

2. Methods

The outputs of climate models listed in the Table 1were processed as follows. The monthly air temperature simulated by a model for the period 1950-2010 over the nodes of the model grid were used to calculate the MAT and AAT values. These values were used to calculate the times series of MAT and AAT 30-year moving averages (MAT_{30} and AAT_{30}) and the coefficient of correlation between them (CMA) at each node of the model grid.

Table 1. The list of climate models, the outputs of which were processed to evaluate the ability of climate models to reproduce the weakening of the annual cycle of surface air temperature over the central part of the Russian Plain.

Model name	Institute	Reference
BCC-ESM1	Beijing Climate Center	[6]
CanESM5	Canadian Centre for Climate Modelling and Analysis	[7]
CESM2	National Center for Atmospheric Research (USA)	[8]
GFDL-ESM4	Geophysical Fluid Dynamics Laboratory (USA)	[9]
HadGEM3-GC31-LL	Hadley Centre for Climate Prediction and Research (UK)	[10]
INM-CM5-0	Institute of Numerical Mathematics (Russia)	[11]
IPSL-CM6A-LR	Institut Pierre-Simon Laplace (France)	[12]
MIROC6	Japanese climate modeling community	[13]
MPI-ESM1-2-LR	Max Planck Institute for Meteorology (Germany)	[14]
MRI-ESM2-0	Meteorological Research Institute (Japan)	[15]

The data of GSWP3 re-analysis [16-17] on the monthly air temperature were processed in the same way to obtain the observation-based estimates of CMA.

3. Results

The values of CMA calculated from the data of GSWP3 re-analysis [16-17] show the negative correlation between MAT_{30} and AAT_{30} over the whole territory of the Russian Plain (Figure 1). The boundary between the regions of positive and negative correlation passes through the western parts of Ukraine and Poland, that is, quite far from the central part of the Russian Plain, represented by the rectangular surrounding Moscow and delineated from the east and west by 31st and 42nd meridians and from the south and north by 51st and 59th parallels.

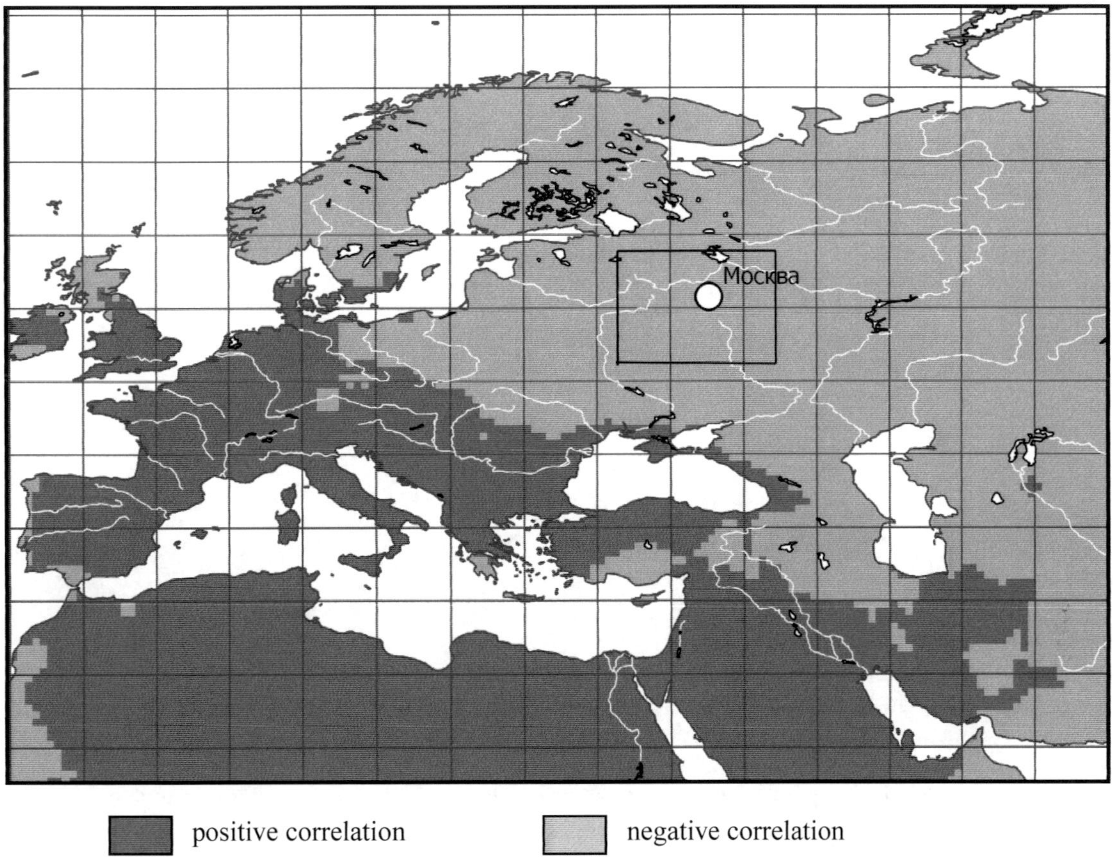

Figure 1. The regions of positive and negative correlation between the time series of 30-year moving averages of mean annual temperature and the amplitude of annual temperature cycle (MAT_{30} and AAT_{30}) derived from the data of GSWP3 re-analysis [16-17].

Nevertheless, not all the climate models listed in Table 1 reproduce the negative correlation between MAT_{30} and AAT_{30} in the nodes of the rectangular (Figure 1) representing the central part of the Russian Plain. CESM2 and INM-CM5-0 do not reproduce it at any of such nodes: the minimal values of CMA are positive (Figure 2). CanESM2 and MIROC6 reproduce it at some nodes; GFDL-ESM4, BCC-ESM1, MPI-ESM1-2_LR and HadGEM3-GC31-LL reproduce it partly; and only MRI-ESM2-0 and IPSL-CM6A-LR reproduce it at each node: the maximal values of CMA are negative.

In summary, the weakening of annual temperature cycle over the central part of the Russian Plain is adequately reproduced by 20% the models, 60% of models reproduce it to some degree, and 20% of models do not reproduce it.

4. Discussion and conclusions

These findings raise a question about the drivers of the negative correlation between MAT_{30} and AAT_{30} over the Russian Plain: are they adequately represented in climate models?

Just like the diurnal asymmetry [18-19], the seasonal asymmetry of global warming can be explained by CO_2 radiative forcing: the higher is the surface temperature the weaker its response to radiative forcing [20]. However, CO_2 radiative forcing is not the only driver of the changes in the annual temperature cycle. AAT depends also on the strength of atmospheric circulation. Since AAT over the oceans is smaller than over the land, the decrease in AAT over the land could be considered as a sign of stronger ocean-to-land heat transfer [21-22].

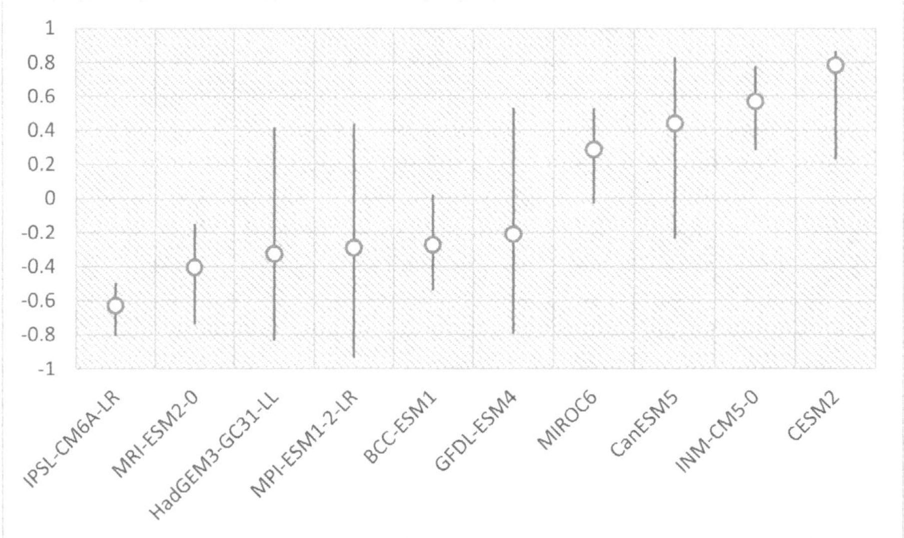

Figure 2. The minimum, maximum, and mean values of the coefficients of correlation between the time series of 30-year moving averages of mean annual temperature and the amplitude of annual temperature cycle (MAT_{30} and AAT_{30}) over the nodes of the geographic grid that falls within the rectangle delineated from the east and west by 31st and 42nd meridians and from the south and north by 51st and 59th parallels.

The strengthening of ocean-to-land heat transfer is seemingly supported by the data on North Atlantic Oscillation index (NAO), that characterizes the variability of the large-scale atmospheric circulation over the North Atlantic and Europe. The annual values of NAO show a tendency to increase during the period from 1950-2010, and therefore one may suggest that a part of AAT decrease could be attributed to the strengthening of zonal atmospheric circulation.

In contrast to CO_2 radiative forcing and NAO, the Arctic warming seemingly increases AAT. Suggesting that Arctic temperature anomalies may reflect a strength of meridional flow, one may suggest that the ability of a climate model to reproduce the negative correlation between MAT_{30} and AAT_{30} over the Russian Plain is determined by the model ability to reproduce the variations in the strength of zonal and meridional circulation [23-24].

The negative correlation between MAT_{30} and AAT_{30} demonstrated by some models allows us to attribute (at least with a low confidence) the weakening of the annual temperature cycle in the central part of the Russian Plain to global warming.

References

[1] Mokhov I I 1985 *Soviet Meteorol. Hydrol.* **5** 14
[2] Stine A R, Huybers P, Fung I Y 2009 *Nature* **457** (228) 435
[3] Eliseev A V and Mokhov I I 2003 *Adv. Atmos. Sci.* **10** 1
[4] Alexandrov G A, Ginzburg A S, Golitsyn G S 2019 *Izvestiya - Atmospheric and Ocean Physics* **55** (5) 407
[5] Eyring V *et al.* 2016 *Geoscientific Model Development* **9**(5) 1937

[6] Wu T, Zhang F, Zhang J, Jie W, Zhang Y, Wu F, Li L, Yan J, Liu X, Lu X, and others 2020 *Geoscientific Model Development* **13** 977
[7] Swart N C, Cole J N, Kharin V V, Lazare M, Scinocca J F, Gillett N P, Anstey J, Arora V, Christian J R, Hanna S, and others 2019 *Geoscientific Model Development* **12** 4823
[8] Danabasoglu G, Lamarque J-F, Bacmeister J, Bailey D, DuVivier A, Edwards J, Emmons L, Fasullo J, Garcia R, Gettelman A, and others 2020 *Journal of Advances in Modeling Earth Systems* **12** e2019MS001916
[9] Dunne J, Horowitz L, Held I, Krasting J, John J, Malyshev S, Shevliakova E, Adcroft A, Dupuis C, Ginoux P, and others 2019 *Journal of Advances in Modeling Earth Systems* **11** 3167–211
[10] Andrews M B, Ridley J K, Wood R A, Andrews T, Blockley E W, Booth B, Burke E, Dittus A J, Florek P, Gray L J, and others 2020 *Journal of Advances in Modeling Earth Systems* **12** e2019MS001995
[11] Volodin E and Gritsun A 2018 *Earth System Dynamics* **9** 1235
[12] Boucher O, Servonnat J, Albright A L, Aumont O, Balkanski Y, Bastrikov V, Bekki S, Bonnet R, Bony S, Bopp L, and others 2020 *Journal of Advances in Modeling Earth Systems* **12** e2019MS002010
[13] Tatebe H, Ogura T, Nitta T, Komuro Y, Ogochi K, Takemura T, Sudo K, Sekiguchi M, Abe M, Saito F, and others 2019 *Geoscientific Model Development* **12** 2727
[14] Mauritsen T, Bader J, Becker T, Behrens J, Bittner M, Brokopf R, Brovkin V, Claussen M, Crueger T, Esch M, and others 2019 *Journal of Advances in Modeling Earth Systems* **11** 998
[15] Yukimoto S, Kawai H, Koshiro T, Oshima N, Yoshida K, Urakawa S, Tsujino H, Deushi M, Tanaka T, Hosaka M, Yabu S, Yoshimura H, Shindo E, Mizuta R, Obata A, Adachi Y and Ishii M 2019 *Journal of the Meteorological Society of Japan Ser. II* **97** 931
[16] Dirmeyer P A 2011 *Journal of Hydrometeorology* **12** 729
[17] Hurk B van den, Kim H, Krinner G, Seneviratne S I, Derksen C, Oki T, Douville H, Colin J, Ducharne A and Cheruy F 2016 *Geoscientific Model Development* **9** 2809
[18] Demchenko P F, Golitsyn G S, Ginzburg A S and Vel'tishchev N N 1994 *Izvestiya, Atmospheric and Oceanic Physics* **30** (5) 595
[19] Davy R, Esau I, Chernokulsky A, Outten S and Zilitinkevich S 2017 *Int. J. Climatol.* **37** (1) 79
[20] Alexandrov G A, Ginzburg A S and Golitsyn G S 2019 *Izvestiya, Atmospheric and Oceanic Physics* **55**(5) 407
[21] Stine A R and Huybers P 2012 *J. Clim.* **25** 7362
[22] McKinnon K A, Stine A R and Huybers P 2013 *Journal of Climate* **26** 7852
[23] Shepherd T 2014 *Nature Geosci* **7** 703
[24] Kononova N K and Lupo A R 2020 *Atmosphere* **11** (3) 255

IOP Conf. Series: Earth and Environmental Science 1040 (2022) 012030 doi:10.1088/1755-1315/1040/1/012030

Aleurite particle concentration profiles in the surface layer of the atmosphere on desertificated areas

R A Gushchin[1,2], G I Gorchakov[1], A V Karpov[1]
[1]A.M. Obukhov Institute of Atmospheric Physics, RAS, 119017, Moscow, Russia.
[2]MIREA - Russian Technological University, 119454, Moscow, Russia.

Email: gengor@ifaran.ru

Abstract The large particle mass flux profile measurement at heights from 0.125 to 16 m and parameters of the size distribution function received by O.E. Semenov on the desertificated areas in the Aral region were analyzed. The technique has been worked out for the large particle height profile characteristic retrieving. The aleurite particle number concentration profiles with sizes less than 50, 73 and 100 μm and sand particle concentration have been retrieved. Power-law approximations of the vertical aleurite and sand particles concentrations profiles are proposed.

1. Introduction

The climate warming in the 21st century is accompanied by the acceleration of desertification and aridization processes, including the process of the drying up of the Aral Sea, an increase in the frequency of weather anomalies, including sand and dust storms, which leads to the spread of the large particles (mineral dust, aleurite) over long distances. To assess the negative consequences for the environment, climate, public health and economic activity of long-range transport of the large particles, it is necessary to know the vertical distributions of large particles near the sources, which are still poorly studied, especially for the aleurite fraction.

A technique for reconstructing the vertical profiles of the concentration of the aleurite particles in the surface layer of the atmosphere on the desertificated areas from the data of measurements of the large particle microphysical characteristics in the Aral Sea region [1] is proposed. Examples of reconstructed vertical profiles of the aleurite particle concentration are given, which are approximated with satisfactory accuracy by power functions. Note that in the near-surface layer of the atmosphere, the concentration profiles of the aleurite and sand particles are approximated by exponents [2].

On the desertificated areas in the Astrakhan oblast the differential particle number concentrations measurements in the layer from 3 to 15 cm were carried out [2, 3]. It is shown that, in addition to the sand particles larger than 100 μm in size, the windsand flux contains the saltating aleurite particles less than 100 μm in size. According to the geological classification [1], the sand particles, the aleurite particles (10-100 μm) and the mineral dust particles with sizes less than about 10 μm are sedimentary rocks.

In the surface layer of the atmosphere, the aleurite particles transport due to turbulent diffusion. With further propagation in the boundary layer of the atmosphere, they can be transported over distances of tens and hundreds kilometers [1].

To assess the aleurite particle content and long-range transport results, it is necessary to have information on the vertical concentration and the particle size parameter distribution function profiles.

In [1], tabular data on the vertical profiles of the large particle mass fluxes with sizes from 50 to 500 μm at heights from 0.125 to 16 m are given, as well as some data on the particle size distribution function parameters. The data [1] were obtained during measurements on the desertificated areas of Kazakhstan, including the Aral region.

This paper presents a method for retrieval the number concentrations of the aleurite particles with sizes less than 50 μm, 73 μm, and 100 μm, as well as the total sand particle number concentration the using data on the mass flux and modal particle size profiles. Vertical concentration profile examples of the aleurite and sand particles are given.

2. Data used

The vertical profile measurement result examples of the mass flux f (kg/m^{2}s) of the large particles with sizes from 50 to 500 μm at heights of 0.125, 0.25, 0.5, 1.0, 2.0, 4.0, 9.0 and 16 m on the desertificated areas are given [1]. At the levels of 0.5, 1.0, 2.0, 4.0, 9.0 and 16 m, wind speed measurements were also carried out. The wind speed profiles were used to determine the friction speed and the roughness parameter z_0 in accordance with the logarithmic wind profile

$$U(z) = \frac{u_*}{\kappa} \ln \frac{z}{z_0}, \quad (1)$$

where $\kappa = 0.4$ is the Karman constant.

Samples of the sand (more precisely, the aleurite and sand particles) have been analyzed. In [1], estimates were obtained for the fractions of particles δ_d with sizes d less than 50, 73, 100, 140, 250, 315, and 500 μm, as well as the modal particle size d_0 and the half-width ν of the observed distribution of large particles. Parameters d_0 and half-width ν of the observed distribution of large particles $g(d)$ are determined when it is approximated by a lognormal distribution

$$g^*(d) = \frac{N}{\sqrt{2\pi}\nu} \exp\left\{-\frac{(\ln d - \ln d_0)^2}{2\nu^2}\right\}, \quad (2)$$

where N is the total concentration of the particles.

This paper analyzes the results of the measurements on the desertificated areas in the Aral region.

3. Vertical profiles of the particle mass flux

The vertical particle mass flux profiles were determined [1] using traps installed on a mast 16 m high. The exposure time varied from several minutes to approximately three hours.

In Figure 1 shows the mass fluxes of the large particles according to measurements near the former Uzunkair Island in the Aral Sea on 06.18.1990 from 19:35 to 20:12 (1) and on 05.30.1992 from 17.30 to 18.30 (2). In logarithmic coordinates, the obtained dependences of the mass flux f on the height z are approximated by linear functions

$$\ln f(z) = b_f + k_f \ln z, \quad (3)$$

which are equivalent to the power approximations

$$f(z) = f_1 (z / z_1)^{k_f}, \quad (4)$$

where z_1 = 1 m, f_1 = 9.1×10^{-3} kg/m^{2}s (1) and 1.5×10^{-3} kg/m^{2}s (2) - mass flux of the particles at the height of 1 m and exponent k_f = -1.1 (1) and -1.28 (2), respectively.

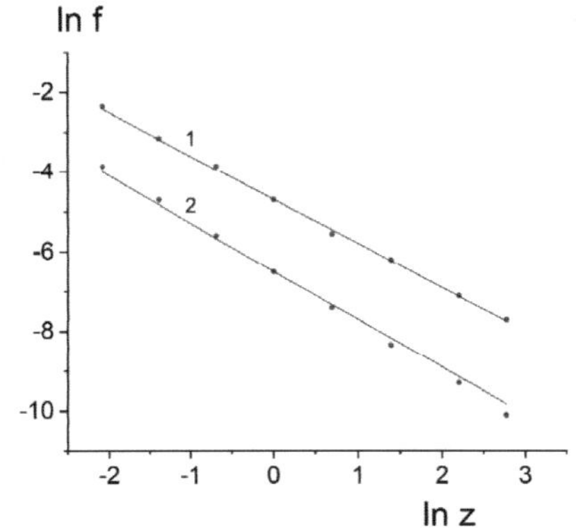

Figure 1. Vertical total mass flux profiles of the sand and aleurite particles on the desertificated areas 06.18.1990 in the period from 19:35 to 20:12 (1) and 05.30.1992 in the period from 17:30 to 18:30 (2).

4. The particle mass concentration profiles

In a windsand flux, where the influence of the saltation on the transport of the sand and aleurite particles prevails, the particle velocities differ markedly from the wind speed [5].

According to the measurement data in the wind channel at a height of about 8 cm [6], the average particles and wind velocities become close. Hence it follows that in the surface layer of the atmosphere, where the diffusion transfer of the large particles predominates [7], the mass flux of the particles is determined by the product of the mass concentration M and the wind speed U:

$$f(z)=M(z)U(z). \tag{5}$$

In Figure 2 shows the profiles of the wind speed measured during the collection of samples of the large particles on a mast with a height of 16 m 06.18.1990 (1) and 05.30.1992 (2).

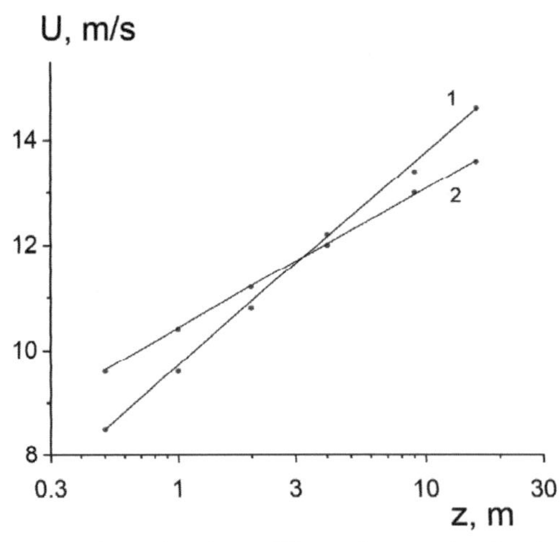

Figure 2. Profiles of the wind speed on the desertificated areas in the Aral region 06.18.1990 (1) and 05.30.1992 (2).

The analysis showed that the obtained dependences U(z) are approximated by logarithmic profiles (1) with parameters $u_* = 0.73$ (1) and 0.45 m/s (2) and $z_0 = 5.5$ (1) and 0.1 mm (2).

According to the $f(z)$ and $U(z)$ data measurements, taking into account the extrapolation $U(z)$ to the height range $z<0.5$ m and using relation (5), the large particle mass concentration profiles $M(z)$ (g/m^3) were obtained. In the logarithmic coordinates $M(z)$ are approximated by linear functions

$$\ln M(z)=b_m+k_m \ln z, \tag{6}$$

which correspond to the power-law approximations

$$M(z)=M_1(z/z_1)^{k_m}, \tag{7}$$

where $M_1 = 0.95$ (1) and 0.0115 g/m^3 (2) and exponents $k_m = -1.28$ (1) and -1.39 (2).

Thus, the profiles of mass concentrations, as well as the large particles mass flux profiles are approximated by power functions.

5. The vertical profiles of the large particle size distribution function parameters

Knowing the vertical profiles of the mass concentration of particles, it is possible to determine the aleurite and sand particles number concentrations profiles if the vertical distribution of the particle size distribution function parameters is known.

The results of determining the modal particle size d_0 and the half-width parameter ν of the approximating log-normal distributions are presented in [1].

The mass concentration of the particles is determined by the distribution of particle volumes by size $V(d)$, where d is the size of the particles.

In the case of a lognormal distribution, the modal size d_v of the volume distribution $V(d)$ and the modal size of the particle mass distribution $d_m = d_v$ are determined by the relation

$$d_v = d_0 \exp\{3\nu^2\}, \tag{8}$$

where d_0 is the modal size for the distribution of the number particle concentration.

We have proposed approximations for the vertical profiles d_v (in μm) (Figure 3)

$$d_v(z)=b_v+k_v \ln z, \tag{9}$$

where $b_v = 87$ (1) and 93 μm (2) and $k_v = 5.2$ (1) and 7.2 (2).

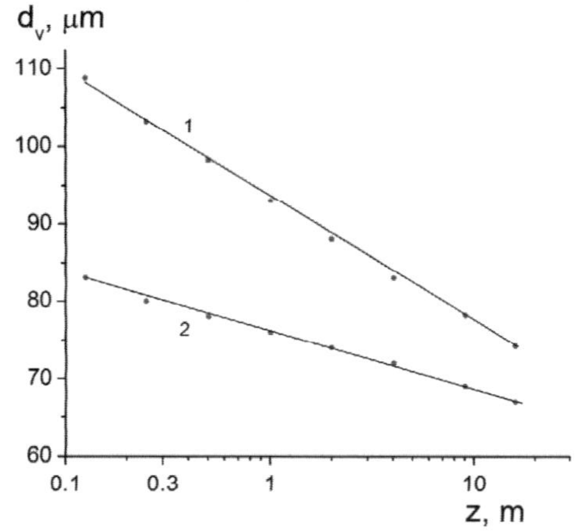

Figure 3. The modal particle volume size distribution dependence on the height according to the measurement data in 1990 (1) and 1992 (2).

The vertical profiles of the parameter ν are also approximated by linear function of ln z.

6. The total large particles number concentration profiles

The substance density ρ of the sand and siliceous aleurite particles equals 2.65 g/cm^3. Since, the particle mass $m=\rho v$ where v is the volume of the particle, it is possible to determine the total number concentration of the particles N through the total volume of particles V using lognormal approximation for the particle size distribution function

$$V = \frac{\pi N d_v^3}{6} \exp\left(\frac{9}{2}\nu^2\right). \tag{10}$$

The results of calculating the $N(z)$ profiles for the cases considered by us are shown in Figure 4.

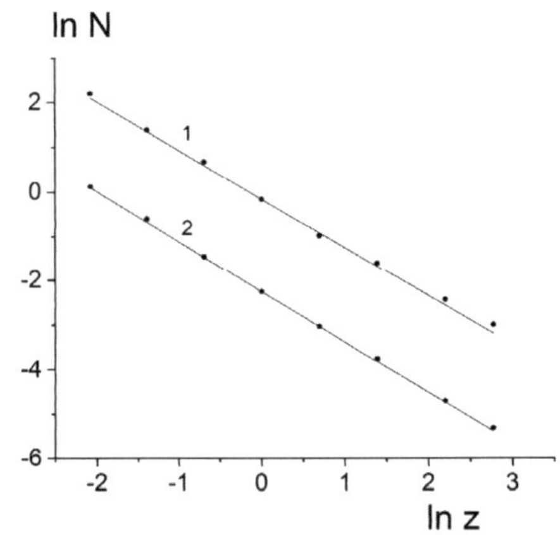

Figure 4. Vertical profiles of the total particles number concentration 06.18.1990 (1) and 05.30.1992 (2).

It is easy to see that the calculated profiles are approximated by power functions

$$N(z) = N_1 \left(z / z_1\right)^{k_N}, \tag{11}$$

where the number counting concentrations at the height of 1 m is $N_1 = 0.86$ (1) and 0.106 cm^{-3} (2) and $k_N = -1.09$ (1) and -1.13 (2).

7. The aleurite particle number concentration fraction evaluations

The fraction estimates for the aleurite particle concentrations with size particles less than 50 μm, 73 μm, and 100 μm were obtained in [1].

In Figure 5 shows the fraction dependences for particles with sizes less than 50 μm (in percent) on ln z, received from the data of measurements in 1990 (1) and in 1998 (2). The indicated dependences are approximated by linear functions ln z

$$\delta_{50}(z)=\alpha_{50}+\beta_{50} \ln z \tag{12}$$

with parameters $\alpha_{50} = 1.3$ (1) and 3.1 (2) and $\beta_{50} = 5.2$ (1) and 12.2 (2).

The dependence α_{50} (ln z) for 1992 is close to the analogous dependence for 1990 (1 in Figure 5).

IOP Conf. Series: Earth and Environmental Science 1040 (2022) 012030 doi:10.1088/1755-1315/1040/1/012030

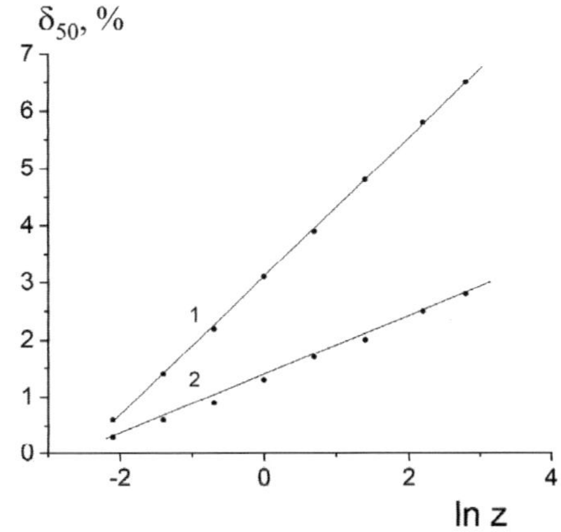

Figure 5. Fraction δ_{50} of the aleurite particles with sizes less than 50 microns according to measurements in 1990 (1) and 1998 (2).

8. Vertical profiles of the aleurite particle number concentrations

With the help of approximation (12), the profiles of the concentration of the aleurite particles $N^{(50)}$ (dm^{-3}) with sizes less than 50 μm were retrieved (Fig. 6). The profiles $N^{(50)}(z)$ are approximated by the power functions

$$N^{(50)} = N_1^{(50)} \left(z / z_1 \right)^{k_{50}}, \tag{13}$$

where $N_1^{(50)} = 11.0$ (1) and 1.4 dm^{-3} (2) and exponent $k_{50} = -0.73$ (1) and -0.80 (2). At the heights of less than 0.5 m, some differences from the power-law approximation are observed (13).

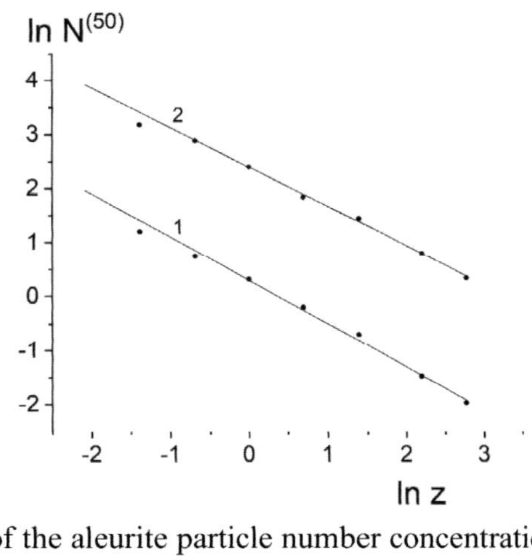

Figure 6. The vertical profiles of the aleurite particle number concentration with sizes less than 50 μm 06.18.1990 (1) and 05.30. 1992 (2).

Approximations of the dependences δ_{73} and δ_{100} on the height were obtained, which made it possible to retrieve the vertical profiles of the aleurite particle concentrations with sizes less than 73 μm and 100 μm. In Figure 7 shows (in logarithmic coordinates) the vertical profiles of the particle concentration with sizes less than 73 μm, which are approximated by power functions

$$N^{(73)} = N_1^{(73)} \left(z / z_1 \right)^{k_{73}}, \tag{14}$$

where $N_1^{(73)} = 0.3$ (1) and 0.027 cm^{-3} (2) and exponent $k_{73} = -0.94$ (1) and -0.91 (2).

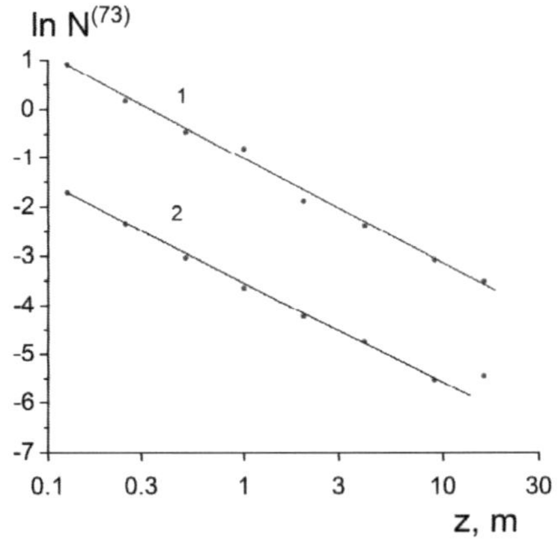

Figure 7. Vertical profiles of the aleurite particle concentrations with sizes less than 73 μm (1-18.06.1990, 2-30.05.1992).

Vertical profiles of the concentration of the aleurite particles $N^{(100)}$ with sizes less than 100 μm and sand particles $N^{(s)}$ with sizes greater than 100 μm (Figure 8) were received from the data measurements in 1992. These profiles are approximated by power function

$$N^{(100)}(z) = N_1^{(100)}(z/z_1)^{k_{100}}, \tag{15}$$

and

$$N^{(s)}(z) = N_1^{(s)}(z/z_1)^{k_s}, \tag{16}$$

where $N_1^{(100)} = 100$ dm^{-3}, $N_1^{(s)} = 11.6$ dm^{-3}, $k_{100} = -1.06$ and $k_s = -1.51$.

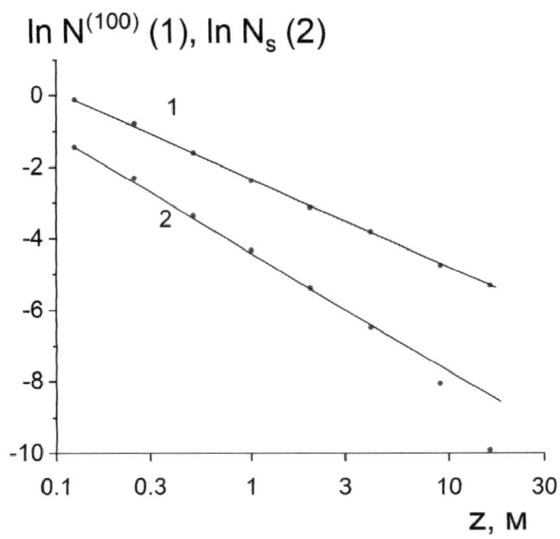

Figure 8. Vertical profile of the total aleurite (1) and sand (2) particle concentration according to measurement data in 1992.

The exponents of power-law approximations of the vertical particle distributions depend on the ratio w_g/u_*, where w_g is the gravitational settling particle velocity and u_* is the friction velocity. According to the measurement data on 05.30.1992 k_N, k_{50}, k_{73}, k_{100} and k_s are equal to -1.13, -0.80, -0.91, -1.06 and -1.51, which qualitatively agrees with the existing data on the dependence of w_g on the particle size.

On the desertified areas, the vertical saltating particle mass flux and concentration distributions are often approximated by exponential profiles [6, 9–12]. When passing from the saltation layer to the surface layer of the atmosphere, where the transport of the particles is determined by turbulent diffusion, deviations from the exponential profiles are observed [6, 10, 12, 13]. In the surface layer of the atmosphere, under stationary conditions, as a rule, these profiles are approximated by power functions [1, 14]. For the range of the heights from about 1 to 30 cm in [15] proposed the approximation

$$f(z) = A\,(z^{-1}-H^{-1})^n, \qquad (17)$$

where A, H and n are parameters.

In general, the problem of the vertical profile concentration and the particle mass flux approximations requires further study.

9. Conclusions

The technique has been worked out for retrieving of the large particle height profile characteristics in the surface layer of the atmosphere.

The vertical profile measurements results of the large particles mass flux in the surface layer of the atmosphere, the profiles of the wind speed, the parameters of the particle size distribution function, including the fractions of particles with sizes less than 50, 73, and 100 μm, are analyzed. Based on the measurement data of the large particle height profiles of the mass flux and the wind speed in the surface layer of the atmosphere on the desertificated areas in the Aral region, the profiles of the total mass concentration of the large particles in the height range from 0.125 to 16 m are received. Power-law approximations of the particle profile mass concentration are proposed.

The technique for retrieving of the total vertical number particle concentration distribution is proposed. The modal size of the large particles height dependences approximations are also obtained. The vertical profiles of the particle number concentrations in the surface layer of the atmosphere on the desertificated areas have been retrieved. Power-law approximations of the total vertical number concentration profiles have been received.

Vertical profiles approximations of the aleurite particles proportions with sizes less than 50, 73, and 100 μm have been proposed.

The profiles of the aleurite particle number concentrations with sizes less than 50 and 73 μm in have been received. Power-law approximations of these profiles are obtained.

According to the measurements in the Aral region in 1992, the profiles of the total aleurite and sand number particle concentrations were retrieved. Power-law vertical profiles approximations of the aleurite and sand particle concentrations are proposed. It is shown that the rate of decrease in the aleurite particles concentration with increasing height is noticeably lower than the rate of the sand particle concentration decrease.

The results obtained are necessary for modeling the processes of the large particle transport from desertificated areas to assess the consequences of desertification and aridization. They are of interest for the theory of climate, ecology, and geomorphology.

Acknowledgments

This work was supported by the Russian Foundation for Basic Research (Grant No. 19-05-00758).

IOP Conf. Series: Earth and Environmental Science 1040 (2022) 012030 doi:10.1088/1755-1315/1040/1/012030

References

[1] Semenov O E 2020 *An Introduction to Experimental Meteorology and Climatology of Sandstorms* (Moscow: Fizmatkniga) p 446 [in Russian]
[2] Gorchakov G I et al 2019 *Doklady Earth Sciences* **488**(1) pp 1103–1106
[3] Gorchakov G I et al 2020 *Atmos Ocean Opt.* **33** pp 198–205
[4] Pettijohn F G 1957 *Sedimentary Rocs* (New York: Harper) p 526
[5] Karpov A V et al 2017 *Atmos Ocean Opt.* **30** pp 456–461
[6] Rasmussen K R and Sorensen M J 2008 *J. Geophys. Res.* **113** FO2S12 pp 1-12
[7] Buttner E K 1978 *Dynamics of the near-surface air layer* (Leningrad: Gidrometeoizdat) p 158 [in Russian]
[8] Barenblatt G I and Golitsyn G S 1973 *Local structure of developed dust storms* (Moscow: Moscow State University) p 44 [in Russian]
[9] Williams G 1964 *Sedimentology* **3** pp 257-287
[10] Greeley R et al 1996 *Sedimentology* **43** pp 41-52
[11] Liu X and Dong Z et al 2004 *Geomorphology* **60** pp 371-381
[12] Gorchakov G I et al 2021 *Doklady Earth Sciences* 496(2) pp 119–124
[13] Liu D et al 2018 *Atmos. Chem. Phys*. **18** pp 7595-7606
[14] Sterk G and Raats PAC 1996 *Science Society of America Journal* **60** pp 1914-1919
[15] Ni J R et al 2003 *Geomorphology* **49** pp 205-218

IOP Conf. Series: Earth and Environmental Science 1040 (2022) 012031 doi:10.1088/1755-1315/1040/1/012031

Application of wavelet analysis methods to study the climate of the Arctic region

DA Solovyev[1], LV Nefedova[2]

[1]Shirshov Institute of Oceanology, Russian Academy of Sciences, Nahimovskiy prospect, 36, Moscow, 117997, Russia
[2]Lomonosov Moscow State University, Faculty of Geography, Leninskie gory, 1, Moscow, 119991, Russia

Email: [1]solovev@ocean.ru, [2]nefludmila@mail.ru,

Abstract. This paper provides an assessment of the possibilities and advantages of using wavelet analysis methods in the study of regional mechanisms of climate formation in the Arctic region by analyzing the relationship between the values of the air temperature anomalies at the Earth's surface and the North Atlantic Oscillation index. It is shown that the methods of wavelet analysis of climatic indices have certain advantages for identifying cyclical changes in climatic characteristics on local time scales and allow an in-depth spectral assessment. The conducted wavelet analysis of the NAO and SAT index (1950-2020) made it possible to estimate the previously known natural quasi-cyclic oscillations with a typical period of 7-15 years and other dominant components of their spectrum.

1. Introduction

Climatic variability in the Arctic is primarily determined by regional and local mechanisms, many of which can play a decisive role. Such tools are associated with a considerable number of feedback regulating climatic processes in the Arctic [1]. Climate changes in the Arctic are amplified by feedbacks, among which the degradation of sea ice in the Arctic Ocean, which is acutely responsive to climate change, attracts particular attention. The removal of freshwater from the Arctic Ocean affects the distribution of sea ice, thermohaline circulation in the adjacent water area of the North Atlantic and through them on the regional and global climate. Several variable heat sources and feedback between them make the Arctic the region of the most significant climate change and fluctuations. Many features of interrelated processes remain insufficiently studied at present [2]. The intensity and direction of the influence of feedback caused by the methods of energy exchange at the ocean-atmosphere interface can be investigated by analyzing the time series of data characterizing climatic fluctuations in the Northern Hemisphere, including the Arctic region. The most significant changes that have a direct impact on the climate of the Arctic region include the North Atlantic Oscillation (*NAO*), Arctic Oscillation (*AO*), and Atlantic Multi-Decade Oscillation (*AMO*) [3–6]. In most cases, the climatic data of these fluctuations are presented in the form of time series set at a finite time interval with a constant time step (day, month, year, etc.). This makes it possible to use spectral and wavelet analysis methods to study the characteristics and assess the potential effects of climatic fluctuations on the climate. It is important to note that it is precisely the role of these fluctuations and their regional impacts that little attention was paid to when predicting climate changes in the Arctic.

At its core, climatic fluctuations are a combination of various details into a full generalized description of the atmosphere or ocean, which can be further used to characterize the factors that affect the global climate system of the Arctic region.

This paper evaluates the possibilities and advantages of using wavelet analysis methods in studying the climate of the Arctic region using the example of the analysis of the North Atlantic Oscillation (NAO) index and its relationship with the interannual variation of the mean annual air temperature anomaly in the Arctic region (SAT) (see Figure 1).

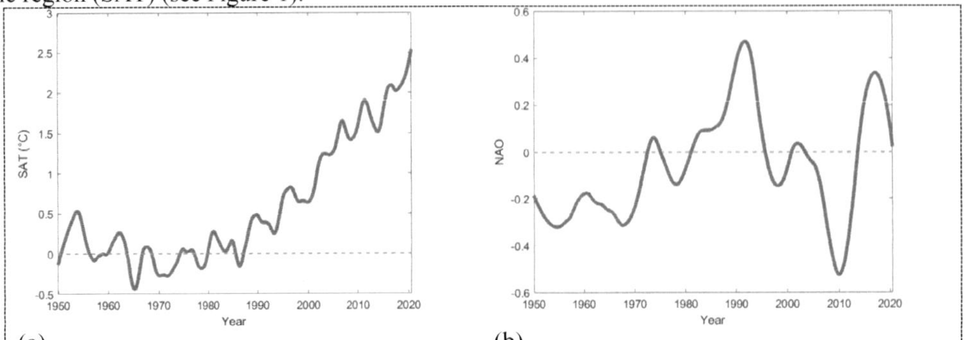

(a) (b)

Figure 1. Interannual variation of the Surface Air Temperature anomaly (*SAT*) (° C) averaged for the latitudinal zone 60–90 ° N. for the period 15.01.1950—15.07.2020 - (a) and the interannual variation of the standardized index *NAO* for the period 01/15/1950 - 07/15/2020. According to the BEST archive [7].

We used the air temperature at the Earth's surface (*SAT*) synthesized at the NASA Goddard Institute of Space Science (NASA GISS) as part of the Berkeley Earth Surface Temperature (BEST) project as the initial information for assessing the interannual variation of the average annual air temperature in the Arctic region (for the latitudinal zone 60-90 ° N) [7]. The period 1951-1980 was used as the base averaging period for assessing air temperature anomalies; the methodology and algorithms for data processing and quality control at all stages are described [8]. In Fig. 1 (a) on the graph for the latitudinal zone 60-90 ° N. for the period 15.01.1950 - 07.15.2020 two pronounced warmings are visible. The first falls on the period 1950 - 1962, the second, more prolonged, - starting from the 1980s of the 20th century. In July 2020, the value of the average annual air temperature anomaly in the Arctic region reached its maximum for the entire history of observations - 2.45 ° C, which confirms the factual data on the unprecedented acceleration of the Arctic climate warming in recent years [9]. It is important to note that warming in the Arctic occurs against the background of a predominantly positive phase of fluctuations in the index *NAO* (warm Arctic) (Figure 1, b) [10].

The North Atlantic Oscillation (*NAO*) is associated with North America, Greenland, and Europe's prevailing weather patterns., the difference in the surface height of 500 hPa between points in Iceland (64 ° N, 24 ° W) and near the Azores (39 ° N, 24 ° W) during the winter-spring season is considered as the index *NAO*. Data source *NAO*: National Oceanic and Atmospheric Administration's (NOAA) Climate Prediction Center. The index values are publicly available on the official website of the organization [11].

During the period when the index values are high, an increase in the subtropical pressure maximum and the deepening of the Icelandic minimum are observed, which leads to large gradients between these atmospheric formations and an increase in winds carrying warm and humid air from the Atlantic Ocean to northern Europe, while in Canada and Greenland is dominated by dry and cold weather. During the period of low values of the index, the intensity of atmospheric formations weakens. The power of winds decreases, and their direction shifts towards the Mediterranean, where warm, humid weather sets in. In northern Europe, in this case, on the contrary, dry and cold weather prevails.

There is reason to believe that the North Atlantic Oscillation affects the Atlantic Ocean and, being part of the global circulation, has a connection with the oscillations of the entire Northern Hemisphere. In particular, the mechanism of its impact on the Arctic region's climatic dynamics was studied using the INM-CM5 climate model with a period of about 15 years [10]. For this, a technique was used to calculate the contribution of various terms to the generation of oscillation energy and phase change. As a result, it was shown that during the positive phase of the oscillation (warm Arctic), the below zero temperature and salinity anomaly occur in the North Atlantic's temperate latitudes. A positive North Atlantic Oscillation Index accompanies these anomalies. A quarter of the period after the warming of the Arctic, warming and salinization occurs in the North Atlantic's temperate latitudes. In the Arctic Ocean, temperature anomalies arise due to increased advection of Atlantic water [12]. The oscillation phase's evolution occurs due to the transfer of heat from the Atlantic and the flow of heat on the surface. Anomalies in currents transferring heat from the Atlantic to the Arctic Ocean are generated mainly by wind friction.

The purpose of this work is to assess the possibilities and advantages of using wavelet analysis methods in the study of regional mechanisms of climate formation in the Arctic region by the example of analyzing the values of the interannual variation of the mean annual air temperature anomaly averaged in the Arctic region near the Earth's surface (SAT) and the North Atlantic Oscillation index (NAO).

2. An object of study and methods

2.1. The advantage of wavelet transforms over Fourier transforms

Wavelet analysis is used in problems related to the study of spatial fields with a complex multiscale structure (turbulent flow) or temporal signals with a spectral composition that changes over time (natural and economic cyclic processes). The wavelet transformation's main idea meets the specifics of many time series, demonstrating the evolution in time of their main characteristics - mean value, variance, periods, amplitudes, and phases of harmonic components. The overwhelming majority of processes studied in various fields of knowledge have the above features [13]. The advantage of the wavelet transform over the Fourier transform is that it allows you to trace the change in the signal's spectral properties with time and indicate which frequencies (scales) dominate the movement [36].

In several works, connection was found between various climatic indices, including the North Atlantic Oscillation (NAO) index (distinct periods of 7-11, 22-40 and 40-80 years, respectively) with the regime of precipitation and temperature, directly affecting the formation of the climatic control of the Arctic [14.15]. To recognize such nonstationary processes, the wavelet analysis method has been increasingly used recently [16]. In studies of this kind, the most interesting is the determination of the nature of a random process and the concept of causes that lead to fluctuations in atmospheric parameters that differ in characteristic periodicities or cycles.

2.2. Results of wavelet analysis of NAO and SAT and ways of their presentation

The one-dimensional Fourier transform also gives one-dimensional information about the relative contribution (amplitudes) of different time scales (frequencies) [16]. The result of the wavelet transform of the one-dimensional series $f(t)$ is a two-dimensional array of wavelet transform amplitudes of the values of the coefficients $\omega(a, b)$:

$$\omega(a,b) = a^k \int_{-\infty}^{+\infty} f(t)\psi^*\left(\frac{t-b}{a}\right)dt,$$ where $\psi(t)$ is a real or complex function.

The distribution of these values in space (a, b) = (time scale, time localization) gives information about the evolution of the relative contribution of components of different scales in time and is called the spectrum of the wavelet transform coefficients (time - scale spectrum, or wavelet spectrum in contrast to the single spectrum Fourier transform).

The spectrum $\omega(a, b)$ *of a* one-dimensional $f(t)$ signal is a surface in three-dimensional space. The way this information is visualized can vary. Instead of images of characters, one often presents their projections onto the plane (a, b) with isolines, which make it possible to trace the change in the intensity of the wavelet transform amplitudes at different scales and in time, as well as pictures of the lines of local extrema of these surfaces (Power spectrum), which reveal the structure of the analyzed process.

Fig. 2 and Fig. 3. show the results of the Wavelet analysis of time fluctuations in *NAO* and the interannual variation of the mean annual air temperature (SAT) anomaly averaged in the Arctic region. In Fig. 2, the lighter areas correspond to positive, and the dark ones, to negative values of $\omega(a, b)$; the ranges of values are highlighted in shades of colour in each of the areas $\omega(a, b)$. It is clear that the value of the wavelet transforms amplitude at the point (a_0, b_0) *is* the more significant (in absolute value), the stronger the correlation between the wavelet of a given scale and the behaviour of the signal $f(t)$ in the vicinity of $t = t_0$. The coefficients' picture demonstrates that the process consists of different scales: the extrema $\omega(a, b)$ are observed at different scales, their intensity changes both with time and with the plate.

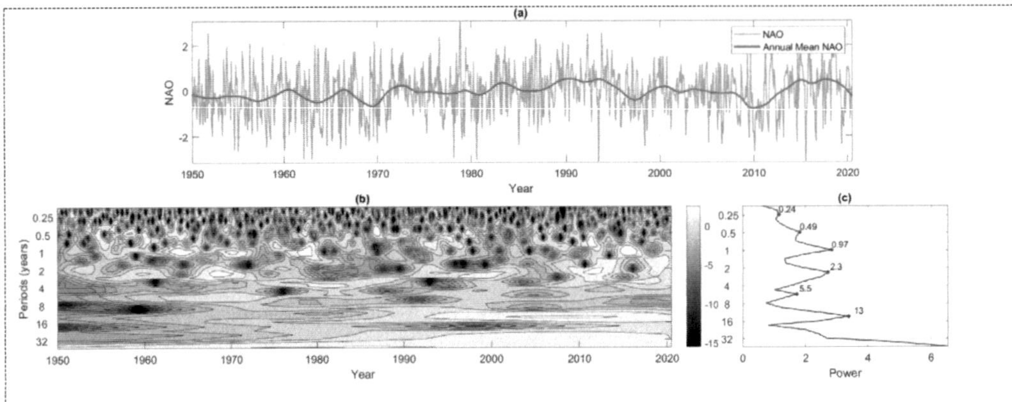

Figure 2. Spectral characteristics of the interannual variation of the smoothed values of the index (NAO) - (a) and its wavelet spectrum for the NAO index - (b and c) for the period 01.1950 - 07.2020.

The wavelet transform decomposes the analyzed process into its constituent waves, components of different scales, and gives information about the process localized in time. Having wavelet spectra, one can calculate valuable characteristics of the process under study and analyze many of its properties. Let us describe the possibilities of exploring the features of the signal and its energy characteristics in more detail.

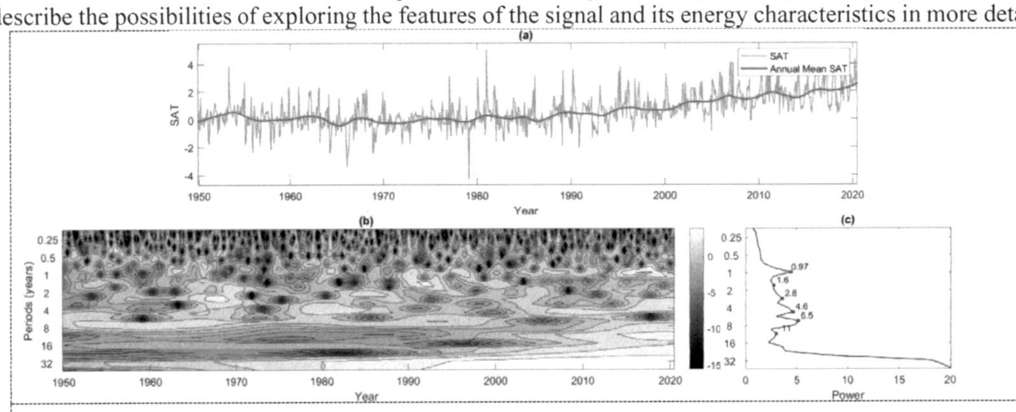

Figure 3. Spectral characteristics of the interannual variation of the mean annual air temperature (SAT) (° C), averaged for the latitudinal zone 60-90 N. - (a) and its wavelet spectrum for SAT - (b and c) for the period 01.1950 - 07.2020. Source: authors' calculations.

The horizontal section of the picture shown in Fig. 2 (c) and 3 (c) (Power spectrum), at a given scale, and demonstrates the change in the component of the selected scale over time. Power spectrum allows you to highlight the distinct periods for the analyzed signal (time series of oscillation data). For NAO and SAT, there are six line maxima each, labelled with digital signatures next to the Power curve. However, in contrast to the Fourier spectrum, we see that both the periods and the amplitudes of these lines change over time. The position of such maxima (peaks) of the Fourier spectrum is usually associated with the frequencies and the corresponding characteristic modes of the analyzed signal, which carry the primary energy of the process. The Power spectrum maxima are interpreted in a similar way - they determine the scale of the process that makes the main contribution to the total energy.

3. Results and discussion

To interpret the obtained results of the wavelet analysis of the NAO time fluctuations and the SAT interannual variation, it is necessary first to consider in detail the mechanisms of the possible NAO impact on climatic variability Arctic region. Since at least the 18th century, it has been known that if Denmark has a harsh winter, then Greenland's winter is milder than usual, and vice versa. This behaviour is part of a climate behaviour pattern known as the North Atlantic Oscillation. The North Atlantic Oscillation is the fluctuation in atmospheric pressure at sea level between the Arctic and the subtropical Atlantic that is most noticeable during the Arctic cold season (November-April) is associated with changes in mean wind speed and direction. Westerly winds blowing across the Atlantic bring warm, humid air to Europe. When westerly winds are strong, summer is cool, winter is moderate, and precipitation is more frequent. If the westerly winds weaken, the temperature in summer and winter reaches its critical levels, leading to the transfer of warm air masses, severe frosts and less precipitation. The constant low-pressure area over Greenland and the Azores' constant high-pressure area govern the westerly winds' direction and strength into Europe. Changes in the North Atlantic Oscillation alter the regular seasonal transport of heat and precipitation between the Atlantic and neighboring continents. The number of storms, their strength and direction of movement also change.

Changes in the North Atlantic Oscillation cause significant changes in ocean surface temperatures, ocean currents, and associated heat transfers. Changes in ice cover in arctic and subarctic regions are also associated with the North Atlantic Oscillation. These climatic changes can impact agricultural yields, water supplies, energy supplies, fish catches, etc. An extensive NAO index is called a positive phase - a small index, respectively, a negative phase. In a positive phase during winter, westerly winds intensify along the North Atlantic, moving relatively warm and moist sea air along much of Europe and further through Asia. In contrast, stronger northerly winds carry cold air along Greenland and northeastern Canada to the south. At the same time, the air temperature and the temperature of the ocean surface in the northeast Atlantic decrease.

Wavelet analysis of the NAO and SAT index (1950 - 2020) revealed the characteristic dominant (11 - 13-year period) harmonic components and smaller ones (0.97 - 6.5 years). The maximum power of the global wavelet spectrum for the NAO is at the 13 - year harmonic, for the SAT - the harmonic at 6.5 years. Simultaneously, there is a statistically significant relationship between the characteristics of the global wavelet spectrum NAO and SAT in the low-frequency region of oscillations (**Table 1.**).

Table 1. Characteristics of the global wavelet spectrum (Power spectrum) for NAO - (a) and for SAT - (b) for the period 01.1950 - 07.2020 for the latitudinal zone 60-90° N. according to the BEST archive [7].

(a)	Power NAO (max)	Periods (y.)	(b)	Power SAT (max)	Periods (y.)
	1.1461	0.24		2.7553	1.60
	1.7155	5.50		3.0113	11.0
	1.8125	0.49		3.5444	2.80
	2.7019	2.30		4.4167	0.97
	2.8252	0.97		4.6527	4.60
	3.3877	13.0		5.1350	6.50

The graphs of the NAO and SAT index changes are a series of peaks, each of which covers a certain period (Fig. 2 (a) and Fig. 3 (a)). It can be seen from the figures that the amplitude of their oscillation cycles is continuously changing. Analysis of the characteristics of the global wavelet spectrum (Power spectrum) for the NAO and SAT index (1950-2020), shown in Table 1, made it possible to estimate not only the previously known natural quasi-cyclic oscillations with a typical period of 7-15 years but also other dominant components of the spectrum. The interannual variability of the NAO index during the warming of the Arctic is complex. The positive phase (growth of the NAO index) means an increase in the western geostrophic transport of air masses in temperate latitudes, an increase in southerly winds in the eastern part of the North Atlantic, an increase in northerly winds in the Labrador Sea, a mixing of cyclone trajectories to the north, which ultimately causes the formation of negative water temperature anomalies in temperate and subpolar latitudes. For the negative phase of NAO, the opposite situation is characteristic: favourable conditions are created for an increase in the SAT temperature anomaly due to the prevailing southwestern winds over the area of the North Atlantic Current. This leads to a rise in the inflow of warm water into the North European Basin and into the Barents Sea, which manifests itself in a decrease in efficiency and an increase in water temperature in the Kola section meridian.

4. Conclusion

Using the methods of wavelet analysis, the contribution of various terms to the generation of the oscillation energy of the NAO index and the interannual variation of the mean annual air temperature (SAT) for the latitudinal zone of 60-90 ° N was estimated. Warming in the Arctic occurs against the background of the positive phase of fluctuations in the NAO index (warm Arctic). This suggests that the generation of temperature anomalies in the Arctic Ocean occurs due to increased advection of Atlantic water. Wavelet analysis data showed that NAO and SAT have characteristic dominant oscillation harmonics close in their values, which confirms a specific contribution of NAO to the regional mechanisms of the currently observed warming of the Arctic. NAO influences ocean currents and associated heat transfer processes. In the positive phase of NAO, westerly winds intensify along the North Atlantic, moving relatively warm and moist sea air along much of Europe and further through Asia. In contrast, stronger northerly winds carry cold air along Greenland and northeastern Canada southward. At the same time, the air temperature decreases in the northeast of the Atlantic.

The performed estimates of the global wavelet spectrum (Power spectrum) for the NAO and the SAT showed a statistically significant relationship between them in the low-frequency region of oscillations. Therefore, climatic changes caused by cyclical fluctuations of the NAO index can be of considerable importance in the formation of the climate of the Arctic region, which is comparable to the current long-term trend of warming of the Earth's atmosphere, in turn, caused by external, mainly anthropogenic factors [17].

The relevance of the use of wavelet analysis methods in the study of the climatic characteristics of the Arctic region is determined by the fact that monitoring, diagnostics and physical description of the leading regional mechanisms of climatic variability (climatic cycles) in high latitudes of the Northern Hemisphere will minimize technological and environmental risks, increase the efficiency and safety of production processes.

Modern global climatic changes will contribute to the fact that the number of permanent residents of the Arctic latitudes will grow steadily. This will be possible due to the ongoing warming, which is most pronounced in the Arctic [18]. Due to the rapid melting of sea ice and permafrost, it will become easier to access the deposits in the coming years. Tourists will be able to access the Arctic. In a warmer climate, it will become easier to conduct economic activities, which will ultimately cause an influx of the region's resident population. This determines the importance of monitoring, diagnosing and physically describing the main regional mechanisms of climate variability (climate cycles) in the Arctic region.

Acknowledgements

This work was acquired within State Assign (0149-2019-0002; AAAA-A16-116032810088-8), with financial support by the Russian Ministry of Science and Higher Education (agreement No. 05.613.21.0090, project RFMEFI61319X0090).

5. References

[1] Sporyshev PV, Katsov VM and Gulev S K 2018 Changes in surface temperature in the Arctic: reliability of model reproduction and probabilistic forecast for the near future *Papers of the Academy of Sciences* 569–73

[2] Sherstyukov BG 2016 Climatic conditions of the Arctic and new approaches to forecasting climate change *Arctic and North* 39–67

[3] Malinin VN and Vainovsky PA 2018 On the causes of the first warming in the Arctic in the 20th centuryRussian State Medical University *Scientific Notes of the* 34

[4] Nesterov ES 2013 The *North Atlantic Oscillation: Atmosphere and Ocean* (Moscow: OOO "Triad")

[5] Gulev SK and Latif M 2015 The origins of a climate oscillation *Nature* **521** 428–30

[6] Hurrell JW and Deser C 2010 North Atlantic climate variability: The role of the North Atlantic Oscillation *J. Mar. Syst.* **79** 231–44

[7] Anon 2021 Data Overview - Berkeley Earth berkeleyearth.org

[8] Rohde R, Muller R, Jacobsen R, Perlmutter S, Rosenfeld A, Wurtele J, Curry J, Wickham C and Mosher S 2013 Berkeley Earth Temperature Averaging Process *Geoinfor Geostat An Overv.* **1**

[9] Yamanouchi T and Takata K 2020 Rapid change of the Arctic climate system and its global influences - Overview of GRENE Arctic climate change research project (2011–2016) *Polar Sci.* 100548

[10] Volodin EM 2020 On the mechanism of climate fluctuations in the Arctic with a period of about 15 years according to the data of the INM RAS climate model *Izvestia of the Russian Academy of Sciences. Physics of the atmosphere and ocean* **56** 139–49

[11] Anon 2021 Climate Indices: Monthly Atmospheric and Ocean Time Series: NOAA Physical Sciences Laboratory *noaa.gov*

[12] Diansky NA 2013 Modeling ocean circulation and studying its response to short-term and long-term atmospheric influences *M. Fizmatlit* **271**

[13] Burnaev E V 2006 Application of wavelet transform for the analysis of economic time series *In*

sb. scientific. Proceedings of the Summer School on Economic and Mathematical Modeling ECOMOD vol 2006 p 95

[14] Mokhov II, Smirnov DA and Karpenko AA 2012 Assessments of the relationship of changes of the global surface air temperature with different natural and anthropogenic factors based on observations *Dokl. Earth Sci.* **443** 381-7

[15] Sutton RT 2005 Atlantic Ocean Forcing of North American and European Summer Climate *Science (80-.).* **309** 115–8NM

[16] Astaf'eva1996 Wavelet analysis: theoretical foundations and examples of application *Advances in physical sciences* **166** 1145–70

[17] Gulev SK, Kattsov VM and Solomina ON 2008 Global warming continues *Vestnik RAN* **78** 20–7

[18] Bushuev VV, Soloviev DA, Morgunova MO, Nefedova LV and Gabderakhmanova T S 2019 *Energy infrastructure of the Arctic zone of the Russian Federation* ed VV Bushuev (Moscow: Publishing House "Shans")

IOP Conf. Series: Earth and Environmental Science 1040 (2022) 012032 doi:10.1088/1755-1315/1040/1/012032

The role of winter net heat fluxes on the modulation of the upper mixed layer temperature and depth in the North Atlantic by the reanalysis data

P A Sukhonos[1] and N A Diansky[2, 3, 4]

[1] Institute of Natural and Technical Systems, Sevastopol, 299011, Russia
[2] Zubov State Oceanographic Institute, Moscow, 119034, Russia
[3] Moscow State University, Moscow, 119991, Russia
[4] Marchuk Institute of Numerical Mathematic, Russian Academy of Sciences, Moscow, 119333, Russia
pasukhonis@mail.ru

Abstract. The role of winter net heat fluxes (NHF) on the upper mixed layer (UML) temperature and depth variability over the North Atlantic is studied using ocean reanalysis data products. The monthly data from the ORA-S3 (1959–2011), GODAS (1980–2020), GECCO3 (1948–2018) datasets were used. The correlation coefficients were calculated between the anomalies of the NHF on the ocean surface and the UML temperature and depth with a shift of 1 month when the former were leading. In most of the North Atlantic, negative correlations between the UML temperature and the NHF preceding 1 month were found. Here, the UML temperature anomalies are probably caused by changes in heat exchange with the atmosphere. Areas of significant positive correlations across all data sets are concentrated in the area of the transition of the Gulf Stream to the North Atlantic Current. Almost the entire water area of the North Atlantic is occupied by positive correlations between the mixed layer depth in winter and the NHF on the ocean surface, with the latter leading for 1 month.

1. Introduction

The upper mixed layer (UML) of the ocean is a layer that forms as a result of turbulent mixing when interacting with the atmosphere. As a result of this interaction, the main hydrophysical parameters (temperature, salinity, density) are distributed uniformly along the vertical within the UML. The mixed layer depth (MLD) depends on the exchange of heat, mass and momentum between the upper layer of the ocean and the atmosphere.

The study of the processes occurring in the UML is of great importance for understanding climate variability. The UML thickness modulates its heat capacity and the ability to accumulate heat from the atmosphere. In addition, the change in the position of the UML lower boundary determines the participation of the deeper layers of the ocean in the exchange of heat with the atmosphere. The study of the UML heat balance plays an important role in understanding the processes that cause changes in the UML temperature [1–5].

In many works, studies of the UML regional variability and its sensitivity to changes in atmospheric forcing have been carried out [6, 7]. The results of model studies [8, 9] showed that the net heat fluxes (NHF) forcing dominates in the UML characteristics variability in the winter season. In this case, wind stress anomalies are ineffective in the formation of sea surface temperature (SST)

IOP Conf. Series: Earth and Environmental Science 1040 (2022) 012032 doi:10.1088/1755-1315/1040/1/012032

variability. Analysis of the cross correlations of the MLD with both the NHF and the wind stress showed that their contribution is mostly at the same level, with a more pronounced correlation for the NHF [7]. However, modern data sets make it possible to consider large water areas over long time periods. This is especially important because the atmospheric forcing is characterized by spatio-temporal heterogeneity caused by the peculiarities of atmospheric circulation and its variability.

The aim of the study is to identify regions in the North Atlantic with a different nature of prevailing atmospheric forcing, expressed by the NHF, on the UML temperature and depth in the winter season by analyzing the ocean reanalysis data.

2. Data and methods

The monthly data on the upper ocean temperature and MLD presented here are taken from the Ocean Reanalysis System 3 (ORA-S3) for the period 1959–2011 [10], the Global Ocean Data Assimilation System (GODAS) for the period 1980–2020 [11] and the German contribution to the Estimating the Circulation and Climate of the Ocean system (GECCO3) for the period 1948–2018 [12]. The monthly data on the NHF on the ocean surface are taken from the atmospheric reanalyzes: ERA-40 [13], NCEP-DOE AMIP-II Reanalysis [14] and NCEP/NCAR Reanalysis [15], respectively. These datasets were used to force the ocean model in the reanalyzes listed above. The North Atlantic water area is limited by coordinates (0–70 °N 8–80 °W). The vertical z-axis is directed upward. Positive values of the NHF on the ocean surface mean the outflow of heat from the ocean to the atmosphere.

Ocean reanalysis is a calculation of one or another numerical model of ocean circulation with the assimilation of available observational data that are non-uniformly distributed in space. Discrepancies may arise between different reanalyzes, both due to different models with their own characteristics, and due to the use of different data assimilation procedures in them [16].

Note that we are using ready-made MLD data. The MLD in the ORA-S3 reanalysis is taken to be equal to the depth at which the Richardson number (*Ri*) reaches the critical value [10]. In the GECCO3 reanalysis the depth-independent critical gradient *Ri* was chosen to control the MLD [12]. Both of these techniques use the *Ri* to determine the position of the UML lower boundary. In the GODAS reanalysis, the MLD is the depth where the buoyancy difference with respect to the surface level is equal to 0.03 (cm/s^2) [https://www.psl.noaa.gov/data/gridded/data.godas.html].

Based on the ocean temperature data, for each of the reanalyzes used, the average temperature was calculated within the space and time variable MLD for each month for the entire available period. The UML temperature is assumed to be constant over the MLD and corresponds to the SST. Further, the UML temperature and depth values in January, February and March and the NHF values with a shift of one month in December, January and February were identified. Then, using the least squares method, the linear trend for the available period was removed from all variables. After that, the values of the correlation coefficients were calculated between the obtained anomalies of the NHF on the ocean surface and the UML temperature and depth with a shift of 1 month when the former were leading. The relationship between the NHF and the UML characteristics is more pronounced when the UML processes are delayed by 1 month. This shift is due to the thermal inertia of the UML [7, 17, 18].

3. Results

The values of the correlation coefficients between the UML temperature in January, February and March and the NHF on the ocean surface with the latter leading for 1 month are shown in figure 1. Most of the North Atlantic water area is occupied by negative correlations. Here, the UML temperature anomalies are probably caused by changes in heat exchange with the atmosphere. Areas of significant negative correlations are located in the Gulf Stream recirculation area and in the Tropical Atlantic, as well as in the Gulf Stream itself before its separation from the continental slope, with the exception of the GODAS data in February and March. This means that the UML cooling (the negative UML temperature anomaly) in the winter season is preceded (with a shift of 1 month) by an increase in heat outflow from the ocean surface (the positive NHF anomaly).

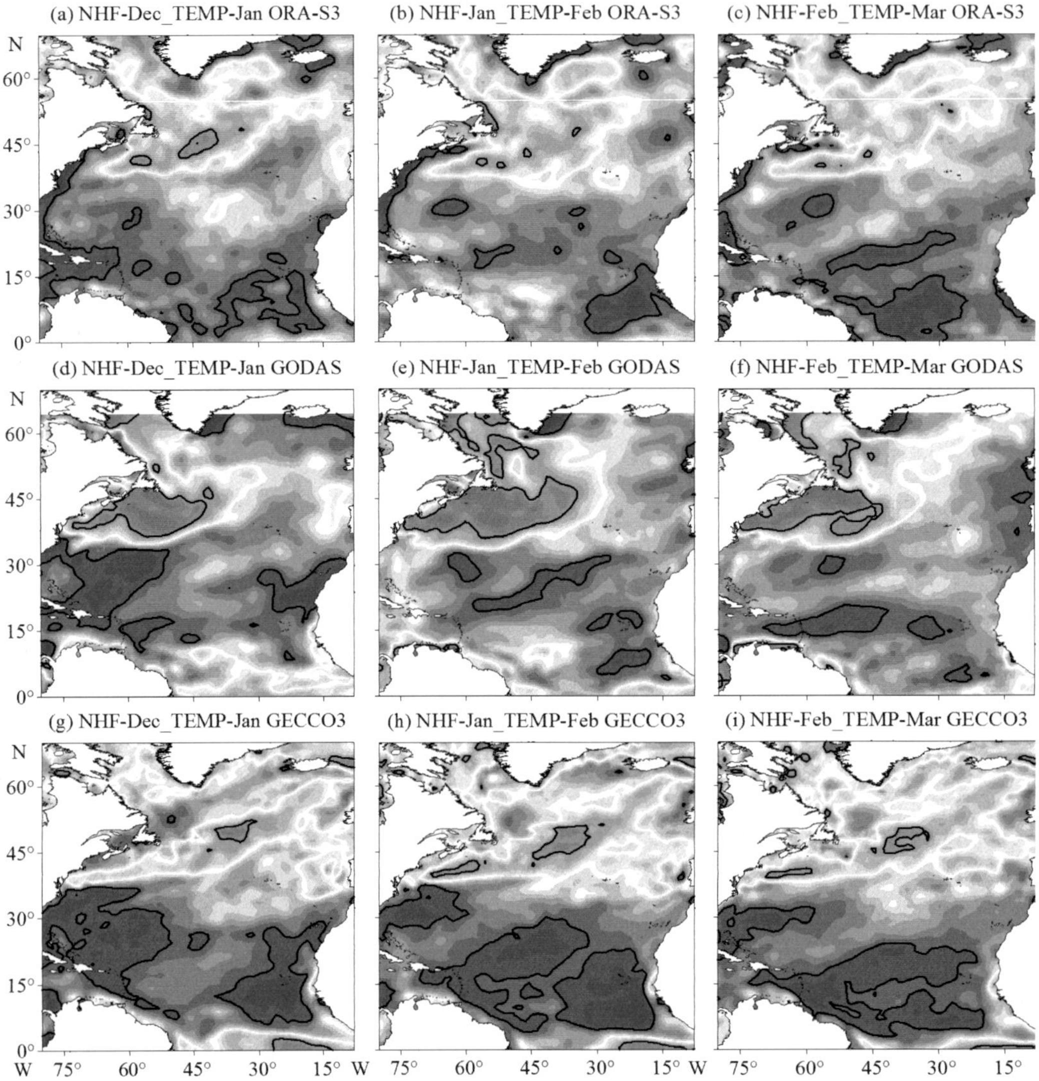

Figure 1. Cross correlation of net heat fluxes (NHF) (positive upward) and UML temperature (TEMP) anomalies from ORA-S3 (a, b, c), GODAS (d, e, f) and GECCO3 (g, h, i) data. All figures represent the corresponding one-month lag correlations, where UML temperature lags the forcing by one month. The left column shows correlation between net heat flux anomalies in December and the UML temperature in January (a, d, g). The middle column shows correlation between net heat flux anomalies in January and the UML temperature in February (b, e, h). The right column shows correlation between net heat flux anomalies in February and the UML temperature in March (c, f, i). Colour code is in such a way that blue shading represents negative correlations and red shading – positive correlations. Black thick line shows the values of the correlation coefficient ±0.5.

The areas of significant positive correlations for all datasets are concentrated in the area of the transition of the Gulf Stream to the North Atlantic Current. Here, an increase in heat outflow from the ocean surface (the positive NHF anomaly) in the winter season is accompanied by an increase in the UML heat content for the next month. In these regions, the formation of temperature anomalies is significantly influenced by inner ocean processes in the UML (see, for example, [1 and 5]). In addition, positive correlations in these areas indicate that here the variability in the UML drive the formation of the variability of the atmosphere, due to the change in the intensity of the transfer of heat to it brought by warm currents. On the whole, it can be stated that, in areas of negative correlations, the NHF between the atmosphere and the ocean affect the formation of the SST anomalies. And, conversely, in areas of positive correlations, the SST anomalies are formed under the influence of inner ocean processes.

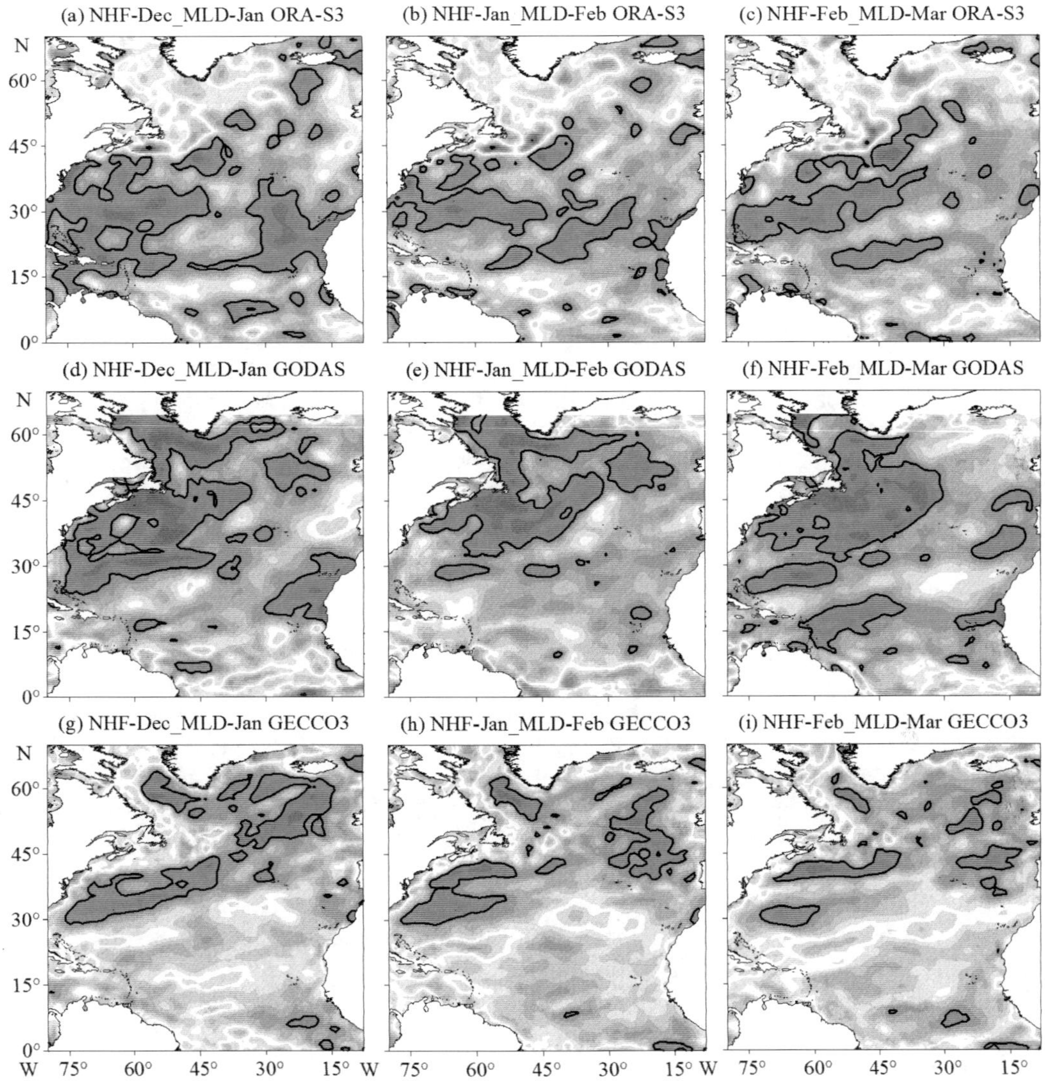

Figure 2. Cross correlation of net heat fluxes (NHF) (positive upward) and mixed layer depth (MLD) anomalies from ORA-S3 (a, b, c), GODAS (d, e, f) and GECCO3 (g, h, i) data. All figures represent the corresponding one-month lag correlations, where mixed layer depth lags the forcing by one month. The left column shows correlation between net heat flux anomalies in December and the mixed layer depth in January (a, d, g). The middle column shows correlation between net heat flux anomalies in January and the mixed layer depth in February (b, e, h). The right column shows correlation between net heat flux anomalies in February and the mixed layer depth in March (c, f, i). Colour code is in such a way that blue shading represents negative correlations and red shading – positive correlations. Black thick line shows the values of the correlation coefficient ±0.5.

However, there are some peculiarities in the fields of the correlation coefficients obtained from the reanalyzes data. The UML temperature anomalies according to ORA-S3 data in a narrow vicinity of the equator (0–5 °N) do not show significant correlation with the NHF preceding 1 month. The UML temperature anomalies according to GODAS data in January and February show a weak positive correlation with the NHF anomalies preceding 1 month, and the UML temperature according to GECCO3 data shows a significant positive correlation. In the vicinity of 10 °N 40 °W the UML temperature anomalies according to GODAS data in January and February show a weak positive correlation with the NHF anomalies preceding 1 month. According to the data of two other reanalyzes, a weak negative correlation is noted here. In the Guiana Current, the UML temperature anomalies, according to GODAS data, show a positive and significant correlation with the NHF anomalies preceding 1 month, but this is not observed according to the data of two other reanalyzes. The UML temperature anomalies, according to the ORA-S3 and GODAS data, in the Labrador Current show a significant positive correlation with the NHF anomalies preceding 1 month. At the same time, the UML temperature anomalies, according to the GECCO3 data, here show a negative correlation in January, and a positive correlation in March. The UML temperature anomalies according to the ORA-S3 data in the East Greenland Current and according to the GODAS data at the southern tip of Greenland show a negative correlation with the NHF anomalies preceding 1 month, and the UML temperature anomalies according to the GECCO3 data show a weak positive correlation.

The values of the correlation coefficients between the MLD in January, February and March and the NHF on the ocean surface with the latter leading for 1 month are shown in figure 2. Almost the entire water area of the North Atlantic is occupied by positive correlations. An increase in heat outflow from the ocean surface (the NHF positive anomaly) in the winter season is accompanied by the UML deepening for the next month (the MLD positive anomaly).

However, there are some peculiarities in the fields of the correlation coefficients between the MLD anomalies and the NHF on the ocean surface. The MLD anomalies according to ORA-S3 and GECCO3 data in the vicinity of 0–10 °N show a weak positive relationship with the NHF preceding 1 month, and according to the GODAS data, there are areas of insignificant negative values of the correlation coefficients. The MLD anomalies according to the ORA-S3 data in the Gulf Stream (42 °N 55 °W) show a significant negative correlation with the NHF preceding 1 month, and according to the data of two other reanalyzes, a significant positive correlation is noted here. The MLD anomalies according to the GODAS data in the Labrador Current show a significant positive correlation with the NHF preceding 1 month, and according to the data of two other reanalyzes, a significant negative correlation is noted here. The MLD anomalies according to GODAS and GECCO3 data in the area of deep convection in the Labrador Sea show a significant positive correlation with the NHF preceding 1 month, and according to ORA-S3 data, there is no significant correlation in this region. The MLD anomalies in February and March in the inner part of the subtropical gyre show a positive correlation with the NHF preceding 1 month according to ORA-S3 and GODAS data, and an insignificant negative correlation according to GECCO3 data.

In this work, different definitions of the MLD are used for the discussed ocean re-analyzes. Since the technique for determining the MLD using the *Ri* is rather crude, there is no fundamental difference between the results obtained from the ORA-S3 and GECCO3 data. The *Ri* takes into account stratification and measures the combined effect of buoyancy forcing and vertical current shear. The MLD according to GODAS data is defined by the difference criterion. This method allows defining the MLD as the depth where the oceanic property (buoyancy) has changed by a critical value from a reference depth near to the surface. Our results show that it is more correct to use the MLD values obtained using the *Ri* criterion. Determining the exact position of the UML lower boundary will make it possible to clarify the impact of winter NHF on the UML temperature and depth variability. Our results are independent of the length of the time series. Since the correlations were obtained both for short time series (GODAS data – 41 years) and for long time series (ORA-S3 data – 53 years and GECCO3 data – 71 years).

4. Conclusion

In most of the North Atlantic water area, negative correlations were found between the UML temperature and the NHF preceding 1 month. Here, the UML temperature anomalies are probably caused by changes in heat exchange with the atmosphere. The UML cooling in the winter season is preceded (with a shift of 1 month) by an increase in heat outflow from the ocean surface. Areas of significant positive correlations across all data sets are concentrated in the area of the transition of the Gulf Stream to the North Atlantic Current. Here, an increase in the heat outflow from the ocean surface in the winter season is accompanied by an increase in the UML heat content for the next month and is probably caused by the inner ocean processes.

Almost the entire water area of the North Atlantic is occupied by positive correlations between the MLD in winter and the NHF on the ocean surface, with the latter leading for 1 month. An increase in heat outflow from the ocean surface in the winter season is accompanied by the formation of unstable stratification in the UML and, as a consequence, the UML deepening for the next month.

Acknowledgments

We thank an anonymous reviewer for friendly and constructive criticism of the first version of the paper. This work was supported by the Russian Science Foundation, project № 17-17-01295.

References

[1] Kelly K A and Qiu B 1995 *J. Phys. Oceanogr.* **25** 2361–73
[2] Alexander M A, Scott J D and Deser C 2000 *J. Geophys. Res.* **105** 16823–42
[3] Dommenget D and Latif M 2002 *Clim. Dyn.* **19** 277–88
[4] Dong S and Kelly K A 2004 *J. Phys. Oceanogr.* **34** 1214–31
[5] Polonsky A B and Sukhonos P A 2017 *Izv. Atmos. Ocean. Phys.* **53** 459–66
[6] Adamec D and Elsberry R L 1984 *J. Phys. Oceanogr.* **14** 1670–76
[7] Pookkandy B, Dommenget D, Klingaman N, Wales S, Chung C, Frauen C and Wolff H 2016 *Clim. Dyn.* **47** 2991–3010
[8] Alexander M A and Penland C 1996 *J. Climate* **9** 2424–42
[9] Zhao B and Haine T W N 2005 *Ocean Modelling* **9** 211–29
[10] Balmaseda M A, Vidard A and Anderson D L T 2008 *Mon. Wea. Rev.* **136** 3018–34
[11] Behringer D W and Xue Y 2004 *Proc. Eighth Symp. on Integrated Observing and Assimilation Systems for Atmosphere, Oceans, and Land Surface* (Washington: AMS)
[12] Köhl A 2020 *Q. J. R. Meteorol. Soc.* **146** 2250–73
[13] Uppala S M, KÅllberg P W, Simmons A J *et al* 2005 *Q. J. R. Meteorol. Soc.* **131B** 2961–3012
[14] Kanamitsu M, Ebisuzaki W, Woollen J, Yang S, Hnilo J J, Fiorino M and Potter G L 2002 *Bull. Amer. Meteorol. Soc.* **83** 1631–43
[15] Kalnay E, Kanamitsu M, Kistler R *et al* 1996 *Bull. Amer. Meteorol. Soc.* **77** 437–71
[16] Kröger J, Müller W and Storch J-S 2012 *Clim. Dyn.* **39** 795–810
[17] Frankignoul C and Hasselmann K 1977 *Tellus* **29** 284–305
[18] Dianskii N A 1998 *Izv. Atmos. Ocean. Phys.* **34** 197–213

IOP Conf. Series: Earth and Environmental Science 1040 (2022) 012033 doi:10.1088/1755-1315/1040/1/012033

Anthropogenic impact on arctic near-surface methane: observations and model simulations

Yu A Shtabkin, K B Moiseenko, A I Skorokhod

A.M. Obukhov Institute of Atmospheric Physics, Russian Academy of Sciences, 3 Pyzhyovskiy Pereulok, Moscow 119017, Russia

E-mail: yuryshtabkin@gmail.com

Abstract. Impact of climatically significant anthropogenic emissions to seasonal methane (CH_4) variations observed at arctic and subarctic background stations in 1999 – 2019 has been quantitatively estimated using GEOS-Chem chemical transport model. It is shown that the formation of a stable continental pollution plume from sources in Western Europe, European Russia and Siberia allows to explain up to 5.5–8.6 % of observed CH_4 surface concentration (~104–165 ppb). These atmospheric response values are several times higher than the of the observed annual methane variability amplitude (22–36 ppb), which allows to conclude that regional anthropogenic methane emissions sources play a significant role in regional CH_4 balance in arctic and subarctic areas.

1. Introduction

According to the fourth synthesis report published in 2007 by the Intergovernmental Panel on Climate Change (IPCCAR4), methane (CH_4) is the second most important (after carbon dioxide, CO_2) greenhouse gas in the atmosphere [1].

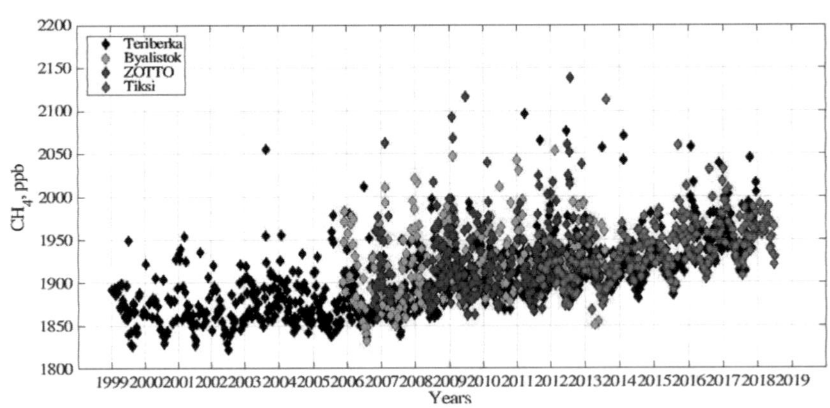

Figure 1. Surface methane concentration according to flask measurements at Teriberka, Bialystok, ZOTTO and Tiksi stations.

Insufficient data on methane content in surface air makes it difficult to quantify methane emissions from major regional anthropogenic and biogenic sources in the Northern Eurasia. This gap is partially filled by long-term observations at stations Teriberka (69.1° N, 35.1° E, 15 m above sea level (m.a.s.l.)), Bialystok (53.1° N, 23.1° E, 160 m.a.s.l.), Zotino (60.8° N, 89.4° E, 300 m.a.s.l.) and Tiksi (71.4° N, 128.5° E, 10 m.a.s.l.). The preliminary analysis of these observations data is given below. Since 2005, according to data at all stations, there has been a significant increase in surface methane concentration (see figure 1). The reasons for this increase are not completely clear. The paper provides quantitative estimates of the most important regional anthropogenic methane emissions sources contribution to observed long-term variability of methane concentration in surface air. Reliable and detailed quantitative estimates of atmospheric methane sources and sinks are needed both to interpret the observed variability and to predict future changes. Background stations observational data, as well as numerical calculations performed by chemical transport models (CTM) is used to analyze surface methane variations. In this paper we use GEOS-Chem model, which is widely applied in chemically active and greenhouse gases fields calculation.

2. Background stations and GEOS-Chem model

GEOS-Chem (http://acmg.seas.harvard.edu/geos/, used version 12-01) - numerical Eulerian global chemical-transport model of atmospheric composition, taking into account all major natural and anthropogenic sources and sinks of chemically active gases and aerosols [2]. Meteorological fields and surface data with a time resolution of 3 (two-dimensional fields) and 6 (three-dimensional fields) hours are taken from the GEOS-GMAO global data assimilation system (Goddard Earth Observing System - NASA Global Modeling Assimilation Office, http://gmao.gsfc.nasa.gov/GEOS/) and are reprojected to model grid. In this paper EDGAR 4.3.2 [3] data were used for anthropogenic CH_4 emissions in the model, WetCHARTs v1.0 (global wetland CH4 emission model ensemble for use in atmospheric chemical-transport models) [4] data - for biogenic emissions and GFED4 – from CH_4 emissions from wildfires [5]. We used 4°×5° model grid, meteorology MERRA2, for comparison with observations were used first level output with a height ~58 m above the ground. Calculations of chemical evolution were carried out in standard mode " NO_x-O_x-hydrocarbon-aerosol" (the so-called "full chemistry" mode for the troposphere, 253 tracers, >500 reactions, time step 60 min.)

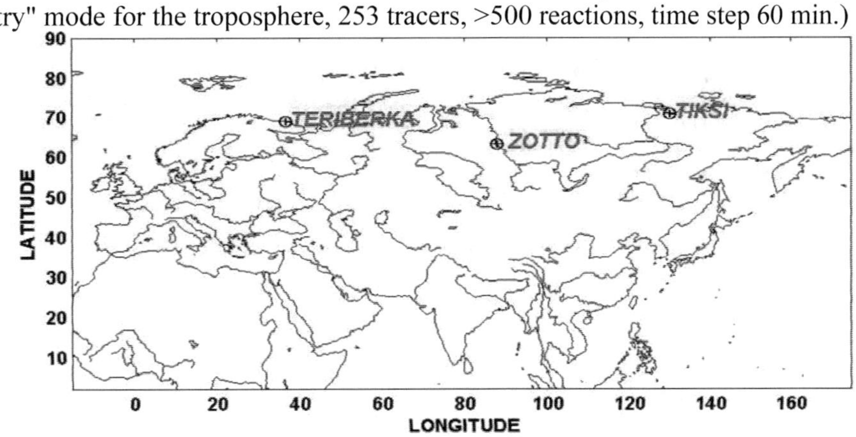

Figure 2. Atmospheric monitoring background stations used for comparison with GEOS-Chem model.

Generally, the results of calculations for all stations (see figure 2), based on monthly averaged data, was in good agreement with the observations despite the limited spatial resolution of the model. The mean bias $\delta = \overline{O} - \overline{C}$, standard error $\mathrm{RMSE} = \left(\overline{(O-C)^2}\right)^{1/2}$, and correlation coefficient $COR = \overline{(O-\overline{O})(C-\overline{C})} / (\sigma_O \sigma_C)$ were used as consistency criteria. O and C are observed and calculated by the model monthly mean concentrations of O_3, σ_O, σ_C - corresponding standard deviations, and the upper line means averaging over all values.

For station Teriberka it was obtained COR = 0.75 (0.65 .. 0.82), RMSE = 19.1 (16.9 .. 22.7) ppb and δ = 2.2 (-1.0 .. 5.6) ppb. The values indicated in parentheses correspond to the limits of the 95% confidence interval calculated by bootstrap method [6]. It should be noted that these values for the station was calculated on basis of flask measurements, carried out 2-10 times a month.

For station Tiksi it was obtained COR = 0.61 (0.44 .. 0.74), RMSE = 22.1 (18.8 .. 26.3) ppb and δ = 5.1 (0.5 .. 10.0) ppb.

For stations ZOTTO it was obtained COR = 0.70 (0.39 .. 0.84), RMSE = 23.9 (18.8 .. 32.1) ppb and δ = -9.6 (-16.1 .. -1.8) ppb.

Generally, calculation results for all three stations, based on monthly averages, was in good agreement with observational data, taking into account limited model spatial resolution. The best agreement is found for Teriberka station, the worst - for Tiksi station. Systematic deviation of calculated methane concentrations from observed in individual seasons was caused by inability of correctly reproducing the high-frequency (synoptic) CH_4 field variations, which make a significant contribution to overall CH_4 variability. Another reason might be the insufficiency of used emission data, especially in Russia. Taking into account the important role of wildfires in middle and high latitudes of Northern Eurasia as a source of large amounts of chemically active gases, including methane, uncertainties in calculation of combustion products emissions might play a major role in observed discrepancies between model calculations and observations. In addition, a significant contribution of total methane variation synoptic component to total measured signal is possible.

3. Influence of regional anthropogenic CH_4 emissions on methane concentration in arctic and subarctic stations

3.1. Model simulations scenario

Quantitative estimates of anthropogenic emissions contribution to CH_4 concentration field were made on the basis of a regional approach: there was identified one macroregion, including Western Europe, European territory of Russia (ETR), Siberia and Russian Far East (see figure 3).

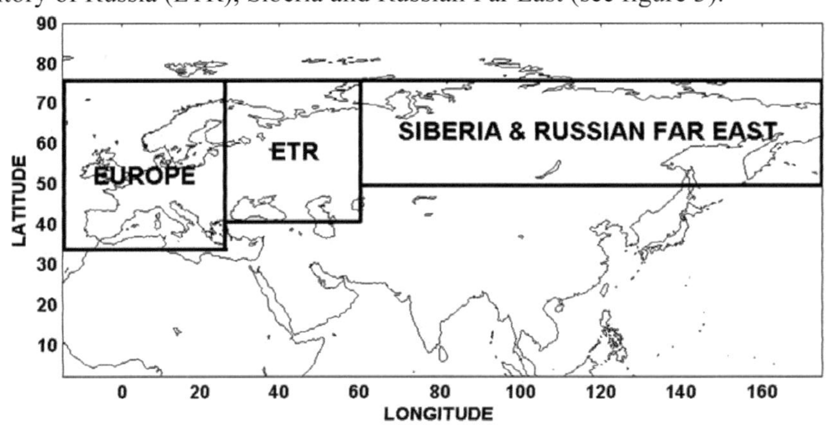

Figure 3. Geographical areas selected for the calculations: Europe (35-75N, 15-27E), European territory of Russia – ETR (41-75N, 27-60E) and Siberia & Russian Far East (49-75N, 60-180E).

The atmospheric response (AR_S) values were calculated as the difference between concentration obtained from baseline model calculation with full emissions, $\chi(CH_4)_0$, and concentration calculated by the model with reduced emissions according to used scenario, $\chi(CH_4)_S$:

$$AR_S = \chi(CH_4)_0 - \chi(CH_4)_S$$

Thus, average seasonal response fields for 2007-2018 were calculated at first model level (~ 58 m from the surface). GEOS-Chem calculations were performed for the period from July 1, 2006 from the model initial startup file with global concentration fields that played the role of initial conditions to December 31, 2018, while the first 6 months were used to accelerate the model (adaptation of atmospheric concentration fields to reduced emission fields) and were excluded from further analysis.

3.2. Calculation results

The average AR values for three stations considered earlier are shown in table 1 and summary diagram based on it in figure 4. Generally, the effect of long-range transport from regions of climatically significant CH_4 emissions in Northern Eurasia on surface methane and AR values is most significant in winter due to several factors: a seasonal increase in emissions from fossil fuels burning, an increase in methane photochemical lifetime due to a decrease in hydroxyl concentration, as well as an increase in of atmospheric tracers residence time in lower troposphere due to a decrease in role of convective transport and a higher static stability of the troposphere as a whole [7]. According to calculations results, AR value in winter period can reach 7-10% of observed methane concentration.

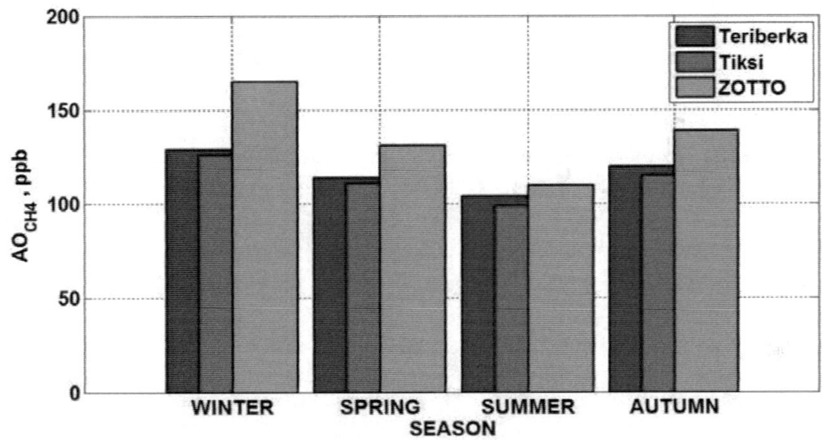

Figure 4. Summary diagram of AR_s for Teriberka, Tiksi and ZOTTO. The average values for all seasons of 2007 - 2018 are given.

Table 1. Seasonally averaged surface methane concentration values for Teriberka, Tiksi and ZOTTO observation stations (ppb). Seasonally averaged atmospheric response values to anthropogenic methane emissions for designated stations (ppb).

	Teriberka		Tiksi		ZOTTO	
	Avg.	AR	Avg.	AR	Avg.	AR
Winter	1934	129	1948	126	1948	165
Spring	1914	114	1927	111	1911	131
Summer	1912	104	1912	99	1917	110
Autumn	1922	120	1939	115	1925	139

The highest AR values (up to 165 ppb in winter) on anthropogenic methane emissions in all seasons are reached for ZOTTO. For most of the year, the station area is located in influence zone of atmospheric pollution sources in Western Europe, ETR and southern Siberia. The continental leeward plume associated with these sources forms an area of high concentrations of pollutants, including CH_4, that extends over almost the entire Northern Eurasia. The stations located on Arctic coast are far from the axis of this plume, so anthropogenic impact on them is even less. The AO values for Teriberka are only slightly (by 3 – 5 ppb) higher than for Tiksi, which is much further to the east.

4. Summary

The impact of climatically significant anthropogenic sources of CH_4 emissions in Northern Eurasia to the observed seasonal methane variability in 1999-2019 on the Kola Peninsula (Teriberka station), in central Siberia (ZOTTO), and in northeast of Eurasia (Tiksi) were quantified, based on GEOS-Chem global chemical transport model calculations. According to results, the contribution of regional emissions to measured methane concentrations is (~104-165 ppbv), or 5.5-8.6 % of average measured at three stations annual value 1926 ppbv. The calculated values of atmospheric response are several times higher than the amplitude of observed annual methane variability (22-36 ppbv), it allows to conclude that regional sources of atmospheric methane emissions in Western Europe, in European territory of Russia and in Siberia play a significant role in regional balance of the surface CH_4 concentration in lower troposphere above the continent. In summer, the impact of regional methane emissions is slightly weaker (105 ppbv) compared to winter (~140 ppb) due to increasing the role of vertical convective exchange in lower troposphere.

This work was supported by the Russian Science Foundation under grants №20-17-00200 (model setup and calculations) and №21-17-00021 (methane observations analysis).

References

[1] Forster P, Ramaswamy V, Artaxo P, Berntsen T, Betts R, Fahey D W, Haywood J, Lean J, Lowe D C, Myhre G, Nganga J, Prinn R, Raga G, Schulz M and Dorland R Van 2007 Changes in Atmospheric Constituents and in Radiative Forcing, in Climate Change 2007: The Physical Science Basis. *Contribution of Working Group I to the Fourth Assessment Report of the Intergovernmental Panel on Climate Change* ed S Solomon, D Qin, M Manning, Z Chen, M Marquis, K B Averyt, M Tignor and H L Miller (Cambridge: Cambridge University Press) pp 129–234

[2] Bey I, Jacob D J, Yantosca R M, Logan J A, Field B, Fiore A M, Li Q, Liu H, Mickley L J and Schultz M 2001 Global modeling of tropospheric chemistry with assimilated meteorology: Model description and evaluation *J. Geophys. Res.* **106** 23073–96

[3] Janssens-Maenhout G, Crippa M, Guizzardi D, Muntean M, Schaaf E, Dentener F, Bergamaschi P, Pagliari V, Olivier J G J, Peters J A H W, van Aardenne J A, Monni S, Doering U, Petrescu A M R, Solazzo E and Oreggioni G D 2019 EDGAR v4.3.2 Global Atlas of the three major greenhouse gas emissions for the period 1970–2012 *Earth Syst. Sci. Data* **11** 959–1002

[4] Bloom A A, Bowman K W, Lee M, Turner A J, Schroeder R, Worden J R, Weidner R, McDonald K C and Jacob D J 2017 A global wetland methane emissions and uncertainty dataset for atmospheric chemical transport models (WetCHARTs version 1.0) *Geosci. Model Dev.* **10** 2141–56

[5] van der Werf G R, Randerson J T, Giglio L, Collatz G J, Mu M, Kasibhatla P S, Morton D C, DeFries R S, Jin Y, van Leeuwen T T 2010 Global fire emissions and the contribution of deforestation, savanna, forest , agricultural, and peat fires (1997-2009) *Atmos. Chem. Phys.* **10** 11707–35

[6] Hanna S R 1989 Confidence limits for air quality model evaluations, as estimated by bootstrap and jackknife resampling methods *Atmospheric Environment* **23** 1385–98.

[7] Elansky N F 2007 Spatial and temporal variations of trace gases surface concentrations over Russia from TROICA observations *Proceedings of the International Symposium on Atmospheric Physics and Chemistry* ed H Wang and G S Golitsyn (Beijin: CAS press) pp 49–56

IOP Conf. Series: Earth and Environmental Science 1040 (2022) 012034 doi:10.1088/1755-1315/1040/1/012034

On the influence of internal gravity waves on the intensity of turbulence in the atmospheric boundary layer

D Zaitseva[1], **M Kallistratova**[1], **V Lyulyukin**[1], **R Kouznetsov**[1,2], **D Kuznetsov**[1]

[1] A.M. Obukhov Institute of Atmospheric Physics Russian Academy of Sciences, Moscow, Russia
[2] Finnish Meteorological Institute, Helsinki, Finland

E-mail: zaycevadv@gmail.com

Abstract. To date experimental data concerning the impact of internal gravity waves on the intensity of turbulence in a stably stratified atmospheric boundary layer are scarce. The paper presents the results of the analysis of the degree of influence on turbulence characteristics of wave-like motions of two classes: Kelvin-Helmholtz billows (KHBs) and horizontally propagating buoyancy waves (BWs). For this purpose the data of long-term sodar measurements carried out in the suburban area of Moscow were used. The passage of 30% of KHB trains and 90% of BW trains was accompanied with increase of turbulent kinetic energy and both heat and momentum fluxes. The relative increase in values of these characteristics was higher during BWs passage. Nevertheless the magnitude of turbulence characteristics were similar for both IGWs classes. In general the difference between characteristics calculated in the presence and in the absence of wavelike motions was smaller than between those in strongly and weakly stable boundary layers.

1. Introduction

The atmospheric boundary layer (ABL) plays an important role in the dynamics of the atmosphere. The exchange of energy, heat and impurities between the earth's surface and the free atmosphere takes place through the ABL. The exchange processes in the stably stratified boundary layer (SBL) have not been sufficiently studied so far, although the SBL is regularly observed during nighttime surface cooling. In cold winter months it can persist for long periods of time.

Non-turbulent motions occur simultaneously on a variety of scales, and their interactions complicate the dynamics of the SBL. Notwithstanding the suppression of the turbulence by the temperature stratification, various processes may result in turbulence intensification. For example, significant vertical wind shear in frequently observed low-level jets results in non-periodic vortex motions that generate turbulence [1]. Internal gravity waves (IGWs) are also supposed to be one of the sources of the turbulence in the SBL [see, e.g., 2]. Field studies of non-turbulent processes and turbulent exchange in the SBL, the development of theoretical schemes and comparison of experimental and model results are of a current interest since they are necessary to improve the results of numerical modeling of the SBL [see, e.g., 3, 4].

IGW modes associated with the shear instability are referred to as Kelvin-Helmholtz billows (KHBs), and those associated with temperature gradients are referred to as buoyancy waves (BWs). [5] theoretically investigated some of the properties of these two classes of waves and the possibility of their separation. They pointed out that these two solutions have different phase speeds if scales of wind shear and density gradient are different. The critical level of KHBs was found near the inflection point in the wind speed profile, and that of BWs in the vicinity of high temperature gradients. Both BW

IOP Conf. Series: Earth and Environmental Science 1040 (2022) 012034 doi:10.1088/1755-1315/1040/1/012034

and KHB classes can reach significant amplitudes within the same general time span. Consequently, their nonlinear interaction is possible and it can result in excitation of disturbances with parameters very different from the initial waves.

The cumulative duration of wavelike motion registration ranges from a few percents of the full measurement time when using visual analysis of sodar echograms [6, 7] and of time series of the near-surface wind speed [8] to nearly full time when using automatic registration of coherent pressure pulsations with the help of microbarographs [9, 10].

Experimental studies of the effect of IGWs on the intensity of turbulent exchange in the SBL are few, but they indicate that wavelike motions influence the overall state of the SBL. The passage of wave trains is accompanied by temperature gradient and wind shear alterations, by bursts of turbulence [11, 12], and also by changes in the low-frequency part of momentum flux [13]. Quantitative estimates of the degree of influence of IGWs on turbulence vary. Some measurements show that the wave contribution to wind speed variances reaches tens of percent [8, 14]. In contrast, [15] found that the vertical wind speed variance in the wave layer was several times higher when the KHB train was passing than when there was no apparent wave activity. Experimental estimates have also shown that the contribution of waves can alter both the value and sign of the heat and momentum fluxes. At the same time, the changes in fluxes during the passage of the BW train turned out to be much higher (more than ten times) than during the passage of the KHB trains (no more than three times) [16].

The published papers partially shed light on the complex picture of the interaction between IGWs and turbulence. However, quantitative estimates of variance and flux changes have been obtained only for a small number of episodes. Our work is aimed at obtaining quantitative estimates of the degree of influence of IGWs of different classes on the turbulence intensity on a base of a fairly extensive array of experimental data. The usage of long-term continuous measurements allowed us to make uniform estimates of changes in turbulent kinetic energy and heat and momentum fluxes for several dozens of cases of wave activity.

2. Measurements and data processing

Continuous year-round measurements of ABL characteristics have been carried out since 2008 at the Zvenigorod Scientific Station (ZSS) of A.M. Obukhov Institute of Atmospheric Physics RAS (IAPh) in the suburban area of Moscow. The ZSS is located 50 km west of Moscow ($55^\circ\ 42'$ N, $36^\circ\ 47'$ E) in a wooded countryside at an elevation of 150 m above sea level. The landscape around the station is mostly flat, sloping gently ($\sim 1^\circ$) to the Moskva River with smooth hills. Registration of IGWs and measurements of wind speed profiles were carried out with the help of the research 3-component Doppler sodar LATAN-3 developed at IAPh [17]. The sodar has been operated at a carrier frequency of 2000 Hz with a vertical resolution of 20 m, and the vertical range from 30 to 150–800 m, depending on noise level and weather. The sounding cycle takes 20 s. The temperature profiles were evaluated with a scanning microwave radiometer MTP-5 [18] and were used for a basic control of the stratification. To calculate the turbulence characteristics we used data from an ultrasonic thermometer-anemometer (sonic) METEK USA-1, which was located next to the sodar on the mast, at a height of 56 m.

Episodes of wave activity were identified by visual analysis of sodar echograms, as is usual [see, e.g., 19]. Despite some subjectivity of this approach, to date visual identification of periodic structures seems to be simpler and more reliable than the published methods of automatic identification [e.g., 20].

Remote acoustic sounding has a number of disadvantages that complicate the registration of waves, among which the main ones are: low time resolution (from a few seconds) and lack of data at low turbulence [see, e.g., 20]. Since periodic structures are identified as oscillations of the temperature structure parameter CT2, rather than that of meteorological quantities, the physical interpretation of the observed motions becomes more complicated. However, ground-based remote sounding has an important advantage: it makes possible to visualize the vertical structure of the wavelike motions and to determine their shape and location.

Since the vertical and time resolution of the sodar we used was not high, only fairly large IGWs were registered. For identification of periodic structures on echograms, we used criteria close to those developed in the work [7]:

1. The minimum period of wavelike patterns T was > 2 min.
2. The depth of the wave layer was > 60 m.
3. The depth of modulation of the background return signal level by a wave structure exceeded 4 dB.

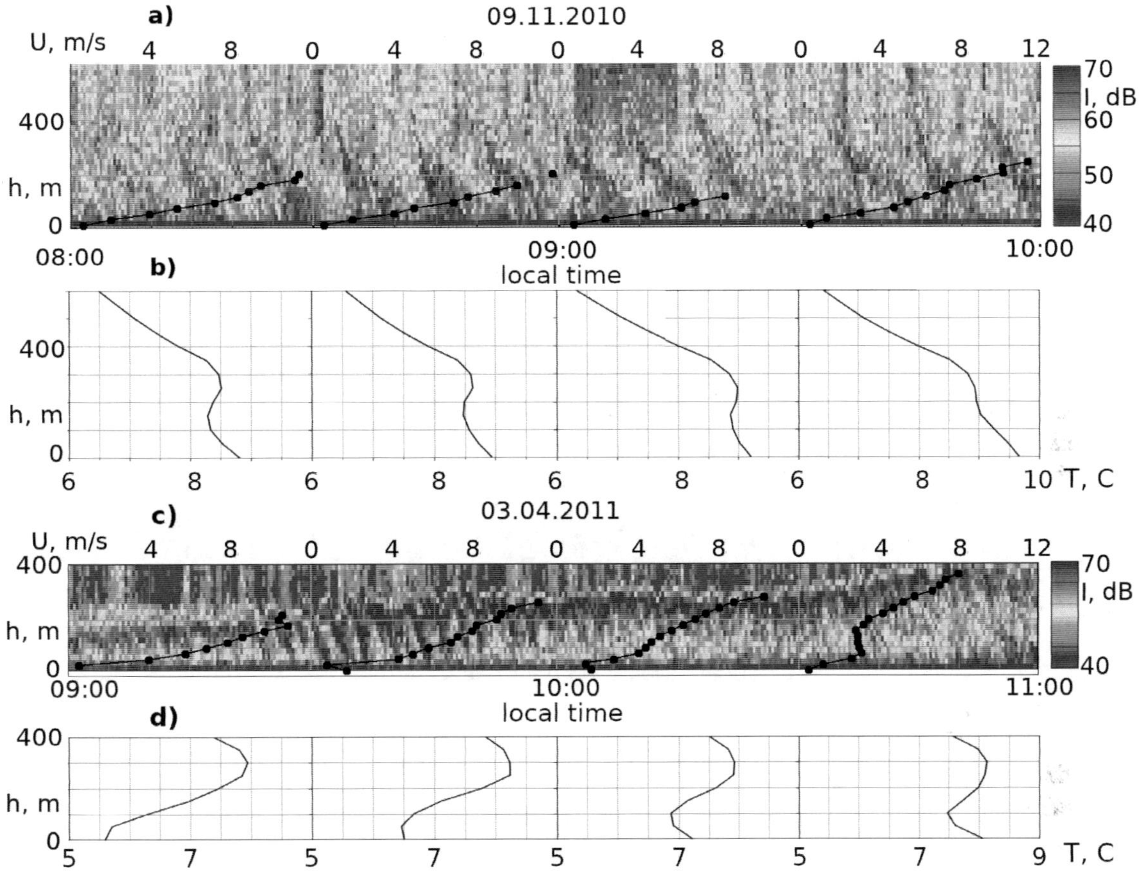

Figure 1. Two examples of KHB observation. (a) and (c) panels present return signal I in height–time coordinates (echograms), colors show the relative intensity of I. Black dots indicate wind speed profiles. The speed scale is shown at the top of each panel. (b) and (d) panels present temperature profiles. Both wind and temperature profiles calculated with 30 minutes averaging

We classified periodic structures into KHBs and BWs. The discrimination was based on the analysis of the vertical structure of observed oscillations of the C_T^2 and of the vertical velocity.

Periodic structures observed in the C_T^2 field in the form of inclined stripes, billows or braids we attributed to the class of KHBs, as, e.g., in [11]. Such waves appear at a strong vertical wind speed shear. Changes in the vertical wind speed at different heights inside the wave layer does not occur synchronously [21]. Examples of sodar echograms with KHBs are shown in figure 1.

Synchronous oscillations of the altitude of one or more turbulized layers or of the boundary of the surface layer we referred to as the BWs. The vertical wind speed oscillates more or less synchronously at different heights inside the layer occupied by BW [e.g. 22]. The presence of a noticeable wind shear is not a necessary condition for the occurrence of BWs. Examples of sodar echogram with BW is shown in figure 2.

IOP Conf. Series: Earth and Environmental Science 1040 (2022) 012034 doi:10.1088/1755-1315/1040/1/012034

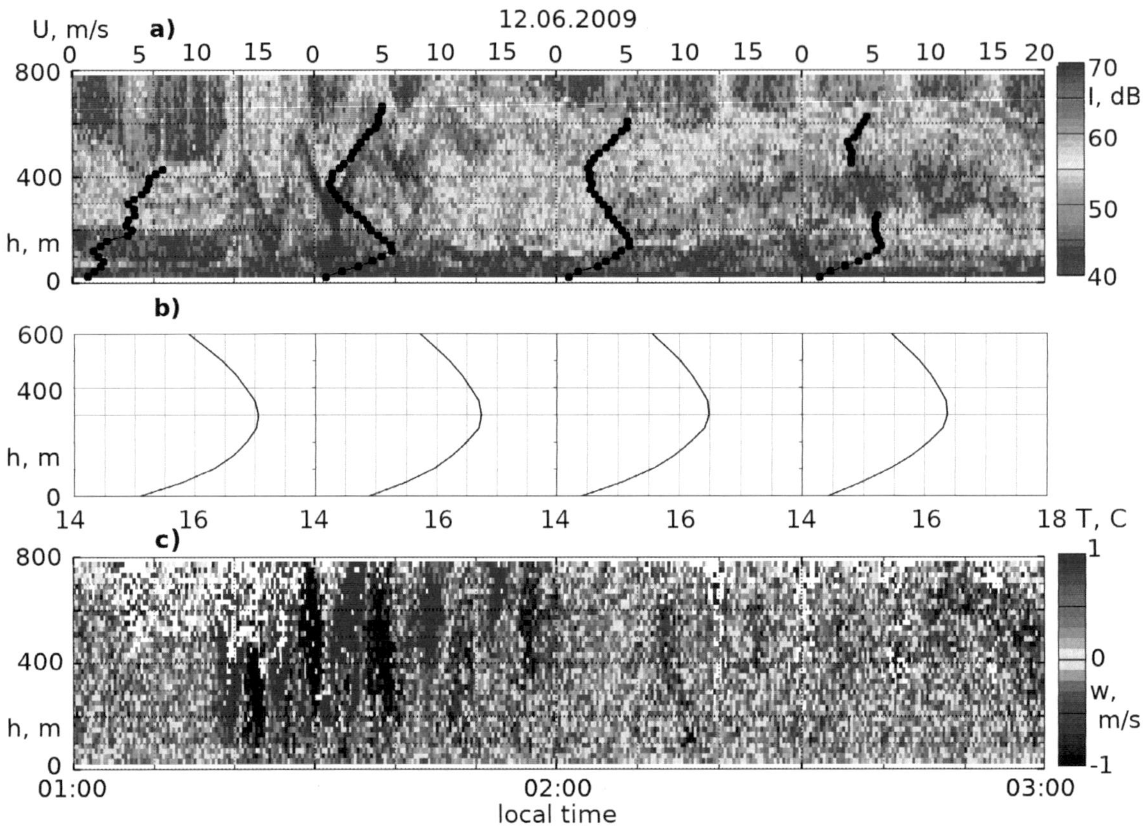

Figure 2. Example of BW observation. (a) and (b) is the same as figure 1. (c) vertical wind speed in time-height coordinates.

In the work we used episodes which could be unambiguously classified. A number of cases were classified as a mixed class and were excluded from the statistics.

We calculated the following characteristics using data obtained with the help of the sonic:

1. turbulent kinetic energy TKE = $< u^2 + v^2 + w^2 >$, where u, v, w are fluctuations of wind speed components;

2. momentum flux MF = $\rho < uw >$;

3. heat flux HF = $c_p\rho < tw >$, where t is fluctations of acoustic temperature measured by the sonic.

A double rotation of sonic axes was applied so that the one of horizontal wind components was aligned with the main flow [23].

For each episode, the ensemble-average of turbulence characteristics were calculated for both period of the passage of the wave train and 30-60 minute period before the episode. After that, the ratio R of these average values was calculated. R $>$ 1 indicates that the characteristic was higher during the episode and R $<$ 1 indicates the opposite.

A minimum of energy in the spectrum of the turbulence (the so-called gap) separates turbulent eddies from larger-scale motions. The position of the gap can vary or be absent in the presence of stable stratification [24]. To asses the position of the cospectral gap we calculated momentum and heat cospectra with the help of multiresolution flux decomposition for each episode [e.g., 25]. These gaps ranged from tens of seconds to tens of minutes. In some cases it was absent. If only true turbulence is required, the use of a constant averaging time is undesirable. To overcome this difficulty,

IOP Conf. Series: Earth and Environmental Science 1040 (2022) 012034 doi:10.1088/1755-1315/1040/1/012034

wave oscillations identified by means of a spectral analysis are usually filtered out of time series of meteorological quantities [e.g., 8, 14].

To make the results comparable with those obtained by means of the traditional approach we calculated the turbulence characteristics for four constant averaging times - 1, 5, 15 and 30 minutes. While the first of these periods in most cases included only small-scale turbulent fluctuations, the others could include both turbulent and wave oscillations. Note that the passage of wave trains recorded by the sodar was not always recorded by the sonic.

In the previous paper [16] only 15-minute averaging was used. Thus both waves and turbulence contributed to calculated characteristics. These results are in qualitative agreement with those presented below.

3. Results

We analyzed 7.5 years of measurements and assessed the frequency of appearance of both KHB and BW trains on sodar echograms. The number of episodes and the fraction of time occupied by wave activity of both classes varied greatly both from month to month and from year to year. Nevertheless, episodes of KHBs, on average, were observed in 25-35% of measurement days regardless of the season. At the same time, BWs were observed mainly from March to August in 4-8% of days. KHBs occupied 1-3% of the total measurement time, and BWs $<1\%$. These data show that observation of IGWs of both classes are not scarce at the ZSS.

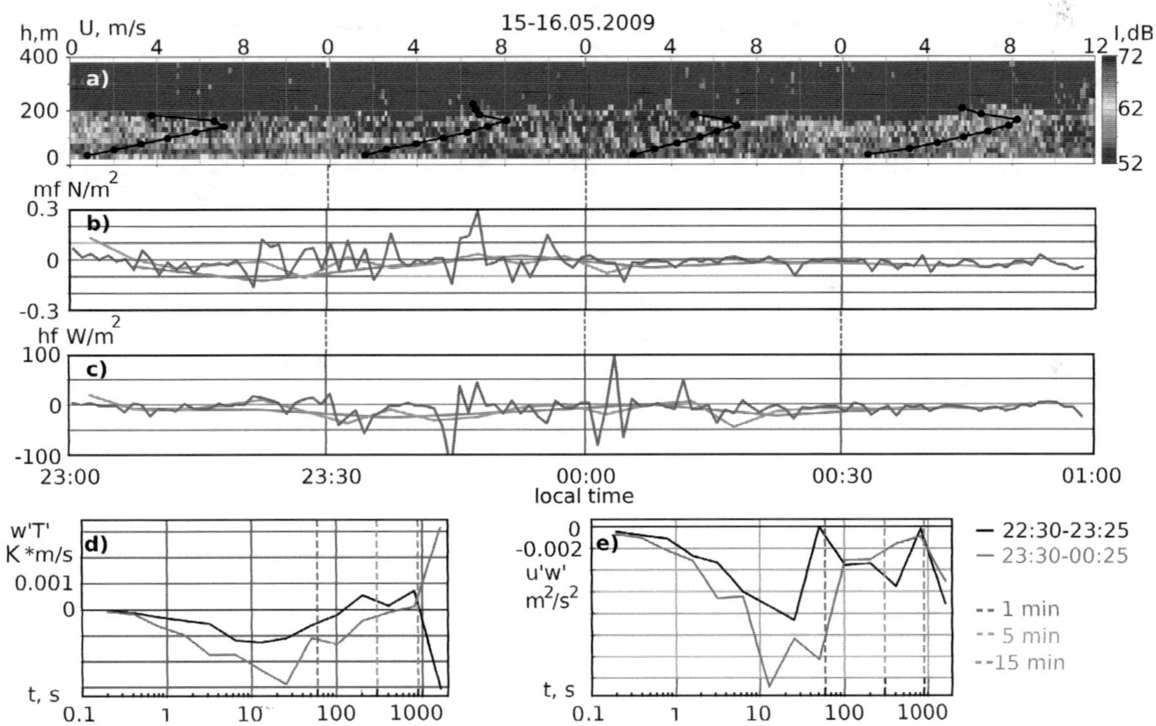

Figure 3. An example of the behavior of turbulent characteristics during the passage of the KHB train. (a) echogram in height-time coordinates. (b), (c) time series of heat and momentum fluxes, calculated with averaging times of 1, 5 and 15 minutes. (d), (e) cospectra of heat and momentum fluxes for time intervals before the KHB passage (22:30-23:25) and during the episode (23:30-00:25).

Figure 3 shows an example of the behavior of turbulence characteristics during the passage of a KHB train. Before the beginning of the episode, the smallest time scales at which the heat and momentum cospectra attained zero were ~100 seconds and ~50 seconds, respectively. During the episode, the momentum cospectrum was negative on all scales, and had increased sharply at ~50-100 s. At the same time the heat cospectrum had undergone a sharp increase at ~25-50 s, but had crossed the zero-line only at ~400 seconds. Before and during the episode, the wind shear under the maximum of the wind speed profile was ~0.05-0.06 s^{-1}. The temperature gradient in the lower 100 meters (not shown) during the hour prior to the passage of KHB train was ~1.0-1.2 K per 100 m, and by the end of the episode it decreased to ~0.2 K per 100 m.

Figure 4 shows an example of the behavior of turbulent characteristics during the passage of a BW train. During the episode heat and momentum fluxes averaged over 1 minute oscillated with a period close to that of the BW. Before the episode both heat and momentum cospectra attained zero-line at ~40-50 s. During the passage of the BW, the heat cospectrum crossed zero-line at ~70 s, and momentum cospectrum had increased to nearly zero at ~40-50 s. With the arrival of the wave, bulk shear decreased from ~0.07 s^{-1} to ~0.05 s^{-1}. By the beginning of the episode, the difference in air temperatures between the surface layer and 100 meters (not shown) was ~ 2.5 K, and by the end ~1.4 K.

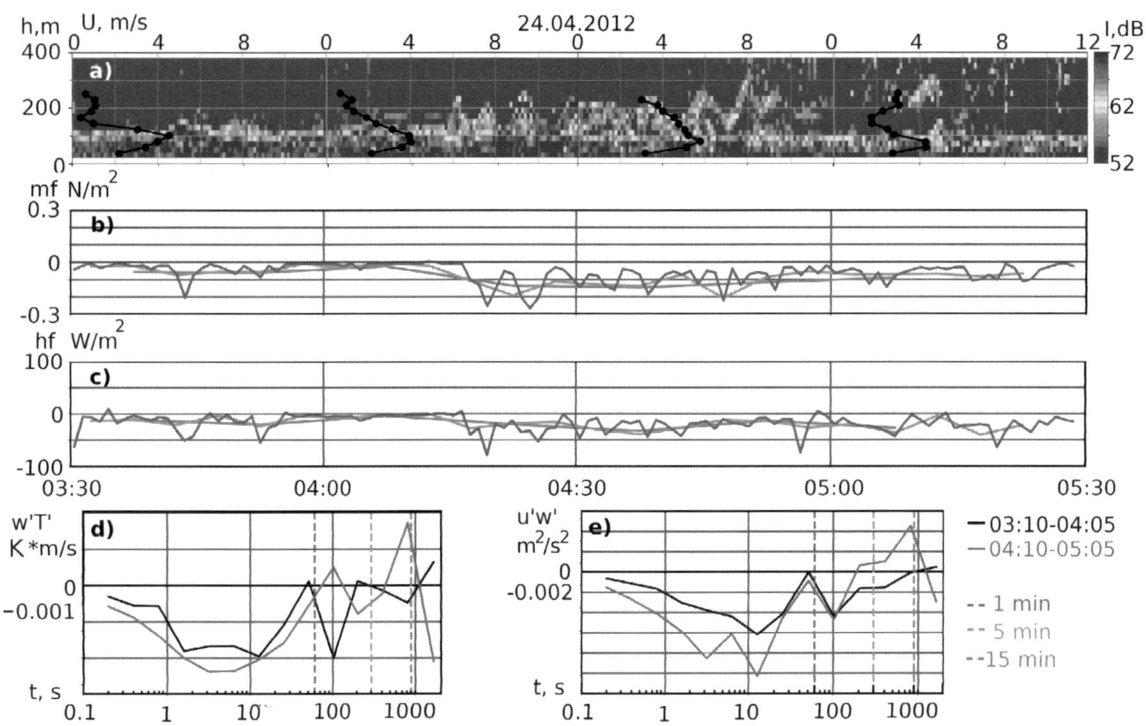

Figure 4. Same as figure 3 for a train of BW.

We have chosen the most distinct episodes of IGWs to analyze their impact on turbulence characteristics. They comprise 28 KHB trains and 10 BW trains. For each episode, the mean values of turbulence characteristics were calculated for three averaging times: 1, 15 and 30 minutes. The largest difference of these values for each KHB episode did not exceed 0.7 m^2s^{-2}, 40 Wm^{-2}, and 0.1 Nm^{-2} for TKE, HF, and MF, respectively. So these changes ranged from tens to hundreds of percent. In general, the differences for TKE and HF are consistent with the data published by [26]. However, differences we calculated for MF are almost double of that presented. And [14] gave even smaller estimates which are 2-3 times lower than ours. During the passage of BWs these differences were even greater: up to 7.3

m^2s^{-2}, up to 300 Wm^{-2}, and up to 0.8 Nm^{-2}. So these changes ranged from hundreds percent to one thousand percent. The large differences are most likely to result from wind speed and temperature oscillations recorded with the help of the sonic during the passage of BWs. As a consequence, turbulence characteristics averaged over 15 and 30 minutes (more than periods of BW trains) were greatly higher [e.g., 26].

Only 10 of 28 KHB trains seemed to increase turbulence characteristics. During other episodes of KHB changes were either absent or masked by other processes, such as gradual alteration of temperature stratification and wind shear. In contrast, 9 of 10 BWs exerted an influence on the turbulence intensity.

For comparison of the degree of intensification of turbulence during the passage of waves of different classes, only episodes that influenced the turbulence characteristics were used. We calculated mean values of the characteristics and of their ratios R for all these cases. They are presented in the table 1. Median values of R are listed in brackets. For KHB trains average and median R generally differed slightly, and for BW trains some values greatly varied. The possible explanation of these differences was mentioned above.

Ratios R show that, in general, BW trains exerted stronger influence than KHB trains. However, 1-minute averages of the turbulence characteristics turned out to be close for both classes of waves. Partly it is the result of episode selection: the most distinct BWs were usually observed on echograms in the SBL with weak winds.

Table 1. Average and median (in brackets) charachteristics of turbulence during the passage of KHB and BW trains and the grade of intensification R.

averaging time 1 minute						
	KE m^2s^{-2}	R	MF $N\,m^{-2}$	R	HF $W\,m^{-2}$	R
KHB	0.5	3.4 (2.4)	-0.07	5.6 (3.6)	-12	2.3 (2.2)
BW	0.7	6.8 (5.7)	-0.06	108.5 (4.6)	-14	4.0 (4.5)
averaging time 15 minutes						
KHB	0.7	2.6 (2.4)	-0.08	9.4 (3.2)	-15	2.1 (2.0)
PBW	3.7	8.7 (6.5)	-0.13	18.5 (4.5)	-8	4.7 (5.7)
averaging time 30 minutes						
KHB	0.8	1.9 (1.9)	-0.07	2.3 (2.2)	-14	1.9 (2.0)
PBW	4.1	8.2 (3.6)	-0.18	14.6 (3.2)	-24	607.5 (1.2)

Turbulence characteristics vary, especially for SBLs with weak and strong turbulence. They can range from very small values up to several m^2s^{-2}, tens of Wm^{-2} and to several of Nm^{-2} for TKE, heat and momentum fluxes, respectively. [e.g., 27, 28]. Thus, turbulence characteristics we obtained fall into the ranges of commonly measured values. Note, that due to the limited resolution of the sodar, we studied only the largest wave motions. It is possible that large differences in values are associated with the passage of waves with amplitude and period being too small to be registered by our sodar.

4. Conclusions

Analysis of long-term continuous sodar measurements carried out in the suburban of Moscow showed that wave-like motions are often registered on sodar echograms in the stably stratified atmospheric boundary layer. It was found that Kelvin-Helmholtz billows were observed on average in 30% of days regardless of the season, and buoyancy waves - on average 4-8% of days from March to August and <1% of days in other months. For 38 episodes of wave activity, the behavior of the turbulence characteristics was analyzed. Only 10 of 28 KHB trains seemed to increase turbulence characteristics. In contrast, 9 of 10 BWs exerted an influence on the turbulence intensity. In general, BW trains exerted a stronger influence than KHB trains. However, the absolute values of the turbulent characteristics turned out to be close during the passage of waves of both classes. This can be partly explained by the choice of episodes: the most distinct episodes of BWs were usually observed in the light-wind SBL.

Acknowledgments

This study was supported by the Russian Foundation for Basic Research (project no. 19-05-01008) and the Russian Science Foundation (project no. 21-17-00021).

References

[1] Kallistratova M A, Kouznetsov R D, Kramar V F and Kuznetsov D D 2013 Profiles of wind speed variances within nocturnal low-level jets observed with a sodar *J. Atmos. Oceanic Technol.* **30** 1970-7
[2] Sun J et al. 2015 Review of wave-turbulence interactions in the stable atmospheric boundary layers *Rev. Geophys.* **53** 956-93
[3] Steeneveld G-J 2014 Current challenges in understanding and forecasting stable boundary layers over land and ice *Front. Environ. Sci.* **2** 1-6
[4] Edwards J M, Beljaars A C M, Holstag A A M and Lock A P 2020 Representation of boundary-layer processes in numerical weather prediction and climate models *Bound.-Layer Meteorol.* **177**51139
[5] Chimonas G and Fua D 1984 Dispersion of small-scale instabilities *J. Atmos. Sci.* **41** 1085-91
[6] Reddy K K K, Kumar T R V, Rao S V B, Kishore P and Rao D 1998 Investigation of gravity waves in the atmospheric boundary layer using sodar and microbarograph *Indian J. Radio Space Phys.* **27** 247-59
[7] Luyluykin V S, Kallistratova M A, Kouznetsov R D, Kuznetsov D D, Chunchuzov I P and Chirokova G Yu 2015 Internal gravity-shear waves in the atmospheric boundary layer from acoustic remote sensing data *Izv. Atmos. Ocean. Phys.* **51** 193-202
[8] Cava D, Giostra U and Katul G 2015 Characteristics of gravity waves over an Antarctic Ice sheet during austral summer *Atmosphere* **6**1271-89
[9] Einaudi F, Bedard A J and Finnigan J J 1989 A climatology of gravity waves and other coherent disturbances at the Boulder Observatory during March-April *J.Atmos. Sci.* **46** 303-29
[10] Rees J M, Denholm-Price J, King J and Anderson P 2000 A climatological study of internal gravity waves in the atmospheric boundary layer overlying the Brunt Ice Shelf, Antarctica *J. Atmos. Sci.* **57** 51126
[11] Blumen W, Banta R, Burns S P, Fritts D C, Newsom R, Poulos G S and Sun J 2001 Turbulence statistics of a KelvinHelmholtz billow event observed in the night-time boundary layer during the Cooperative AtmosphereSurface Exchange Study field program *Dynam. Atmos. Oceans* **34** 189-204
[12] Sun J et al. 2004 Atmospheric disturbances that generate intermittent turbulence in nocturnal boundary layers *Bound.-Layer Meteor.* **110** 255-79
[13] Newsom R K and Banta R M 2003 Shear-flow instability in the stable nocturnal boundary layer as observed by Doppler lidar during CASES-99 *J. Atmos. Sci.* **30** 1633
[14] Durden D J, Nappo C J, Leclerc M Y, Duarte H F, Zhang G, Parker M J and Kurzeja R J 2013 On the impact of wave-like disturbances on turbulent fluxes and turbulence statistics in nighttime conditions: a case study *Biogeosciences* **10** 843343
[15] Coulter R L 1990 A case study of turbulence in the stable nocturnal boundary layer *Bound.-Layer Meteorol.* **52** 75-91
[16] Zaitseva D V, Kallistratova M A, Lyulyukin V S, kouznetsov R D and Kuznetsov D D 2018 The effect of internal gravity waves on fluctuations in meteorological parameters of the atmospheric boundary layer *Izv. Atmos. Ocean. Phys.* **54** 17381
[17] Kouznetsov R D 2006 The new PC-based sodar LATAN-3 *Proc. 13-th Int. Symp. Acoust. Rem. Sens. (Garmisch-Partenkirchen)* pp 978
[18] Westwater E R, Han Y, Irisov V G, Leuskiy V, Kadygrov E N and Viazankin S A 1999 Remote sensing of boundary layer temperature profiles by a scanning 5-mm microwave radiometer and RASS: comparison experiments *J. Atmos. Ocean. Technol.* **16** 805-818
[19] Petenko I, Mastrantonio G, Viola A, Argentini S and Pietroni I 2012 Wavy vertical motions in the ABL observed by sodar *Bound.-Layer Meteorol.* **143** 125-41
[20] Bull G 1997 Sodar investigations of gravity waves by cross spectral analysis *Acoust. Rem. Sen. Appl. Lecture Notes in Earth Sci.* **69** 275-92

[21] Lyulyukin V, Kouznetsov R and Kallistratova M 2012 The Composite Shape and Structure of Braid Patterns in KelvinHelmholtz Billows Observed with a Sodar *J. Atmos. Oceanic Technol.* **30** 2704-11

[22] Banakh V A, Smalikho I N and Falits A V 2020 Windtemperature regime and wind turbulence in a stable boundary layer of the atmosphere: case study *Remote Sens.* **12** 955

[23] Kaimal J C and Finnigan J J 1994 *Atmsopheric Boundary Layer Flows. Their Structure and Measurements* (New York, Oxford: Oxford University Press) p 282

[24] Viana S, Yagüe C and Maqueda G 2009 Propagation and effects of a mesoscale gravity wave over a weakly-stratified nocturnal boundary layer during the SABLES2006 field campaign *Bound.-Layer Meteor.* **133** 16588

[25] Vickers D and Mahrt L 2003 The cospectral gap and turbulent flux calculations *J. Atmos. Oceanic Technol.* **20** 660-72

[26] Nappo C J, Miller D R and Hiscox A L 2008 Wave-modified flux and plume dispersion in the stable boundary layer *Bound.-Layer Meteor.* **129** 211-23.

[27] Mahrt L and Vickers D 2006 Extremely weak mixing in stable conditions *Bound.-Layer Meteorol.* **119** 1939

[28] Sun J, Mahrt L, Banta R M and Pichugina Y L 2012 Turbulence regimes and turbulence intermittency in the stable boundary layer during CASES-99 // *J. Atmos. Sci.* **69** 338-51

IOP Conf. Series: Earth and Environmental Science 1040 (2022) 012035 doi:10.1088/1755-1315/1040/1/012035

Temporal variations of downward solar radiation, aerosol optical thickness and cloudiness over North of the European part of Russia based on surface data records (during 40 years)

I N Plakhina[1], I A Repina[1, 2] and E L Makhotkina[3]

[1]A.M. Obukhov Institute of Atmospheric Physics RAS, 119017, Moscow, Russia
[2]Research Computing Center, Lomonosov Moscow State University, 119234, Moscow, Russia
[3]A.I. Voeikova, Main Geophysical Observatory , 194021 St. Petersburg, Russia

E-mail: inna@ifaran.ru, repina@ifaran.ru

Abstract. The results of the analysis of temporal changes in the aerosol optical thickness of the atmosphere, the factor of turbidity, cloudiness and incoming solar radiation near the Earth's surface in the north of the European Territory of Russia (ETR) based on observations of the Roshydromet actinometric network for 1976—2014 are presented. Trends of synchronous changes in the above parameters have been revealed, both for the entire observation period and in the post-volcanic period (1995—2014). Under conditions of the aerosol optical thickness (AOT) decreasing (an increasing of the transparency of the atmosphere) for average cloudiness, a neutral time dependence of the incoming radiation is observed. Comparison with the results at actinometric stations in the south of ETR was carried out.

1. Introduction

Solar short-wave radiation arriving at the Earth's surface is the main component of the radiation balance in the surface layer of the atmosphere and an important factor influencing climatic processes [1-3]. Variations in the incoming solar radiation cause changes in other components of the radiation (and heat) balance of the underlying surface and are obviously the object of climate monitoring. For example, in works [4-10], data on changes in the total incoming radiation in Europe, China, India, Africa, and North America have been analyzed. At the same time, it was found that until the mid-1980s, it decreased by 0.07—0.03 W/m^2 per year; in subsequent years, an increase of 0.08—0.02 W/m^2 per year was noted for 75% of stations [11]. At a quarter of the stations (part of China and India), since the mid-1980s, the total radiation continued to decrease by 0.08 W/m^2 per year. The background station Mauna Loa showed a neutral course of total radiation over time. In [11], it was assumed that until the mid-1980s, the ubiquitous decrease in the incoming total solar radiation Q was due to rapid industrial growth, accompanied by powerful industrial emissions. 1980 to 2000 The observed increase in Q is associated with the introduction of technologies for cleaning emissions in developed countries of Europe and America and the partial transfer of production to Asia (India and China). The last decade has been characterized by a decrease in Q in India and in parts of China, reflecting industrial growth. In publications [11-13], the trends of changes in total solar radiation are summarized according to the data of ground-based observing systems BSRN (BASELINE SURFACE

IOP Conf. Series: Earth and Environmental Science 1040 (2022) 012035 doi:10.1088/1755-1315/1040/1/012035

RADIATION NETWORK) and AERONET (AEROSOL ROBOTIC NETWORK). The tendencies (sign) and rates of change in the flux of incoming solar radiation are the same both in cloudy and cloudless conditions, which indicates the possible role of aerosol in changes in total radiation. Atmospheric aerosol is constantly considered as a factor influencing solar radiation in the atmosphere. The scattering aerosol weakens due to the reflection the solar radiation arriving at the upper boundary of the aerosol layer, causing local and sometimes regional cooling of the lower layers of the atmosphere; the absorbing aerosol causes heating of the aerosol layer itself, which leads to the subsequent heating of the surface atmospheric air due to mixing.

Most reliable way to obtain information about the AOT (aerosol optical thickness) of the atmosphere is through the combined use of satellite and ground-based observations. The data of actinometric observations carried out on the Roshydromet network can be used to obtain additional and regular information on the spatial and temporal AOT variability. Such information is necessary primarily for improving models of optically active atmospheric components and increasing the reliability of calculations of radiation transfer in the atmosphere. Russia is characterized in IPCC reports and in publications [1-2, 14-15] as a territory of decreasing AOT in the direction from south to north and is a kind of "white" spot; AOT estimates based on ground-based network observations are an advantageous supplement to the data from AOT satellite monitoring over Russia [16]. The merit of these estimates is the length of the considered series of actinometric observations (40 years), the versatility of the measurement technique and the processing of results for all stations, as well as the vast coverage area of the Russia, which is little considered, but large in area. Atmospheric aerosol is constantly considered as a factor influencing solar radiation in the atmosphere. The scattering aerosol weakens due to the reflection the solar radiation arriving at the upper boundary of the aerosol layer, causing local and sometimes regional cooling of the lower layers of the atmosphere; the absorbing aerosol causes heating of the aerosol layer itself, which leads to the subsequent heating of the surface atmospheric air due to mixing.

2. Results: the AOT changes over the Russian Federation, based on the data from 75 stations

Actinometric observations performed at the network of Roshydromet stations provide an information basis for monitoring the components of the radiation balance, cloudiness, transparency aerosol optical thickness (AOT) of the atmosphere [14-15, 17-19]. In these studies, the AOT was estimated using a technique developed by specialists from the Meteorological Observatory of Moscow State University [20]. A typical picture of the change in the annual AOT values in the period 1961-2013 gives the presented in Fig. 1 for the northern station Ust-Vym (62.2 ° N, 50.4 ° E), which clearly shows the AOT responses to stratospheric disturbances caused by the consequences of the four powerful volcanic eruptions: Agung (8 ° S, 116 ° E) – 1963, Fuego (14 ° N, 91 ° W) – 1974, El Chichon (17 ° N, 93 ° W) – 1982, Pinatubo (15 ° N, 120 ° W) – 1991. Obviously, there is also a fairly stable decrease in the AOT value since 1995.

This character of changes in annual AOT is typical for most stations. Average long-term and extreme values of annual AOD, and estimates of trends for the territory of Russia as a whole for three periods (based on data from 75 stations) are presented in table 1.

The mean AOT value for the analyzed dataset is 0.15 and varies from 0.26 to 0.09, which is consistent with the range of spatial changes in AOT (0.30—0.05) for the Russia according to the IPCC reports. The spatial distribution of AOT over the territory of Russia is consistent with the modeling results presented in the IPCC reports. Over the territory of Russia AOT decreases from the southwest to the northeast. Since 1995 year, at most stations, the process of “cleaning” the atmosphere from aerosol has been manifested. Thus, for the Russia the trend of AOT changes is negative; the absolute values of the rate of AOT changes over 10 years vary from (- 0.06) in the south to (+ 0.01) in the north. The relative trend is on average (- 13%) for 10 years.

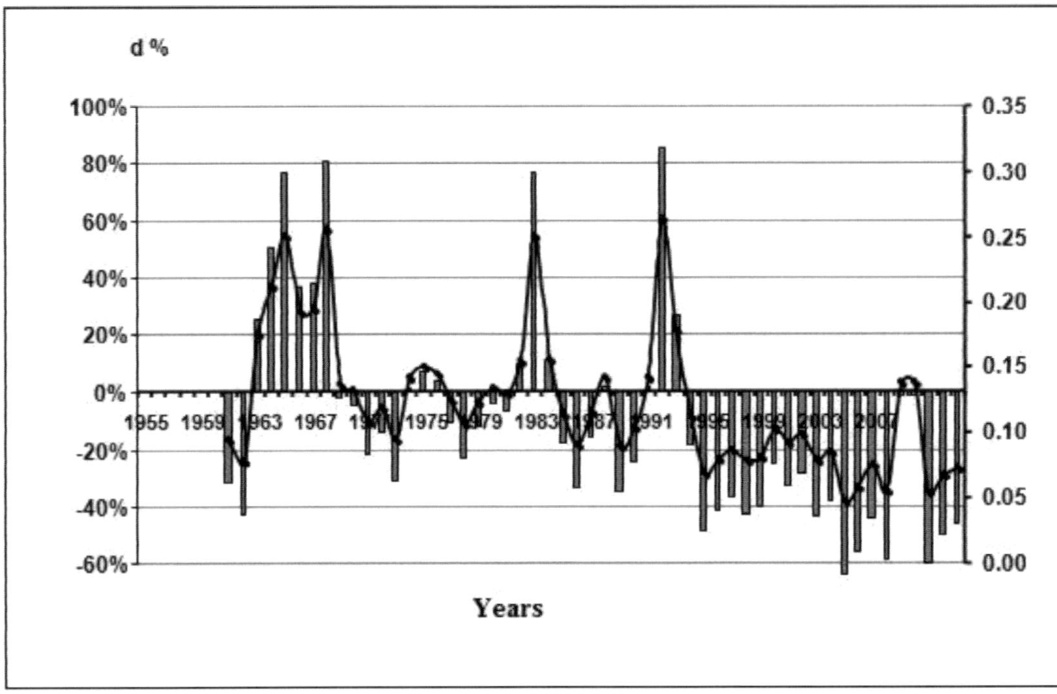

Figure 1. An example of long-term changes in the annual values of aerosol optical thickness (AOT) and their deviations from the average d (%) = 100% (AOT_i - AOT_{av}) / AOT_{av} at Ust-Vym station for the period 1961—2014 (base period adopted by WMO, 1961-1990).

Table 1. Average annual AOT and their changes in the indicated years by the ground-based measurements over the Russian Federation (75 stations).

Time interval	AOT			Trends of AOT (for 10 years)		
	Average	Maximal	Minimal	Average	Maximal	Minimal
1976-2014	0,15	0,26	0,09	-0,03	0,01	-0,06
1976-1994	0,18	0,26	0,12	-0,02	0,02	-0,11
1995-2014	0,11	0,20	0,06	-0,02	0,05	-0,12

3. Analysis of changes in AOT, cloud cover and incoming radiation for the north of the ETR

Analysis of data on atmospheric turbidity and trends in its change in the northern polar region makes it possible to obtain additional data on the state of the atmosphere in this region, which is most important from the point of view of its influence on the Earth's climate [21]. Our paper presents the results of changes in AOT and total solar radiation Q for stations in the Russian north in the ETR. To characterize the conditions of the European sector of the Arctic, we used data on the stations of the Northern UGMS: Bugrino, Kotkino, Arkhangelsk, Kargopol, Eletskaya, Ust-Vym, Irael for the period 1961-2013, as well as data from Umba and Apatity stations located on the Kola Peninsula since 1976. In the north of the ETR (in the region north of 62° N), reliable AOT values can be obtained for the period from March to October. The region as a whole is characterized by high transparency. The annual variation of the AOT in the north of the ETR has a spring maximum in April and a second less pronounced summer maximum in July. Differences are observed in the annual course of AOT for two observation periods: for a 30-year period (1976-2005) and a period of relatively high transparency (1994-2013), which average 16%. The average over a 30-year period under real and "undisturbed" conditions diverge within 8%. The data series for "undisturbed" conditions are formed by removing from the real observation series 2 years following the years of powerful volcanic eruptions El Chichon (1982 year) and Pinatubo (1991year). To characterize long-term changes in turbidity, we analyzed the series of annual values of AOT and turbidity factor T_2 for the period 1976-2013. Turbidity factor T_2 was calculated using standard methods. The linear approximation of the time course indicates a decrease in the atmospheric turbidity, while during the last 5 years both the rate of change in AOT and the sign of changes have been steadily retained. At the same time, the statistical significance of the trends in T_2 and AOT changes for "undisturbed" conditions increases in comparison with real conditions.

a)

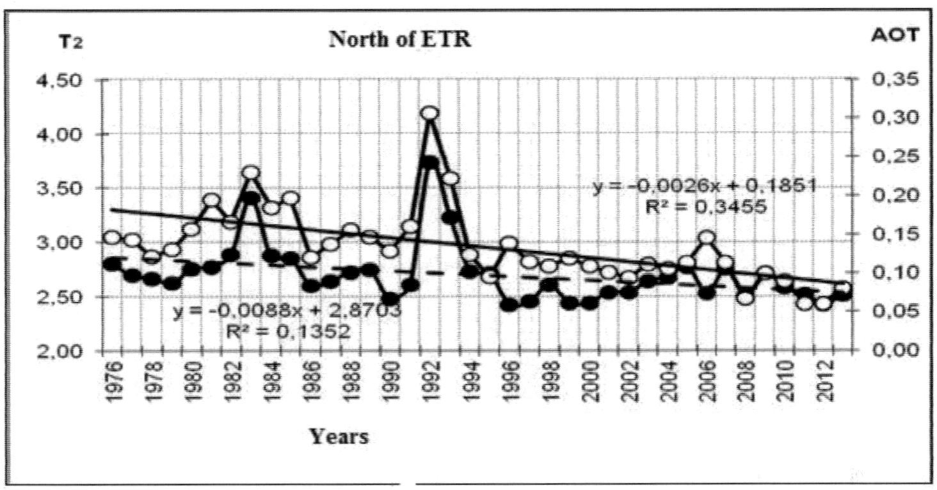

b)

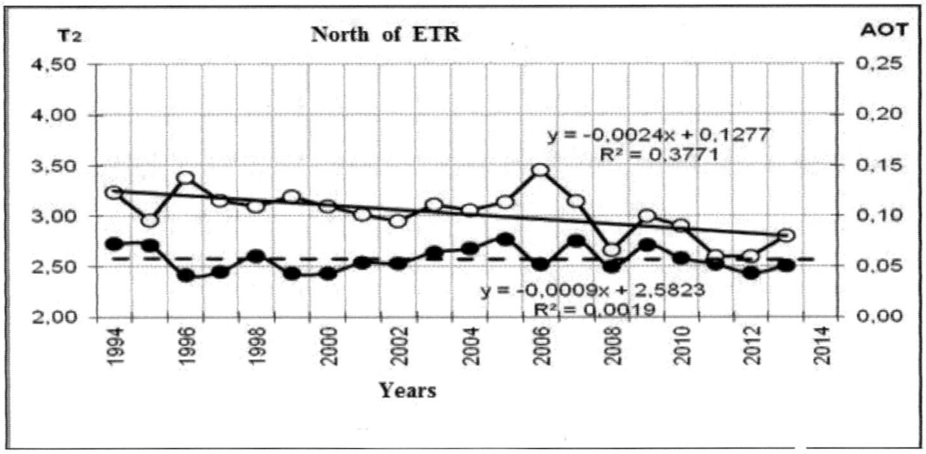

Figure 2. Change in the annual values of the turbidity factor T2 (•) and the aerosol optical thickness of the atmosphere AOT (○) in the North of the ETR during the period 1976-2014 (a) and the period 1994-2013 (b).

IOP Conf. Series: Earth and Environmental Science 1040 (2022) 012035 doi:10.1088/1755-1315/1040/1/012035

Long-term changes in the annual values of T_2 and AOT for the north of the ETR are presented in figure 2 (a, b). The duration of the period of relatively high transparency of the atmosphere over the territory of Russia is currently 20 years. Changes in T_2 and AOT occurring within this period in the north of the ETR indicate that conditions of relatively high atmospheric transparency remain.

Generally, long-term changes in AOT both throughout the period under consideration and in the last 20 years are negative, with a low rate of change (several % per year). Nevertheless, the relatively long length of the time series and the annual averaging of daily observations allow us to conclude that the rate of change in AOT has retained its magnitude and sign in recent years.

Figure 3 shows the temporal changes in the total solar radiation for the Umba station. There are insignificant and small positive (for average annual daily values) and negative (for average July daily values) tendencies of changes. The results were obtained for practically unchanged (with annual averaging) cloud conditions. Figure 4 shows the tendencies of changes in the coverage of the sky with clouds (annual values) for the stations Umba (1), Tsimlyansk (2). The degree of coverage of the sky with clouds was downloaded through a specialized server https://giovanni.gsfc.nasa.gov/giovanni, monthly values were used " cloud fraction from cloud mask", TERRA / MODIS.

As follows from the graphs shown in figure 5, the relationship between the aerosol content in the atmosphere (more precisely, AOT) and the incoming solar radiation flux is not observed in the presented time interval. However, for pronounced abrupt changes in AOT (powerful volcanic eruptions El Chichon – 1982 y. and Pinatubo – 1991y.), negative anomalies of total radiation Q are expressed in the next 1-2 year after the entry and formation of the stratospheric aerosol layer. The norm is calculated for 1981-2001 in accordance with the recommendations made in the Assessment Report on the peculiarities of the climate in the territory of the Russian Federation [22].

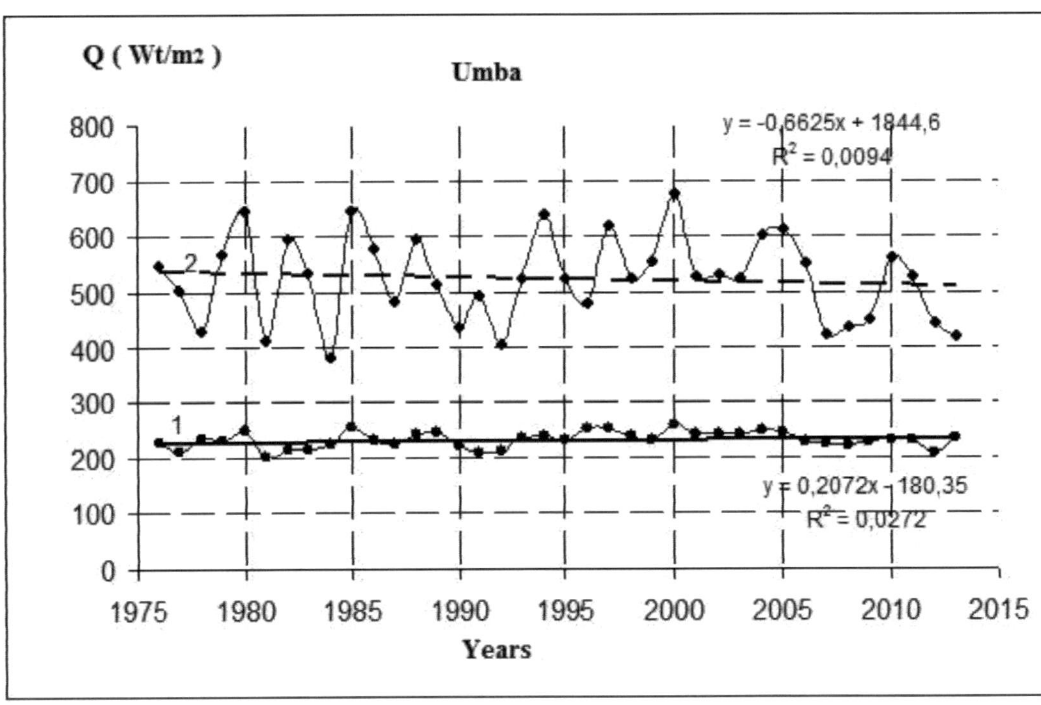

Figure 3. Long-term changes in total radiation (1 - mean annual daily values, 2 - mean July daily values) for Umba station (66.7 ° N; 34.3 ° E).

IOP Conf. Series: Earth and Environmental Science 1040 (2022) 012035 doi:10.1088/1755-1315/1040/1/012035

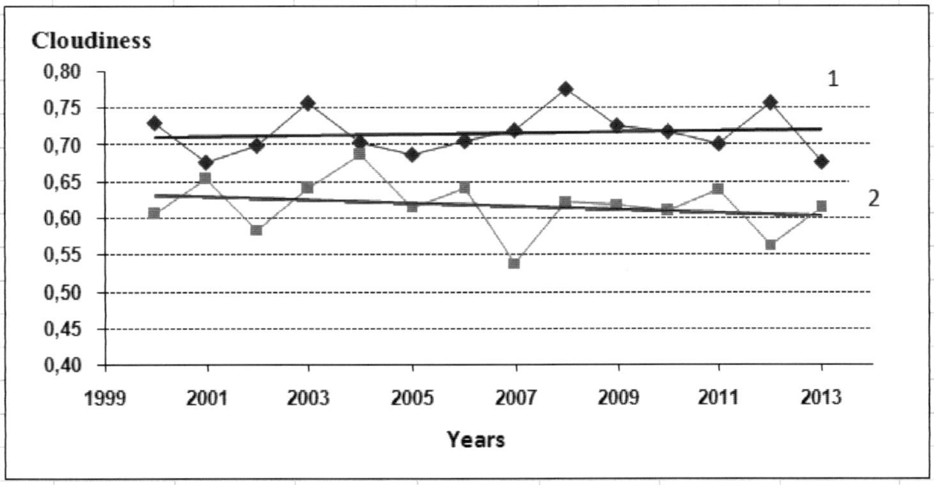

Figure 4. Trends in cloud coverage (annual values) for stations Umba (1), Tsimlyansk (2).

Table 2 shows the absolute and relative anomalies of Q and AOT, as well as their average values for the "volcanic" 1976-1994. and "post-volcanic" 1995-2013 periods. The rate is calculated as average values for 1981-2001. AOT is characterized by positive anomalies of 10% in the period 1976-1994 and negative anomalies of 40% for the 1995 period. At the same time, the total radiation in the north of the EPR is characterized by a "correct" tendency of Q changes: positive AOT anomalies correspond to negative values of Q anomalies and, conversely, for the last 20 years.
Table 2. Anomalies and average values of AOT and Q.

4.

Time interval	Anomalies of AOT Umba		Average values AOT Umba	Anomalies of AOT Tsimlyansk		Average values AOT Tsimlyansk	Anomalies Q (W/m^2) Umba		Average values Q(W/m^2) Umba	Anomalies of Q (W/m^2) Tsimlyansk		Average value Q(W/м^2) Tsimlyansk
	abs	rel %		abs	rel %		abs	rel %		abs	rel %	
1976 - 1994	0,01	10	0,15	0.04	15	0,25	-4	-2	229	0	0	365
1995 - 2013	-0.07	-45	0,08	-0.08	-40	0,13	4	2	237	1	0	366
1976 - 2013			0.13			0,19			233			365

In the south of the EPR, the annual Q values have a neutral time course, despite the pronounced negative AOD anomalies in the last 20 years, which is possibly related to the cloudiness trends in the southern region and (or) the more pronounced influence of the cloud factor on the incoming solar radiation flux.

The obtained estimates of interannual changes in the AOT and Q series, as well as their annual variation, are consistent with the data of [23 - 24], in which the analysis of data on the total incoming radiation and integral transparency of the atmosphere at the Russian Arctic stations was carried out and the nature of long-term, interannual, and seasonal changes to these parameters was presented. In these works, it was shown that for more than 50 years of observations of the total radiation, no statistically significant trends in the change in its annual sums were observed in both polar regions, as well as the fact that in the polar regions the manifestation of powerful volcanic eruptions in the AOT atmosphere is expressed, solar radiation for one and a half years or more after the eruption. In the spring, the AOT of the atmosphere is formed under the influence of the transfer of anthropogenic emissions to the Arctic from sources located in the temperate latitudes of the Northern Hemisphere.

a)

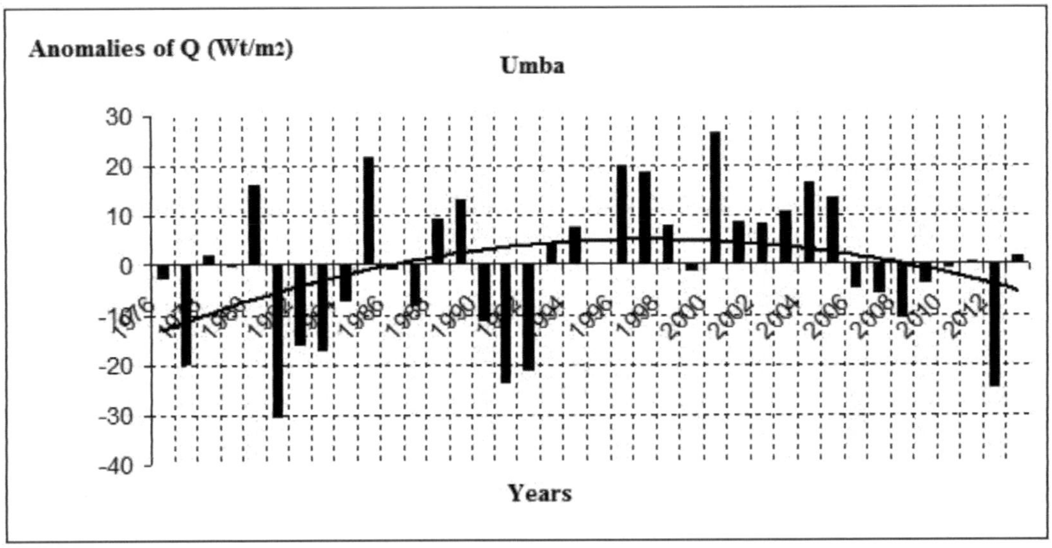

b)

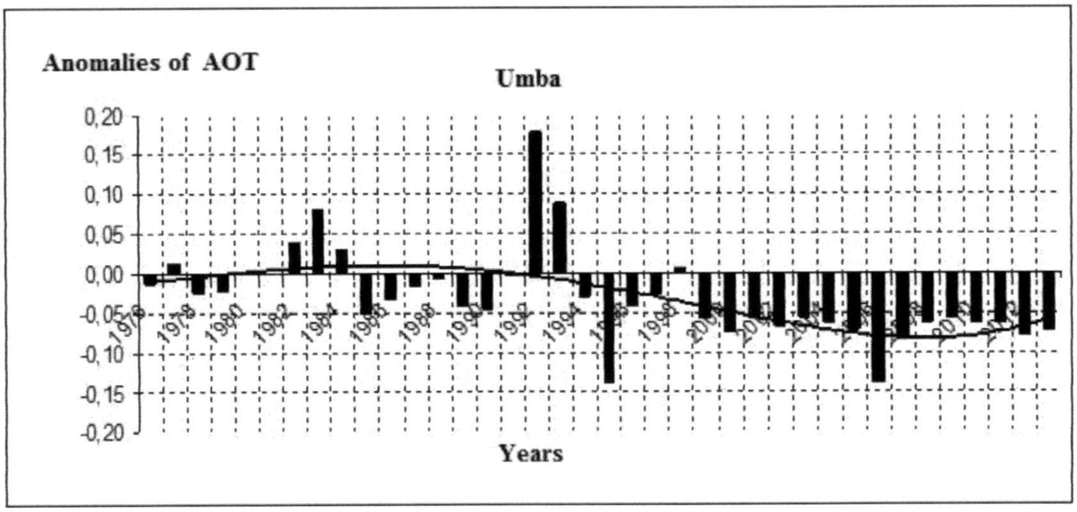

Figure 5. Anomalies of the annual values of the total radiation Q (a) and corresponding AOT anomalies (b) for Umba station. Quadratic approximation of the trend line has been carried out.

5. Conclusion

Quantitative estimates of changes in the aerosol optical thickness of the atmosphere and incoming solar radiation are presented based on the data of observations of the Russian actinometric network over 40 years for several stations in the north and south of the ETR. For AOT, there are quite pronounced anomalies, positive (1976-1994) and negative (after 1994). Negative AOD anomalies in the last 20 years are caused both by the absence of large volcanic eruptions with a global manifestation and increasing the AOT of stratospheric aerosol, and by the regional manifestation of a decrease in anthropogenic emissions on the territory of Russia. For the incoming radiation fluxes, a neutral time dependence is observed under the conditions of aerosol "clearing" of the atmosphere and average cloudiness. At the same time, in 1–2 years, following the powerful volcanic eruptions of El Chichon and Pinatubo, stable negative anomalies of the annual average daily value of incoming radiation Q of 20 Wt/m^2 are observed relative to the norm calculated for 1981–2001.

References

[1] Intergovermental panel on climate change 2018 Special Report «Global Warming of 1.5°C» WMO UNEP
[2] IPCC. Climate Change 2013 Working Group I. Contribution to the Intergovernmental Panel on Climate Change. Fifth Assessment Report of Climate Change: The Physical Science Basis Ch.7 Cambridge University Press 1535
[3] Gorbarenko E.V. 2020 An empirical model of the variability of the radiation balance of the underlying surface *Bulletin of Moscow University* series 5 Geography **2** 23 - 35
[4] Alpert P, Kirsha P. 2005 Global dimming or local dimming: Effect of urbanization on sunlight availability *Geophysical Research Letter.* **32** L17802
[5] Kishcha P, Starobinets B 2008 Quantification of the effect of urbanization on solar dimming *Geophysical Research Letters* **35** L08801
[6] Kishcha P, Starobinets B 2009 Variations of meridional aerosol distribution and solar dimming *Journal of Geophysical Research* **114** D00D14
[7] Norris J R, Wild M 2007 Trends in aerosol radiative effects over Euorope inferred from observed cloud cover, solar "dimming", and solar "brightening" *Journal of Geophysical Research* **112** D08214
[8] Norris J R., Wild M 2009 Trends in aerosol radiative effects over China and Japan inferred from observed cloud cover, solar "dimming", and solar "brightening" Journal of Geophysical Research 114 D00D14. DOI: 10.1029/2008JD010975
[9] Ohmura A 2006 Observed long-term variations of solar irradiance at the earth surface *Space Science Reviews* **125** 111-128
[10] Wang K, Dickinson R, Liang S 2009 Clear sky visibility has decreased over land globally from 1973 to 2007 *Science* **323** 1468-70
[11] Wild M 2012 Enlightening global dimming and brightening *American Meteorological Society* **1** 27-37
[12] Wild M, Trussel B 2009 Global dimming and brightening : An update beyond 2000 *Journal of Geophysical Research* **114** D00D13
[13] Wild M 2009 Global dimming and brightening : A review *Journal of Geophysical Research* **114** D00D16
[14] Plakhina I N, Pankratova N V, Makhotkina E L 2009 Variations of the aerosol in the atmospheric thickness according to the Russian actinometric network (1976–2006) *Izvestiya RAS, Atmospheric and Oceanic Physics* **45(4)** 456-66
[15] Plakhina I N 2012 Variations in the aerosol optical depth above the Russia from the data obtained at the Russian actinometric network in 1976-2010 years Atmospheric aerosols – Regional characteristics - Chemistry and physics 2012. Chapter 1 3-28
[16] Rakitin V S, Elansky N F, Pankratova N V and Skorokhod A I 2018 Study of changes in the composition of the atmosphere over Eurasia Turbulence, dynamics of the atmosphere and climate, Moscow RAS IAP named by A.M. Obukhov Fizmatbook 393-402

[17] Khlebnikova E I, Makhotkina E L 2014 Clouds and radiation regime The second assessment report of Roshydromet on climate change and its consequences in the Russian Federation Moscow ch 1.5 111- 25
[18] Samukova E A, Gorbarenko E V and Erokhina A E 2014 Long-term changes in solar radiation in Europe *Meteorology and hydrology* **8** 15-24
[19] Plakhina I N, Pankratova N V and Makhotkina E L 2011 Spatial variations in the air turbidity factor above the European part of Russia under conditions of abnormal summer of 2010. *Izvestiya, Atmospheric and Oceanic Physics* **47(6)** 708-13
[20] Yarkho E V 1994 Temporal variability of the aerosol optical thickness of the atmosphere in different climatic zones *Izvestiya, Atmospheric and Oceanic Physics* **30(3)** 417- 24
[21] Uttal T, Starkweather S, Drummond J R, Vihma T, Makshtas A P, Darby L S, Burkhart J F, Cox C J, Schmeisser L N, Haiden T, Maturilli M, Shupe M D, de Boer G, Saha A, Grachev A A, Crepinsek S M, Bruhwiler L, Goodison B, McArthur B, Walden V P, Dlugokencky E J, Persson P O G, Lesins G, Laurila T, Ogren J A, Stone R, Long C N, Sharma S, Massling A, Turner D D, Stanitski D M, Asmi E, Aurela M, Skov H, Eleftheriadis K, Virkkula A, Platt A, Fuгland E J, Iijima Y, Nielsen I E, Bergin M H, Candlish L, Zimov N S, Zimov S A, O'Neill N T, Fogal P F, Kivi R, Konopleva-Akish E A, Verlinde J, Kustov V Y, Vasel B, Ivakhov V M, Viisanen Y and Intrieri J M 2016 International Arctic systems for observing the atmosphere: an international polar year legacy consortium *Bull Am Meteorol Soc* **97(6)** 1033–56
[22] Second estimation report of Roshydromet on climate change and their consequences on the territory of the Russian Federation territory 2014 Moscow Roshydromet 225
[23] Radionov V F, Rusina E N and Sibir E A 2017 Long-term variability of annual sums of total and absorbed solar radiation in the Arctic *Problems of the Arctic and Antarctic* **3(113)** 39-50
[24] Sakerin S M., Chernov D G 2012 Preliminary results of studies of aerosol characteristics of the atmosphere in the region of Barentsburg (Spitsbergen) *Problems of the Arctic and Antarctic* **1(91)** 20-31

IOP Conf. Series: Earth and Environmental Science 1040 (2022) 012036 doi:10.1088/1755-1315/1040/1/012036

Sodar observation of the breeze return currents over the coastal zone of the Black Sea

V Lyulyukin[1], **D Zaitseva**[1], **D Kuznetsov**[1], **I Repina**[1,2]

[1] A.M. Obukhov Institute of Atmospheric Physics, Russian Academy of Sciences, Moscow, Russia
[2] Research Computer Center, M.V. Lomonosov Moscow State University, Moscow, Russia

E-mail: lyulyukin@ifaran.ru

Abstract. The breeze circulation was studied with the help of sodars in 2 short-term expeditions on the northern coast of the Black Sea. The measurements were carried out at a stationary oceanographic platform located 400 m from the coast. The main attention was paid to the vertical structure of breeze flows with the accompanying return currents aloft. The existence of turbulent exchange between the near-sea breeze flow and the return current at an altitude of several hundred meters is shown. Hence, the return current must be taken into account when determining the height of the mixing layer above the sea surface in the offshore zone.

1. Introduction

The physical mechanism for the formation of an "ideal" breeze is well known. However, in real conditions, there are many factors impacting breezes, and in each specific place the breeze circulation has its own characteristics [1]. Breezes significantly affect the weather, as well as air quality and the wind energy resource of coastal regions, where 40 to 60% of humanity live according to various sources. Therefore, a huge number of scientific publications have been devoted to the study of breezes over the past century. At the same time, experimental studies of the vertical structure of breezes are few in number, since they are very laborious. Such studies are possible using aircrafts [2] or remote sensing techniques such as sodars [3], lidars [4] and radars [5]. Sometimes (but not often) these tools are able to register not only main breeze flows, but also return currents above [5]. Knowledge of the vertical structure of the breezes, including return currents, is necessary to advancement of the mixing layer height H_{mix} determination over water areas. The mixing layer height is generally understood to mean the height of the layer adjacent to the ground over which any contamination emitted into this layer become vertically dispersed due to convection or mechanical turbulence. The H_{mix} value is a key parameter for air pollution models, and is also important for the marine ABL parameterization in weather and climate models (see, e.g. [6]). The H_{mix} is not measured by standard meteorological practices and its estimation is often not straightforward.

This work is devoted to sodar studies of the vertical structure of breeze flows with a return currents above. The main task was to study the turbulent exchange between the surface breeze flow and the return current and estimate the height of the mixing layer H_{mix} under such conditions. Due to the large temporal variability of the breeze circulation, in order to fully understand its characteristics in a specific place, year-round long-term measurements at stationary points are desirable. Such approach is implemented, for example, in the works [7, 8]. As a palliative, multiple short-term measurements at

IOP Conf. Series: Earth and Environmental Science 1040 (2022) 012036 doi:10.1088/1755-1315/1040/1/012036

the same site are possible. During this study the measurements were carried out for 1–2 weeks in the summer of 2015 and in autumn of 2016.

2. Measurements site and equipment

The measurements were carried out on a stationary oceanographic platform located 400 meters from the coast on the shelf slope of the southern coast of the Crimean Peninsula (figure 1(a)). The wind regime in this area is largely determined by the proximity of the mountain plateau, which blocks the southerly wind directions and leads to the occurrence of slope winds that enhance the night breeze [9]. The main wind directions in the near-sea layer are west and east in the daytime and north - at night.

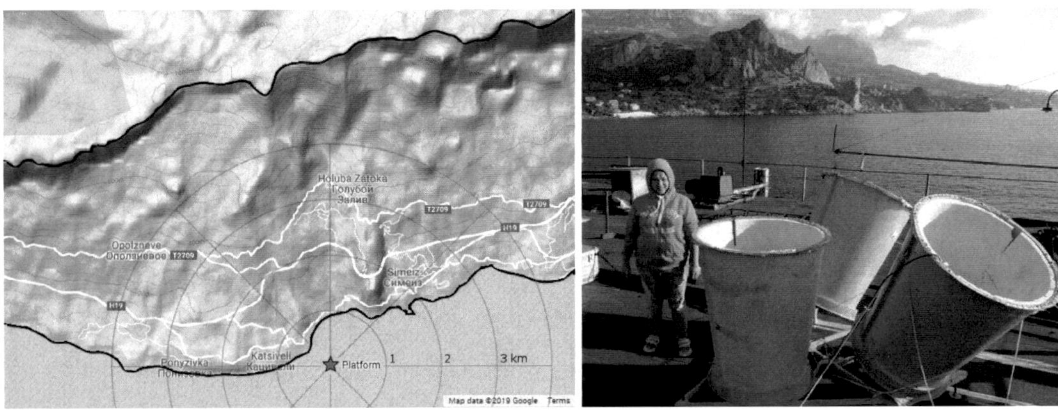

(a) (b)

Figure 1. Black Sea expedition, October 2016. (**a**) Coastline topography near the platform. (**b**) Minisodar LATAN-3M installed on the upper deck of the oceanographic platform.

Remote sensing of the ABL parameters was carried out using a three-antenna Doppler minisodar LATAN-3m (Institute of Atmospheric Phisics: Moscow, Russia [10]) and a meteorological temperature profiler MTP-5 (Attex: Dolgoprudny, Russia [11]). The minisodar was operated in a synchronous mode (simultaneous sounding with three antennas) with an altitude resolution of 10 m and a temporal resolution of 6 s in 2015 and 3 s in 2016. The altitude range of sounding was from 30 to 400 meters above sea level, taking into account the "dead" zone of sounding and the height of the sodar location. The study used instantaneous echo and radial velocity profiles as well as average wind speed and direction profiles. A photograph of the minisodar installed on the upper deck of the platform is shown in figure 1(b). The vertical temperature profiles up to 600 m were measured by the temperature profiler MTP-5 placed at 15 m above sea level (a.s.l.) and facing the open sea. The MTP-5 provided data on the vertical temperature distribution with a height resolution of 50 m and a 5 min measurement cycle.

The accompanying standard meteorological measurements were provided by the WXT536 weather transmitter (Vaisala: Helsinki, Finland), monitoring of the radiation balance and measuring the temperature of the water surface were carried out using radiometers (Kipp & Zonen: Delft, The Netherlands).

3. Breeze activity in different seasons

The expeditions were carried out in early summer and autumn. This work uses data from two expeditions: June 12–21, 2015 and October 3-10, 2016. In June, the water surface temperature was below air temperature at 15 m a.s.l., stratification was mostly stable, convection was observed less than 10 percent of the measurement time, mainly with wind from the coast. There was a pronounced diurnal cycle of wind speed and direction in the near-sea level, with dominant direction from the north for night hours (from 19:00 to 7:00 local time) and from the west (along the coast) for the daytime (from 7:00 to 19:00). The mean wind speed time course had two maxima: about 6 m s^{-1} at around 03:00 and about 2.5 m s^{-1} at around 15:00; and two minima: at 08:00 and at around midnight. Breeze flows were observed mainly at night (land breeze), sea breezes were observed rarely and for a short time. In the autumn season, the water surface temperature most of the time exceeded the air temperature, convection was observed by sodar for 60% of the time, including at night. In the autumn season, episodes of a steady breeze, both sea and land, were observed; however, the short duration of measurements and bad weather do not allow judging the regularity of the breeze circulation in autumn (the expedition was interrupted by the start of the storm season). In the wind field for the autumn season, a greater contribution from the eastern wind directions was observed compared to the summer.

The return currents were determined by the characteristic shape of the vertical profiles of wind speed and direction: the presence of a minimum in the speed profile, accompanied by a change in direction by 90–180 degrees at the same height. The directions of the main breeze flows and the return currents, as a rule, did not correspond to the situation of the classical breeze circulation [9]. Return breeze currents within the sounding range of the sodar (up to 400 m) were observed regularly during the summer season (about one episode per day), but they were characterized by instability of direction and short duration (as a rule, no longer than an hour). All episodes were observed with a stable stratification; the total duration of the episodes was about 5% of the total observation time.

4. Vertical structure of the ABL in breeze currents with a return flow

Figures 2–4 present data on the vertical structure of the ABL from the remote sensing for three cases of observing the return current with a 180 degree wind direction turn. The duration of each episode was 30 minutes, which approximately corresponds to the characteristic time of the change in the ABL structure (the shape of vertical profiles of meteorological quantities and turbulent structures on the echograms) under the observed conditions. The profiles of wind speed and direction with a characteristic turn at heights from 100 to 200 m are shown in figures 2–4(a). The wind speed at the maximum of the jets reached 4 m s^{-1}, which corresponds to the typical values for the breeze flows in the given area. Figures 2–4(b) show the vertical profiles of the temperature structure parameter C_T^2 and vertical velocity variance σ_w^2. The C_T^2 parameter derived from sodar echo-signal intensity is directly related to the temperature fluctuations and its profile can be used to determine the mixing layer height H_{mix}. The σ_w^2 profile can be used to estimate the intensity of vertical turbulent exchange. Sodar echograms are shown in figures 2–4(c). In the field of the scattered signal, in most cases, two turbulized layers located at heights with maximum wind shear were observed. Sometimes a layer of the local minimum of the echo signal was seen at the height of the minimum of the wind speed. In the turbulized layers, Kelvin-Helmholtz billows (KHBs) were regularly observed, with an inclination to the right in the lower layer and an inclination to the left in the return current. In the presented cases, stable stratification with temperature inversion was observed (vertical temperature profiles are shown in figures 2–4(d)). A bulk Richardson number Ri vertical profile calculated from the wind speed and potential temperature gradients is shown in figures 2–4(e). The Ri profiles for all episodes had two minima at the heights of the maximum wind speed gradient and a local minimum at the height of the direction change. The dotted line marks the value Ri_{cr} = 0.25, which is a critical value for the occurrence of the Kelvin-Helmholtz instability in the layer. Corresponding fields of vertical velocity obtained by the sodar are presented in figures 2–4(f).

Figure 2 shows the episode of the daytime breeze on June 18, 2015. The echogram shows two turbulized layers corresponding to the main and return currents. The C_T^2 profile does not have a sharp step and smoothly decreases with height. KHBs are observed in both layers, which corresponds to the two minima in the Ri profile. Ascending and descending motions with a vertical length of up to 150 m are observed in the vertical velocity field (figure 2(f)), penetrating both layers, which may indicate the presence of mixing between the direct and return currents. This hypothesis is also indirectly confirmed by the monotonicity of the σ_w^2 profile (figure 2(b)). The height of the mixing layer H_{mix} can be estimated as 300 m.

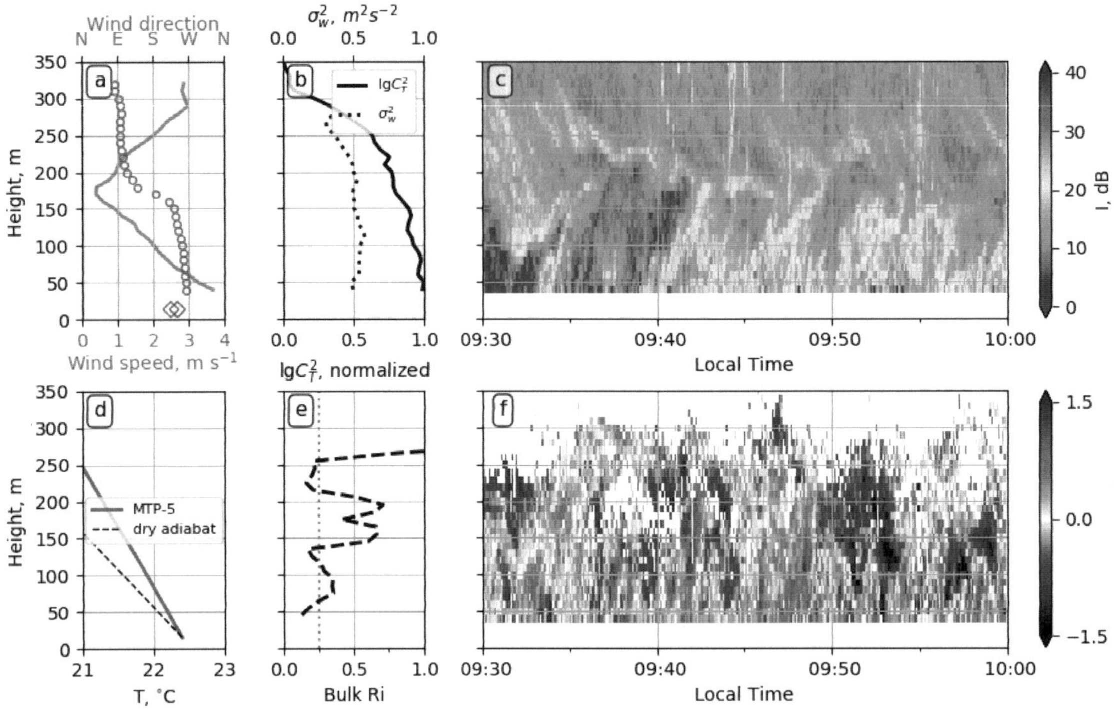

Figure 2. Vertical ABL structure in the presence of a return current, June 18, 2015. The case represents strong turbulent exchange between two turbulized layers with KHBs. (**a**) Vertical profiles of wind speed and direction. The values from the weather station are marked with diamonds. (**b**) Vertical profiles of the vertical velocity variance σ_w^2 and the logarithm of the structure parameter C_T^2, proportional to the intensity of the sodar echosignal. (**c**) Sodar return signal in height–time coordinates (echogram). (**d**) Temperature profile according to the temperature profiler data. (**e**) Vertical profile of the bulk Richardson number calculated from the sodar and temperature profiler data, dotted line marks the critical value Ri_{cr} = 0.25. (**f**) Vertical velocity field obtained by the sodar. The averaging time for all profiles is 30 min.

Figure 3 shows the episode of the daytime breeze on June 17, 2015. The KHBs are observed only in the upper layer, which corresponds to the Richardson number profile, and the C_T^2 increases to the height of the return current core. The wave-like movements corresponding to the KHBs are also observed in the vertical velocity field (figure 3(f)). According to the echo signal intensity, the height of the mixing layer can be estimated as 200 m. However, this value is most likely greatly overestimated, since figure 3(c) shows a layer with a minimum of the echo signal in the lower part of the sounding range (at the height of the maximum of the lower jet or below), which indicates the absence of turbulent mixing between the near-surface layer and the raised turbulent layer. In this case, it is most likely that the mixing layer has a height of less than 30 m and lies below the sodar sounding range.

Figure 4 shows the episode of the night breeze on October 4, 2016, in which clear KHBs are observed only in the lower layer, and the height of the mixing layer approximately corresponds to the height of the minimum wind speed between the direct and return flows – about 100 m.

5. Conclusions

Sodar sounding of the ABL in a coastal zone makes it possible to study the vertical structure of breeze cells, including the return currents. Return breeze currents at altitudes of 200-300 meters were observed in summer and autumn seasons under stable conditions for the sea and land breezes in the form of short-term and unstable episodes. The observed monotonous profiles of the vertical velocity variance

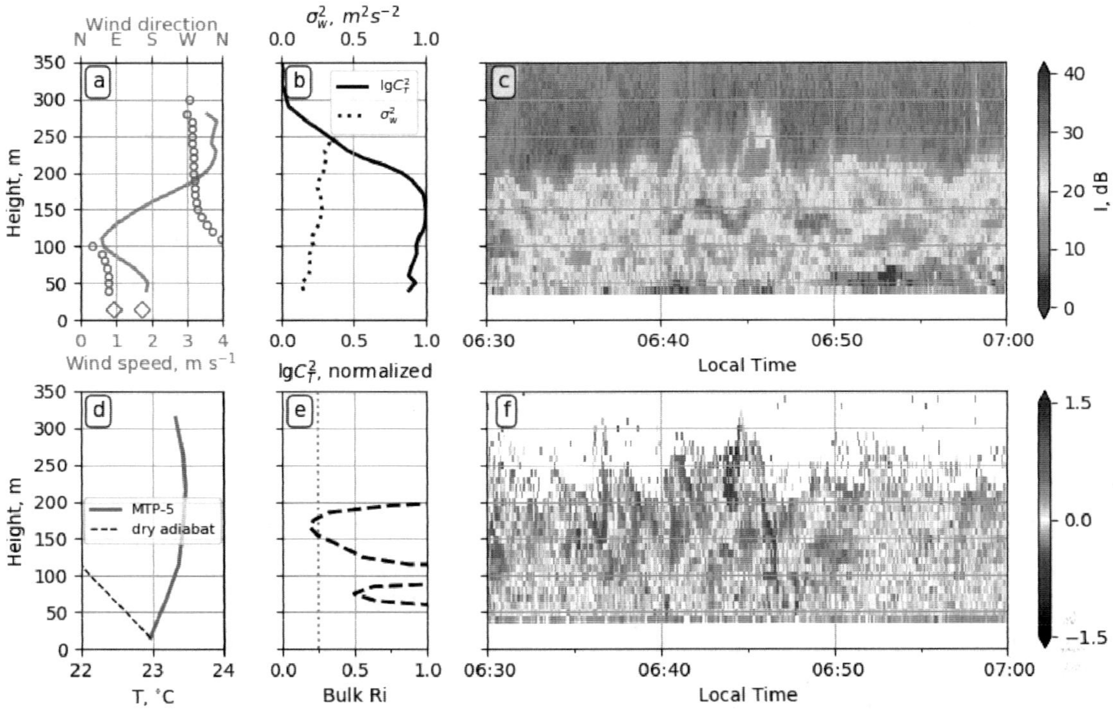

Figure 3. Vertical ABL structure in the presence of a return current, June 17, 2015. The case represents strong KHBs in the return current. All notations are the same as in figure 2.

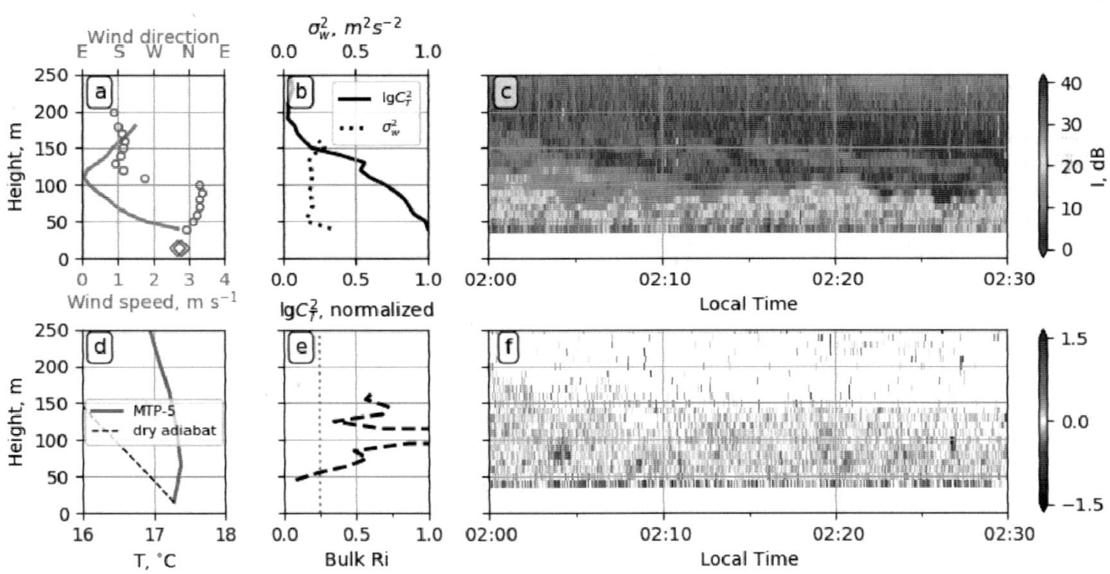

Figure 4. Vertical ABL structure in the presence of a return current, October 4, 2016. The case represents strong KHBs in the lower layer and a low level of the signal scattered in the return current. All notations are the same as in figure 2.

σ_w^2 and the temperature structure parameter C_T^2 indicate the presence of turbulent exchange between the direct and return flows in breezes and the need to take into account the return flow when determining

H_{mix}. However, the question remains as to how much the values of σ_w^2 and C_T^2 measured by sodar are determined by turbulent motions, and whether there is a contribution of wave motions that do not lead to vertical mixing of air masses (see [12]).

In general, in the case of breeze circulation, the height of the mixing layer H_{mix} is determined ambiguously and in various cases can approximately correspond to either the height of the near-surface breeze current core, or the height of the minimum wind speed between the direct and return flows, or the height of the return current core. In addition, it is important to note that under these conditions it is difficult to accurately determine the height of the core of the near-surface current from the sodar sounding data, since its lower part (below the maximum of the wind speed) is below the minimum sounding height of the sodar in all the cases described. Long-term experimental studies of the ABL by means of remote sensing and local methods in the near-surface layer are required to solve the problem of parametrization of the mixing layer in the coastal zone taking into account the return currents.

Acknowledgments

This study was supported by the Russian Foundation for Basic Research (project no. 19-05-01008, 20-05-00834, 19-05-00547) and the Russian Science Foundation (project no. 21-17-00021).

References

[1] Miller S T K, Keim B D, Talbot R W and Mao H 2003 *Reviews of Geophysics* **41** 1011
[2] Finkele K, Hacker J M, Kraus H and Byron-Scott R A 1995 *Boundary-layer meteorology* **73** 299
[3] Lyulyukin V S, Kallistratova M A, Kouznetsov R D, Kuznetsov D D, Chunchuzov I P and Chirokova G Y 2015 *Izvestia, Atmospheric and Oceanic Physics* **51** 193
[4] Banta R M 1995 *Mon. Weather Rev.* **123** 3614
[5] Tijm A, Holtslag A and Van Delden A 1999 *Monthly Weather Review* **127** 625
[6] Zeng X, Brunke M A, Zhou M, Fairall C, Bond N A and Lenschow D H 2004 *Journal of Climate* **17** 4159
[7] Batchvarova E, Barantiev D and Novitzky M 2012 *Proceedings of the 9th International Symposium on Tropospheric Profiling*
[8] Petenko I, Casasanta G, Bucci S, Kallistratova M, Sozzi R and Argentini S 2020 *Atmosphere* **11** 28
[9] Lyulyukin V, Kallistratova M, Zaitseva D, Kuznetsov D, Artamonov A, Repina I, Petenko I, Kouznetsov R and Pashkin A 2019 *Atmosphere* **10** 811
[10] Kouznetsov R D 2009 *Meteor. Zeitschrift* **18** 169
[11] Kadygrov E N, Shur G N and Viazankin A S 2003 *Radio Science* **38** 8048
[12] Zaitseva D V, Kallistratova M A, Lyulyukin V S, Kouznetsov R D and Kuznetsov D D 2018 *Izvestiya Atmospheric and Oceanic Physics* **54** 173

AUTHOR INDEX

Institute of Physics
Dirac House, Temple Back
Bristol BS1 6BE UK

ISSN: 1755-1307
ISBN 978-1-7138-5818-8

9 781713 858188

CURRAN ASSOCIATES INC.
proceedings.com